Hot-Carrier Effects in MOS Devices

Hot-Carrier Effects in MOS Devices

Eiji Takeda

Cary Y. Yang

Akemi Miura-Hamada

ACADEMIC PRESS

San Diego New York Boston

London Sydney Tokyo Toronto

Academic Press, Inc.
A Division of Harcourt Brace & Company
525 B Street, Suite 1900, San Diego, California 92101-4495

United Kingdom Edition published by
Academic Press Limited
24-28 Oval Road, London NW1 7DX

Library of Congress Cataloging-in-Publication Data

Takeda, Eiji, date.
Hot-carrier effects in MOS devices / by Eiji Takeda, Cary Y. Yang, and Akemi Miura-Hamada.
p. cm.
Includes index.
ISBN 0-12-682240-9
1. Metal oxide semiconductors. 2. Hot carriers. I. Yang, C. Y.-W. (Cary Y. -W.), date. II. Miura-Hamada, Akemi. III. Title.
TK7871.99.M44T35 1995
537.6'225-dc20 95-30713
CIP

CONTENTS

CHAPTER 3

HOT-CARRIER DEVICE DEGRADATION

CHAPTER 4

AC AND PROCESS-INDUCED HOT-CARRIER EFFECTS

CHAPTER 5
HOT-CARRIER EFFECTS AT LOW TEMPERATURE AND LOW VOLTAGE

CHAPTER 6
DEPENDENCE OF HOT-CARRIER PHENOMENA ON DEVICE STRUCTURE

CHAPTER 7

As–P Double Diffused Drain (DDD) Versus Lightly Doped Drain (LDD) Devices

CHAPTER 8

Gate-to-Drain Overlapped Devices (GOLD)

PREFACE

This book represents a compilation of recently published results and analyses on the subject of hot carriers in metal–oxide–silicon devices. It is an updated and substantially expanded version of the book on the same subject written in Japanese by the first author and published by Nikkei in 1987. It is intended to be a reference for practicing device engineers as well as researchers in academia and industry. Knowing full well that the subject matter is exceedingly dynamic, the authors designed the contents in such a way that the findings would still be useful to the reader even after they have been supplanted by results subsequent to the book's publication.

The book is organized using a "bottom-up" scheme, where the early chapters contain fundamental scientific knowledge and later ones technology development. The first chapter is a comprehensive review of MOS device physics to a level equivalent to what is covered in a first-year graduate course on semiconductor devices. No prior exposure to MOS device physics is required, even though some knowledge of electrostatics and modern physics would be helpful. The next two chapters treat the origins of hot carriers and how they affect device operations (DC), respectively. AC response and process-induced effects are introduced in Chapter 4, including mechanical effects. Chapter 5 presents a survey of phenomena at low operating temperatures and at low applied voltages. The final three chapters examine the details of device structures and their relationship with various hot-carrier phenomena, leading to what is popularly known as drain engineering. Specifically, Chapter 7 contains an in-depth comparison between the two standard drain structures, DDD and LDD, while Chapter 8 introduces the GOLD structure.

A reference list with almost two thousand articles is included. Although not all of them are cited in the text, the list nonetheless represents an attempt by the authors to include most, if not all, publications related to the subject matter published prior to the completion of the manuscript.

A task of this magnitude would not be possible without the support, guidance, advice, and valuable time and effort of many individuals. First and foremost, the authors are indebted to the management teams of the Central Research Laboratory of Hitachi, Ltd., especially Dr. Yasutsugu Takeda, Senior Executive Managing Director, who provided support for the first author's research on this subject over the past decade and a half and who generously hosted the visits of the second author and his students under the Hitachi Research Visit Programs (HIVIPS). The vision of Professor Takashi Tokuyama in making the initial suggestion to produce such a volume and his graciousness in hosting the second author's three-month visit to Tsukuba University during the planning and development phase are greatly appreciated.

The authors are deeply grateful to their numerous collaborators over the past fifteen years, who have contributed to the advancement of this important field. They are also grateful to the many authors and publishers who have granted permissions to reproduce figures and text from their publications. Special thanks are extended to Dr. Bruce E. Deal, who co-taught a course on semiconductor surfaces and interfaces with the second author, for using draft versions of this book as text, and for his valuable suggestions. The valuable assistance of the library staff, especially Edward Wladas, at Santa Clara University during the compilation of the reference list is gratefully acknowledged. Outstanding technical assistance was provided by Dr. Jianmin Qiao, Dr. Raymond Li, and Ashawant Gupta during various stages of manuscript preparation. Finally the second author expresses his heartfelt gratitude to all of his students who were first exposed to the subject of MOS devices in his classes and are applying that knowledge in their current professions, and to his research students, from whom he has derived much insight into the field.

EIJI TAKEDA
Hitachi Ltd.

CARY Y. YANG
Santa Clara University

AKEMI MIURA-HAMADA
Hitachi Ltd.

CHAPTER

1

MOS Device Fundamentals

The object of this chapter is twofold. It provides the reader with a brief but self-contained treatment of metal-oxide-semiconductor (MOS) device physics. While minimal background of semiconductor device physics is assumed in this treatment, the device specialist may find it useful to review the more advanced topics contained in this chapter before proceeding with subsequent chapters. The second purpose is to orient the reader to the approach used throughout this book. The approach is based on the assumption that understanding of empirical phenomena can be attained to some extent with appropriate modeling of the physical system. We hope that the content in this chapter will serve to provide the reader with the background information and understanding necessary for studies of hot-carrier effects.

This chapter begins with a short historical narrative on MOS devices and proceeds with a complete treatment of the MOS diode. The MOS field-effect transistor (MOSFET) operation is discussed next, with emphasis on short-channel effects. The chapter concludes with an introductory survey of key issues surrounding the subject of hot carriers.

1.1 From Discrete to ULSI

The first MOSFET device structure was proposed by Lilienfeld in 1928 [see the detailed historical account in Sah, 1988d]. But it was not until the late 1950s that the first stable working device was fabricated and tested. The subsequent successful applications of MOS capacitors and transistors hinged upon the understanding and control of the properties of the oxide layer and the oxide–silicon interface.

Through the legendary innovations of Noyce [1961] and Kilby [1976], the first monolithic integrated circuits consisting of resistors, capacitors, diodes, and transistors were fabricated on a single silicon substrate. With support from the U.S. government in the early 1960s, the development of integrated circuits blossomed and by 1970, large-scale integration (LSI) based on MOS technology started to emerge. The first LSI chip measured 3 mm × 3 mm and contained several thousand components. In the 1980s, very large-scale integration (VLSI) became the standard technology for memory and logic applications. The VLSI chip is usually less than 2 cm × 2 cm in size and typically contains several million transistors. In the early 1990s, advanced lithography, etching, and deposition techniques allowed further decrease in transistor dimensions and the corresponding increase in density of each silicon chip. The era of ultra LSI (ULSI) was born.

Despite such phenomenal advances in silicon integrated circuit technology, some basic problems remain as far as the oxide and oxide–silicon interface are concerned. For example, understanding of the physical mechanisms governing the initial stage of thermal oxidation is still lacking. Also the technology of silicon surface preparation is largely empirical and no unified scientific basis has been established. Lastly, and certainly not the least significant, the exact nature of the oxide–silicon interface traps and their creation and annihilation are far from being well understood.

1.2 Physics of the MOS Diode

The operation of MOSFET, which is a four-terminal device, is based on the properties of the two-terminal MOS structure. The MOS diode (or capacitor) is also important in its own right, in terms of actual applications as well as its utility as a test structure. The device is usually fabricated by oxidizing a crystalline silicon substrate, depositing a conducting film on the resulting amorphous SiO_2 layer, forming the gate, and finally metallizing the gate and the substrate, forming the necessary ohmic contacts. The fabrication process, no matter how well controlled, introduces defects at the SiO_2–Si interface, which critically affects the device characteristics of both the diode and the transistor. Moreover, the amorphous SiO_2 layer is highly susceptible to various charges, both fixed and mobile, and the defects present in the layer can serve as trap centers for carriers. A

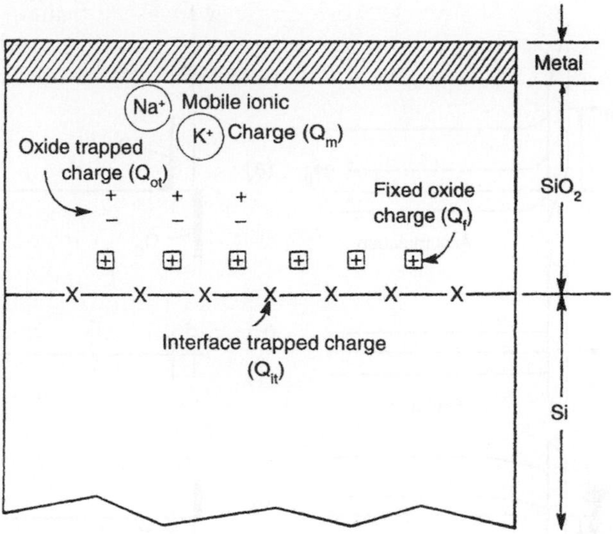

FIGURE 1.1 Schematic of oxide and interface charges in a MOS diode.

schematic description of the charges in the MOS structure is shown in Figure 1.1 [Deal, 1980].

1.2.1 THE IDEAL MOS DIODE

An idealized model would be to assume that all charges shown in Figure 1.1 are zero. In addition, the work function difference between the gate metal and the Si substrate is assumed to be zero. This model simplifies the analysis and serves as the basis for nonideal considerations.

1.2.1.1 Energy Band Diagrams

When a DC voltage V_G is applied across the device, charges are induced in the semiconductor near the oxide interface. An electric field results and the energy bands bend accordingly. This phenomenon is illustrated in Figure 1.2 for a device with a *p*-type substrate. When no gate voltage is applied, the device is in the *flat-band* state (Figure 1.2a). For negative V_G, holes are *accumulated* near the interface (Figure 1.2b). This region becomes *depleted* of carriers when V_G is positive (Figure 1.2c). Further increase in V_G results in the formation of an *n*-type region (Figure 1.2d), known as the *inversion* layer. The semiconductor is assumed to be in thermal equilibrium throughout the gate voltage sweep. In addition, the

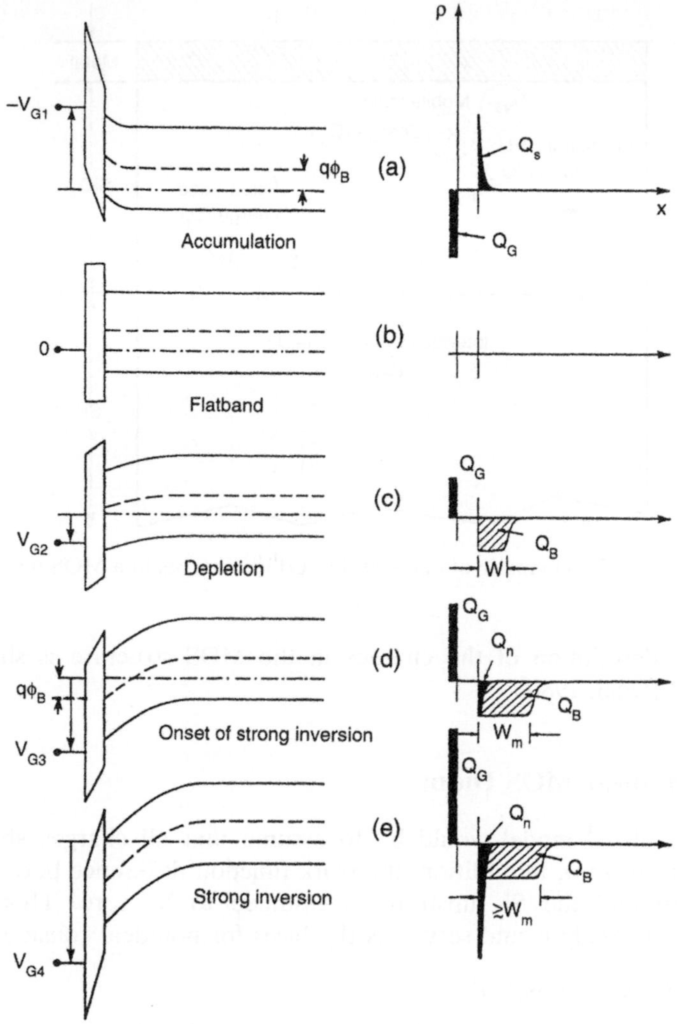

FIGURE 1.2 Energy band diagram of a *p*-substrate MOS diode in (a) accumulation, (b) flat band, (c) depletion, and (d and e) inversion states.

semiconductor is assumed to be nondegenerate so that carriers obey Boltzmann statistics.

The band bending parameter, sometimes known as surface potential, ψ_s, serves as a simple indicator of the state of the MOS diode. When $\psi_s = 0$, the condition corresponds to flat band. Negative ψ_s gives rise to hole "accumulation," while positive ψ_s results in carrier "depletion" and eventually "inversion." When the semiconductor surface becomes as n-type as the substrate is p-type, i.e.,

$$n_s = n(0) = p(\infty) \equiv p_{\text{bulk}}, \tag{1.1}$$

where n and p denote electron and hole concentrations, respectively, the condition is known as the "onset of strong inversion." Under this condition,

$$\psi_s = 2\phi_B \equiv 2\frac{kT}{q}\ln\frac{p_{\text{bulk}}}{n_i}, \tag{1.2}$$

where n_i is the intrinsic carrier concentration, T the absolute temperature in degrees Kelvin, k the Boltzmann constant, and q the unit electronic charge. The following summarizes the above description for a p-substrate MOS diode.

$$\begin{aligned} \psi_s &< 0 && \text{accumulation}, \\ \psi_s &= 0 && \text{flat band}, \\ 2\phi_B > \psi_s &> 0 && \text{depletion/weak inversion}, \\ \psi_s &\geq 2\phi_B && \text{strong inversion}. \end{aligned} \tag{1.3}$$

It should be noted that this description is generic for a p-type substrate and is completely independent of the ideal MOS assumptions.

The MOS diode in accumulation resembles a parallel-plate capacitor, with charge buildup on either side of the oxide. The situations in depletion and inversion are somewhat more complicated and require closer scrutiny. We now proceed to present two solutions of this electrostatics problem.

1.2.1.2 Solution with the Depletion Approximation

The charge density in the semiconductor substrate is given by

$$\rho_s(x) = q[p - n + N_D^+ - N_A^-], \tag{1.4}$$

where N_D^+ and N_A^- are the ionized donor and acceptor impurity concentrations, respectively. Using the depletion approximation for the MOS diode in the depletion state, Eq. (1.4) becomes

$$\begin{aligned} \rho_s &= q[N_D^+ - N_A^-] \equiv qN_B && W \geq x \geq 0, \\ \rho_s &= 0 && x > W, \end{aligned} \tag{1.5}$$

where W is the depletion width and N_B is negative for p-type substrate. For uniform doping, N_B is a constant of position. Solution of the one-dimensional Poisson equation

$$\frac{d\mathscr{E}}{dx} = \frac{\rho_s}{\epsilon_s} \tag{1.6}$$

yields the electric field

$$\begin{aligned} \mathscr{E}(x) &= q\frac{N_B}{\epsilon_s}(x - W) \qquad && W \geq x \geq 0, \\ \mathscr{E}(x) &= 0 && x > W, \end{aligned} \tag{1.7}$$

where ϵ_s is the semiconductor permittivity. From simple electrostatics the band-bending parameter can be expressed as

$$\psi_s = \frac{1}{2}\mathscr{E}(0)W = -\frac{qN_B W^2}{2\epsilon_s}. \tag{1.8}$$

Conversely, the depletion width can be written as

$$W = \left|\frac{2\epsilon_s \psi_s}{qN_B}\right|^{1/2}. \tag{1.9}$$

The treatment thus far is applicable to either p-type or n-type substrates.

The total charge per unit area in the p-type semiconductor substrate in depletion/inversion is given by

$$Q_s = Q_n + Q_B, \tag{1.10}$$

where Q_n is the electron charge in the inversion layer and Q_B is the space charge in the depletion region. Integration of Poisson's equation [Eq. (1.6)] yields

$$Q_s = -\epsilon_s \mathscr{E}(0). \tag{1.11}$$

In depletion, Q_n is assumed to be negligibly small and $Q_B = qN_B W$ under the depletion approximation. Together with Eqs. (1.7) and (1.10), Eq. (1.11) is thus verified. However, in order to extend the depletion approximation to treat inversion, additional assumptions have to be made. In view of the charge screening provided by the inversion layer, one assumes that when strong inversion is reached, the depletion width is maximized so that further increase in V_G would not increase W. Thus this maximum or saturated width is given by

$$W_m = \left(\frac{2\epsilon_s 2\phi_B}{q|N_B|}\right)^{1/2} \tag{1.12}$$

Beyond the onset of strong inversion, Q_B is independent of V_G and Q_n is no longer negligible.

The gate voltage consists of an oxide component and ψ_s. The drop across the oxide is given by

$$V_{ox} = \mathscr{E}_{ox} t_{ox} = \frac{\epsilon_s \mathscr{E}(0)}{\epsilon_{ox}} t_{ox} = -\frac{Q_s}{C_{ox}}, \tag{1.13}$$

where ϵ_{ox}, t_{ox}, and C_{ox} are the oxide permittivity, thickness, and capacitance, respectively. In depletion, using Eqs. (1.9) and (1.10), one obtains

$$V_{ox} = \frac{|Q_B|}{C_{ox}} = \frac{q|N_B|W}{C_{ox}} = \frac{1}{C_{ox}}[2\epsilon_s q|N_B|\psi_s]^{1/2}. \tag{1.14}$$

At the onset of strong inversion, $\psi_s = 2\phi_B$ and V_G becomes the so-called threshold voltage

$$\begin{aligned} V_{th} &= V_{ox}(\psi_s = 2\phi_B) + 2\phi_B \\ &= \frac{1}{C_{ox}}[2\epsilon_s q|N_B|2\phi_B]^{1/2} + 2\phi_B \qquad p \text{ substrate}. \end{aligned} \tag{1.15}$$

For an n-type substrate, identical arguments lead to

$$V_{th} = -\frac{1}{C_{ox}}[2\epsilon_s q N_B 2\phi_B]^{1/2} - 2\phi_B \qquad n \text{ substrate}, \tag{1.16}$$

where

$$\phi_B \equiv \frac{kT}{q} \ln \frac{n_{bulk}}{n_i}. \tag{1.17}$$

Despite the serious conceptual shortcomings of this simplified description, it is the basis of all analyses for the MOSFET. For the MOS diode, it is sometimes necessary to relax some of these assumptions and treat the electrostatics problem a bit more rigorously. This analysis will be presented following a brief description of the ideal capacitance–voltage characteristics in the next section.

1.2.1.3 Capacitance–Voltage Characteristics

Based on the depletion approximation, one can deduce the capacitance for the MOS diode in accumulation, depletion, and inversion. To simplify the present discussion, two additional assumptions are made. First, the DC gate voltage sweep rate is such that sufficient minority carriers are generated in the inversion layer during the sweep. This prevents "deep depletion" from occurring, which would result in a nonequilibrium situation. In practice, this can be accomplished by proper illumination of device to create the necessary carriers in the inversion layer. Second, at this stage we confine ourselves to AC gate voltage signals having frequencies sufficiently high that the charge fluctuation is dominated by the space charge near the edge of the depletion region. In other words, minority carrier generation in the inversion layer cannot keep up with the AC signal. Under these constraints, the MOS diode can be represented as a series combination of two capacitor as shown in Figure 1.3. The oxide capacitance, C_{ox}, was given previously in Eq. (1.13), while the semiconductor capacitance, C_s, represents charge fluctuations in the substrate.

The capacitance in accumulation is simply given by C_{ox}, since the device in this state resembles a parallel-plate capacitor. Alternatively, one can view that as a result of majority carrier buildup near the interface, C_s is so large that C_{ox} dominates the series combination. Under the depletion approximation, this domination persists up to flat band. This is illustrated

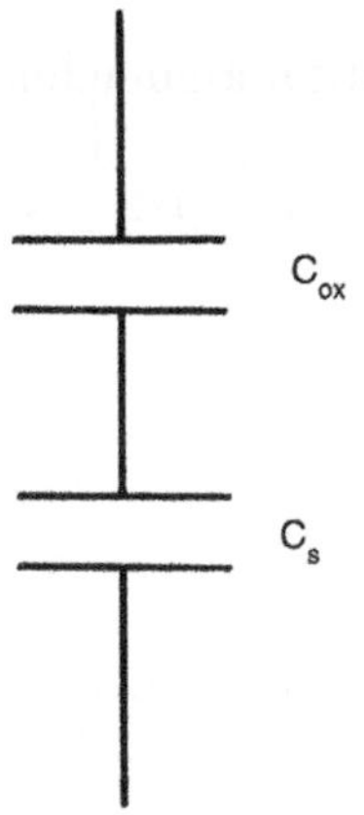

FIGURE 1.3 High-frequency circuit model of an ideal MOS diode.

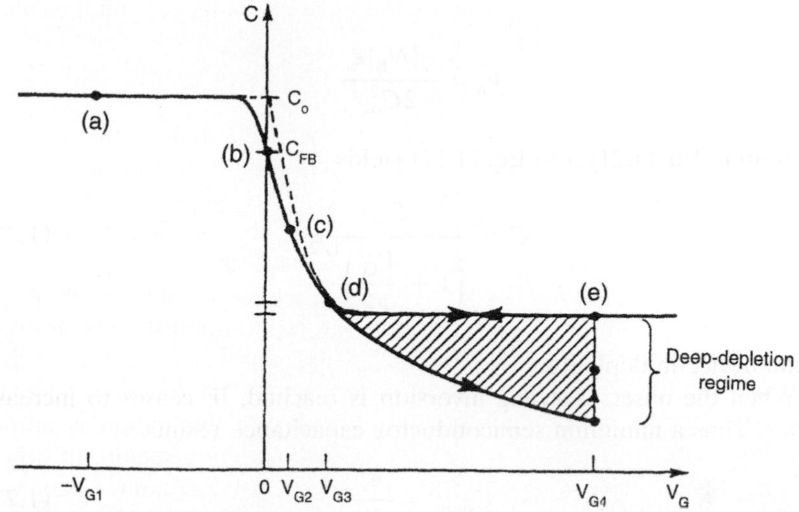

FIGURE 1.4 High-frequency capacitance–voltage characteristics of an ideal MOS diode. The solid curve indicates exact solution, while the broken curve is derived from the depletion approximation.

in Figure 1.4 by the broken curve. The existence of a depletion region when the device is in depletion gives rise to a semiconductor capacitance per unit area

$$C_s = \frac{\epsilon_s}{W}. \tag{1.18}$$

The capacitance of the series combination becomes

$$C = \left(\frac{1}{C_{ox}} + \frac{1}{C_s}\right)^{-1} = \left(\frac{1}{C_{ox}} + \frac{W}{\epsilon_s}\right)^{-1}. \tag{1.19}$$

Recognizing from Eqs. (1.8) and (1.14) that

$$V_G = V_{ox} + \psi_s = \frac{q|N_B|W}{C_{ox}} + \frac{q|N_B|W^2}{2\epsilon_s}, \tag{1.20}$$

W can be expressed as

$$W = \frac{\epsilon_s}{C_{ox}}\left[\left(1 + \frac{V_G}{V_\delta}\right)^{1/2} - 1\right], \tag{1.21}$$

where

$$V_\delta = \frac{q|N_B|\epsilon_s}{2C_{ox}^2}.$$

Substituting Eq. (1.21) into Eq. (1.19) yields

$$C = \frac{C_{ox}}{\left(1 + \frac{V_G}{V_\delta}\right)^{1/2}} \tag{1.22}$$

for the device in depletion.

When the onset of strong inversion is reached, W ceases to increase further. Thus a minimum semiconductor capacitance results.

$$C_{smin} = \frac{\epsilon_s}{W_m}. \tag{1.23}$$

This leads to a minimum MOS capacitance

$$C_{min} = \left(\frac{1}{C_{ox}} + \frac{1}{C_{smin}}\right)^{-1}. \tag{1.24}$$

The broken curve in Figure 1.4 reveals the entire C–V_G behavior under the depletion approximation.

One important utility of this approximation is the determination of substrate doping, N_B. From Eqs. (1.23) and (1.24), one obtains

$$W_m = \frac{\epsilon_s}{C_{ox}}\left(\frac{C_{ox}}{C_{min}} - 1\right). \tag{1.25}$$

Thus W_m can be determined from the maximum and minimum values of the C–V_G curve. But W_m is also given by Eq. (1.12). Combining it with Eq. (1.2) and assuming extrinsic behavior for the bulk substrate, namely,

$$p_{bulk} = |N_B|, \tag{1.26}$$

one arrives at

$$|N_B| = \frac{4\epsilon_s kT}{q^2 W_m^2} \ln\left|\frac{N_B}{n_i}\right|. \tag{1.27}$$

Equation (1.27) can be solved iteratively to yield N_B.

It is important to note that the depletion approximation is most severe near flat band. When the device approaches flat band from accumulation, the capacitance starts to decrease as a result of the series combination of C_{ox} and C_s. This behavior is illustrated by the solid curve in Figure 1.4. In fact, this result can be obtained directly from an exact solution of Poisson's equation. A sketch of this solution will be presented next.

1.2.1.4 General Solution

One must recognize at the outset that the semiconductor capacitance, C_s, is defined as charge fluctuations in the substrate as a result of changes in band bending. Expressed in a more compact form,

$$C_s \equiv \frac{dQ_s}{d\psi_s}. \tag{1.28}$$

In view of Eq. (1.11), if $\mathscr{E}(0)$ can be determined, then the MOS capacitance is obtained.

To facilitate the solution of this electrostatics problem, we examine the MOS diode in the depletion state with a schematic band diagram shown in Figure 1.5. Poisson's equation in the semiconductor region is written as

$$\frac{d\mathscr{E}}{dx} = -\frac{d^2\psi}{dx^2} = \frac{\rho_s}{\epsilon_s} = \frac{q}{\epsilon_s}[p - n + N_B] \qquad x \geq 0, \tag{1.29}$$

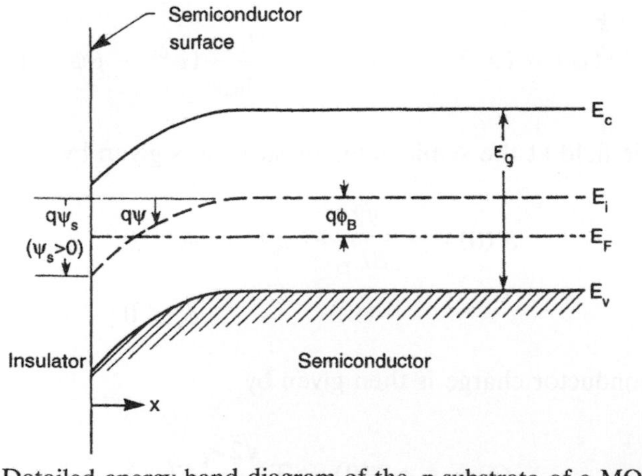

FIGURE 1.5 Detailed energy band diagram of the p-substrate of a MOS diode in depletion/weak inversion.

where ψ is the electrostatic potential with the reference in the semiconductor bulk, i.e., $\psi(\infty) = 0$. Since $\rho_s = 0$ in the bulk, $N_B = n_{bulk} - p_{bulk}$. Simple deduction using Figure 1.5 results in

$$n = n_{bulk} e^{q\psi/kT}, \qquad p = p_{bulk} e^{-q\psi/kT}. \tag{1.30}$$

Thus Eq. (1.29) can be written as

$$\frac{d\mathscr{E}}{dx} = \frac{q}{\epsilon_s}\left[p_{bulk}(e^{-\beta\psi} - 1) - n_{bulk}(e^{\beta\psi} - 1)\right], \tag{1.31}$$

where $\beta \equiv q/kT$ is used. Multiplying Eq. (1.31) by $\mathscr{E}$ and integrating, one obtains

$$\mathscr{E}^2 = 2\left(\frac{1}{\beta L_D}\right)^2 F^2(\psi), \tag{1.32}$$

where L_D is the so-called extrinsic Debye length

$$L_D \equiv \left(\frac{\epsilon_s kT}{q^2 p_{bulk}}\right)^{1/2} \tag{1.33}$$

and

$$F^2(\psi) = (e^{-\beta\psi} + \beta\psi - 1) + \frac{n_{bulk}}{p_{bulk}}(e^{\beta\psi} - \beta\psi - 1). \tag{1.34}$$

The electric field at the semiconductor surface is given by

$$\mathscr{E}(0) = \pm\frac{\sqrt{2}}{\beta L_D}F(\psi_s) \qquad \psi_s > 0, \quad \psi_s < 0. \tag{1.35}$$

The semiconductor charge is then given by

$$Q_s = -\epsilon_s \mathscr{E}(0) = \mp\frac{\sqrt{2}\,\epsilon_s}{\beta L_D}F(\psi_s). \tag{1.36}$$

Thus the semiconductor capacitance becomes

$$C_s = \left|\frac{\partial Q_s}{\partial \psi_s}\right| = \frac{\epsilon_s}{\sqrt{2}L_D F(\psi_s)}\left|1 - e^{-\beta\psi_s} + \frac{n_{\text{bulk}}}{p_{\text{bulk}}}(e^{\beta\psi_s} - 1)\right|. \quad (1.37)$$

In the limit $\psi_s \to 0$, or flat band, Eq. (1.37) reduces to

$$C_{\text{sFB}} = \frac{\epsilon_s}{L_D}. \quad (1.38)$$

This leads to the MOS flat-band capacitance

$$C_{\text{FB}} = \left(\frac{1}{C_{\text{ox}}} + \frac{1}{C_{\text{sFB}}}\right)^{-1} = \left(\frac{1}{C_{\text{ox}}} + \frac{L_D}{\epsilon_s}\right)^{-1}. \quad (1.39)$$

The gate voltage corresponding to C_{FB} is known as the flat-band voltage, V_{FB}. For the ideal MOS diode, it follows from definition that $V_{\text{FB}} = 0$. It is apparent that with a room-temperature value of several hundred angstroms for L_D, C_{sFB} is comparable to C_{ox}, resulting in a significant reduction of the MOS capacitance when flat band is reached from accumulation. The solid curve in Figure 1.4 represents a solution based on Eq. (1.37) and $C = (1/C_{\text{ox}} + 1/C_s)^{-1}$ for accumulation and depletion. The band bending has been converted to gate voltage through

$$V_G = -\frac{Q_s}{C_{\text{ox}}} + \psi_s = \frac{\epsilon_s \mathscr{E}(0)}{C_{\text{ox}}} + \psi_s.$$

It is instructive to examine the behavior of the semiconductor charge as a function of band bending. This behavior for a *p*-substrate MOS diode as shown in Figure 1.6 reveals that in depletion/weak inversion ($2\phi_B > \psi_s > 0$), Q_s varies roughly as $\psi_s^{1/2}$, which supports the depletion approximation. However, the spatial variation of the charge density is not well represented by the depletion approximation in the vicinity of the depletion region edge, as shown in Figure 1.7. The drop-off of the true charge density is not abrupt and occurs over a distance comparable to L_D around the depletion region edge. In strong inversion, Q_s varies roughly as

$$\left[\psi_s + \frac{kT}{q}\left(\frac{n_{\text{bulk}}}{p_{\text{bulk}}}\right)e^{(q\psi_s/kT)}\right]^{1/2},$$

where the second term, which represents the inversion charge, dominates for large ψ_s as shown in Figure 1.6. Comparing this behavior with that in depletion, this variation lends some credibility to the assumption that the depletion width "saturates" in strong inversion.

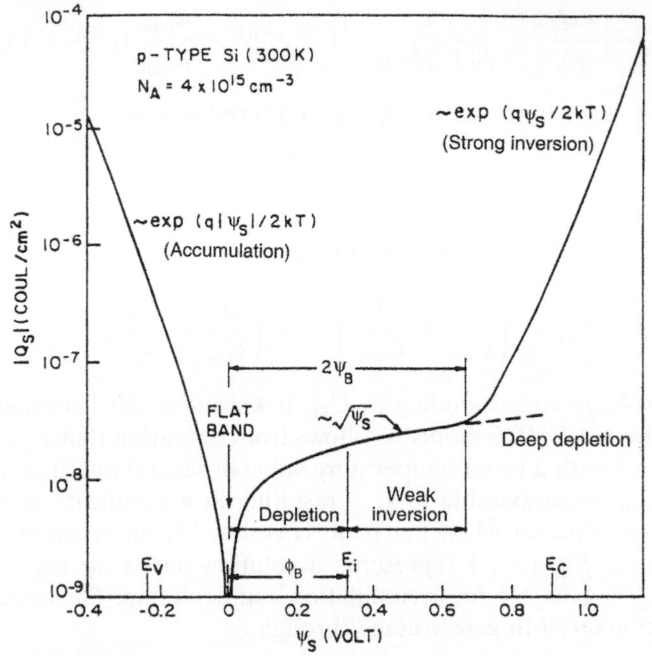

FIGURE 1.6 Semiconductor charge as a function of band bending. (Reprinted by permission from John Wiley & Sons, Inc., from S. M. Sze, *Physics of Semiconductor Devices*, Wiley, New York (© 1988 John Wiley & Sons, Inc.).)

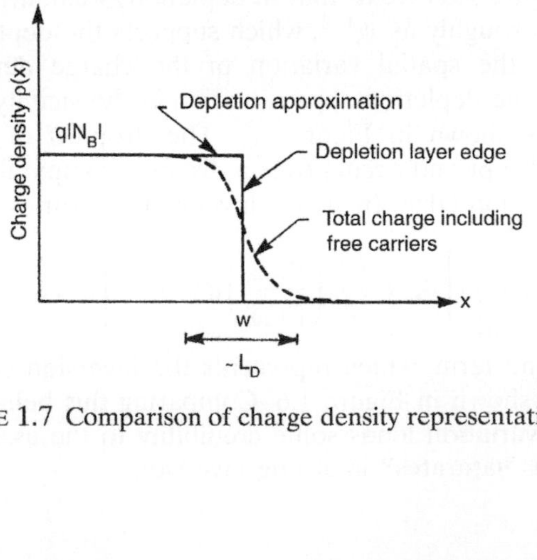

FIGURE 1.7 Comparison of charge density representations.

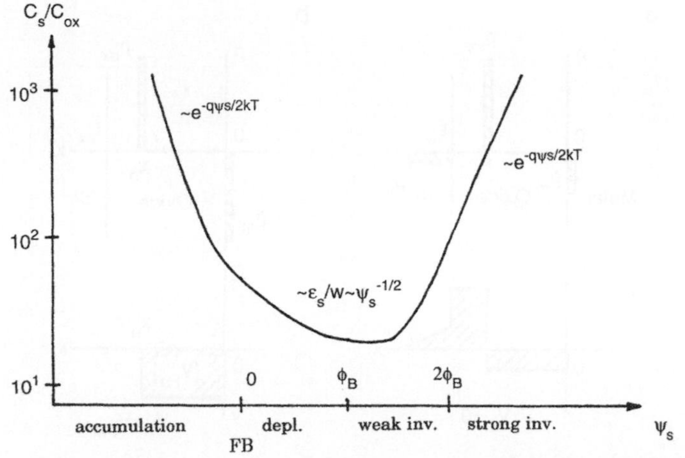

FIGURE 1.8 Semiconductor capacitance for a p-substrate MOS diode.

The semiconductor capacitance versus band bending behavior given by Eq. (1.37) is shown in Figure 1.8. As in the charge case, the behavior in depletion resembles that given by the depletion approximation.

1.2.2 NONIDEAL CONSIDERATIONS

A real MOS capacitor generally differs from the ideal one. As shown in Figure 1.1, the device contains oxide and interface charges, which give rise to changes in C–V_G behavior. The gate–substrate work function difference is usually not zero. Further, substrate dopant concentration can vary with position.

This section is organized as follows. Work function difference, fixed oxide charge, oxide trapped charge, and mobile ions are treated together, since they result in voltage-independent flat-band voltage shifts from the ideal behavior. The interface trapped charge, on the other hand, varies with the surface potential (band bending), which in turn varies with gate voltage. Thus interface charges give rise to to voltage-dependent flat-band shifts as well as changes in ψ_s vs. V_G behavior. The latter causes a "stretch-out" of the C–V_G curve. Following the two subsections on oxide and interface charges, a discussion of the effect of nonuniform substrate doping is presented.

1.2.2.1 Voltage-Independent Flat-Band Shifts

A difference is gate–substrate work function, $\phi_{ms} \equiv \phi_m - \phi_s$, results in band bending in the substrate to accommodate a uniform Fermi level in

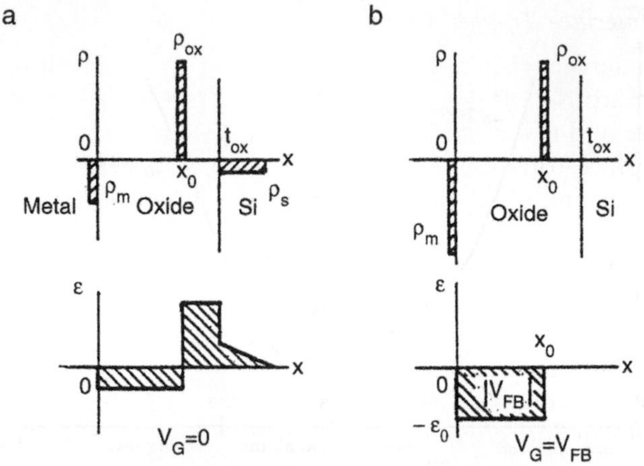

FIGURE 1.9 Electrostatics of oxide charges.

equilibrium. To return to flat-band condition, a gate voltage of $V_G = \phi_{ms}$ must be applied. Thus the flat-band voltage shift due to work function difference is simply ϕ_{ms}.

The oxide charges, regardless of their origins, are distributed across the entire insulating layer. A schematic of the electrostatics is shown in Figure 1.9. The oxide charge distribution, $\rho_{ox}(x)$, includes fixed oxide charge, oxide trapped charge, and mobile ions. In principle, these are quite different physical entities with very different distributions (see Figure 1.1). In practice, these charges manifest themselves in similar fashions electrically and thus are not easy to distinguish [Nicollian and Brews, 1982]. As in work-function difference, the band bending resulting from these charges can be made to vanish with a suitable gate voltage. From the electrostatics scheme shown in Figure 1.9, using superposition across the oxide layer, one can deduce that $\rho_{ox}(x)$ requires a gate voltage of ΔV to achieve flat band [Sze, 1981]. ΔV is given by

$$\Delta V = -\frac{Q_0}{C_{ox}} = -\frac{1}{C_{ox}}\left[\frac{1}{t_{ox}}\int_0^{t_{ox}} x\,\rho_{ox}(x)\,dx\right], \tag{1.40}$$

where $x = 0$ defines the gate and Q_0 is the total effective oxide charge per unit area at the Si–SiO_2 interface. Combining with the effect of work function difference, the total flat-band voltage shift is given by

$$V_{FB} = \phi_{ms} - \frac{Q_0}{C_{ox}}. \tag{1.41}$$

This excludes interface trapped charges, which will be discussed next.

1.2.2.2 Interface Trapped Charge

This is a subject that will be discussed at great length throughout this book. The purpose of this introductory discussion is to establish the nomenclature and review some basic understanding of energy states at the SiO_2–Si interface (often known as interface traps). Another purpose is to examine the extent to which interface trapped charge affects the MOS capacitance–voltage behavior.

The voltage-independent nonideal considerations discussed in the previous subsection simply produce a parallel shift of the C–V_G curve along the voltage axis. The interface trapped charge, on the other hand, depends on the Fermi energy position and thus is a function of band bending and gate voltage.

We shall confine our discussion to two types of interface traps: a donor-like interface trap is one that is neutral when filled with an electron and positively charged when empty; an acceptor-like trap is negatively charged when filled with an electron and neutral when empty. In general, a donor-like trap is neutral when it is a few kT below E_F and positive above, while an acceptor-like trap is negative a few kT below E_F and neutral above.

Figure 1.10 serves to illustrate the behavior of the interface traps and their effect on the MOS characteristics for a *p*-type and an *n*-type substrate. For simplicity, let us assume these traps are all donor-like with a density of states, $D_{it}(E)$, as shown. In Figure 1.10a, the system is at flat band; the traps above E_F are positively charged and below are neutral. Figure 1.10b–d depict the system in accumulation, depletion, and inversion, respectively. It is clear that the amount of the interface trapped charge (positive for donor-like traps) depends on band bending. For an *n*-type substrate, it is the smallest in accumulation and continues to increase as the device goes to flat band, depletion, and inversion; for *p*-type, it is the largest in accumulation and continues to decrease as the device goes to flat band, depletion, and inversion.

Since a positive sheet charge at the SiO_2–Si interface causes a negative shift of the C–V_G curve according to Eq. (1.41), the resulting flat-band shift can be written as

$$\Delta V = -\frac{Q_{it}(\psi_s)}{C_{ox}}, \tag{1.42}$$

where $Q_{it}(\psi_s)$ is the interface-trapped charge for the given surface band bending ψ_s. For the case that the interface traps are all donor-like,

$$Q_{it}(\psi_s) = q\int_{E_F}^{E_{CB}-q\psi_s} D_{it}(E)\,dE, \tag{1.43}$$

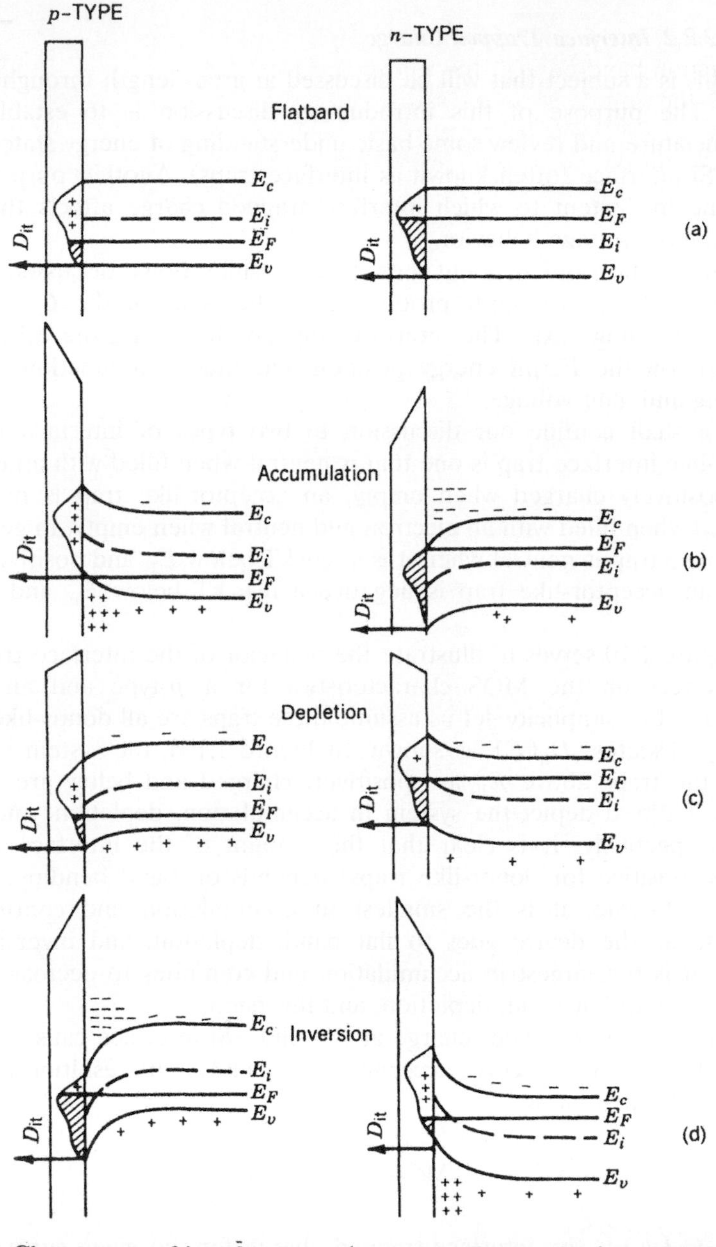

FIGURE 1.10 Charge state of interface traps (assuming all donor-like) in a MOS capacitor under (a) flat band, (b) accumualtion, (c) depletion, and (d) inversion. (Reprinted by permission from John Wiley & Sons, Inc., from T. P. Ma and P. V. Dressendorfer, *Ionizing Radiation Effects in MOS Devices*, Wiley, New York (© 1989 John Wiley & Sons, Inc.).)

where E_{CB} is the conduction band edge in the bulk, and $E_{CB} - q\psi_s$ is the conduction band edge at the silicon surface.

If the interface traps are acceptor-like instead, then the trapped charge, $Q_{it}(\psi_s)$, will be negative and the voltage shift ΔV_G in Eq. (1.42) would be in the positive direction. Equation (1.43) now becomes

$$Q_{it}(\psi_s) = -q \int_{E_{VB}-q\psi_s}^{E_F} D_{it}(E)\,dE, \tag{1.44}$$

where E_{VB} is the valence band edge in the bulk, and $E_{VB} - q\psi_s$ is the valence band edge at the silicon surface. Both Eqs. (1.43) and (1.44) ignore the variation of Fermi function with temperature. This assumption is fairly accurate for room temperature and below, and if $D_{it}(E)$ does not vary rapidly over a few kT.

Figure 1.11 compares C–V_G curves for three MOS capacitors having otherwise identical parameters, except that one is free of interface traps, one contains interface traps that are all donor-like, and one contains interface traps that are all acceptor-like. One can see that the influence of

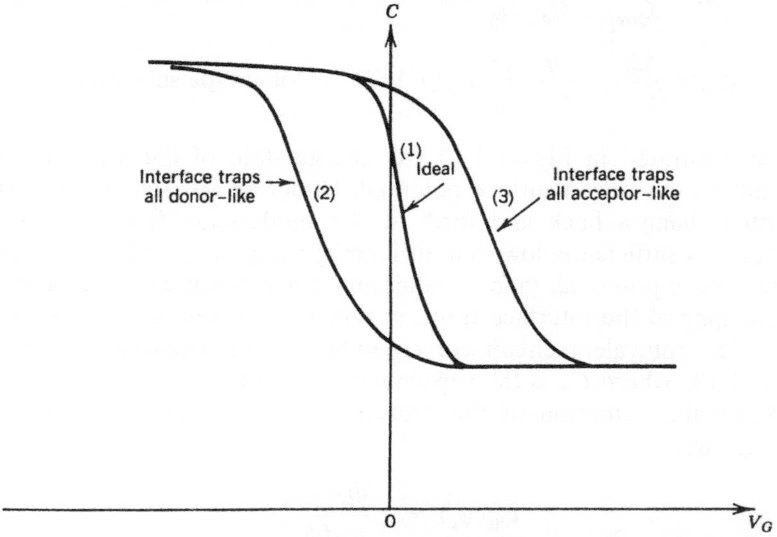

FIGURE 1.11 Influence of interface traps on the high-frequency MOS C–V curve: curve 1, with no interface traps; curve 2, with only donor-like interface traps; curve 3, with only acceptor-like interface traps. (Reprinted by permission from John Wiley & Sons, Inc., from T. P. Ma and P. V. Dressendorfer, *Ionizing Radiation Effects in MOS Devices*, Wiley, New York (© 1989 John Wiley & Sons, Inc.).)

the donor-like interface traps is to stretch out the curve to the left [see curve (2)], in accordance with Eqs. (1.42) and (1.43).

In actual devices, there has been strong evidence that the interface traps below the middle of the silicon band gap, E_{mg}, are donor-like and those above acceptor-like [Gray and Brown, 1966]. Therefore, at a gate voltage when the Fermi level coincides with E_{mg} (essentially E_i) at the silicon surface, called the midgap voltage, V_{mg}, the charge state of the interface traps is practically neutral, because most donor-like traps are filled and most acceptor-like ones are empty. Thus, the presence of the interface traps at this gate voltage does not contribute to Q_{it}, and the flat band shift in midgap voltage, ΔV_{mg}, is determined only by the oxide charge and work function difference:

$$\Delta V_{mg} = \phi_{ms} - \frac{Q_0}{C_{ox}}. \tag{1.45}$$

Assuming the same interface trap distribution, the total flat-band voltage is then given by

$$V_{FB} = \phi_{ms} - \frac{Q_0}{C_{ox}} + \frac{q}{C_{ox}} \int_{E_i}^{E_F} D_{it}(E)\, dE \qquad \text{for } n\text{-type substrates}, \tag{1.46}$$

$$V_{FB} = \phi_{ms} - \frac{Q_0}{C_{ox}} - \frac{q}{C_{ox}} \int_{E_F}^{E_i} D_{it}(E)\, dE \qquad \text{for } p\text{-type substrates}. \tag{1.47}$$

As illustrated in Figure 1.10, the charge state of the interface traps depends on the silicon surface potential. Under an AC signal, the surface potential changes back and forth at the modulation frequency. If the frequency is sufficiently low that all interface traps respond to the change in the surface potential, then an additional capacitance component, due to the charging of the interface traps, is added to the semiconductor capacitance. The equivalent circuit corresponding to this situation is shown in Figure 1.12, where C_{it} is the capacitance due to the interface traps.

From the definition of the capacitance due to the charging of the interface traps

$$C_{it}(\psi_s) = -\frac{dQ_{it}(\psi_s)}{d\psi_s} \tag{1.48}$$

and using Eqs. (1.43) and (1.44), one can deduce that the density of interface traps is simply given by

$$D_{it}(\psi_s) = C_{it}(\psi_s)/q. \tag{1.49}$$

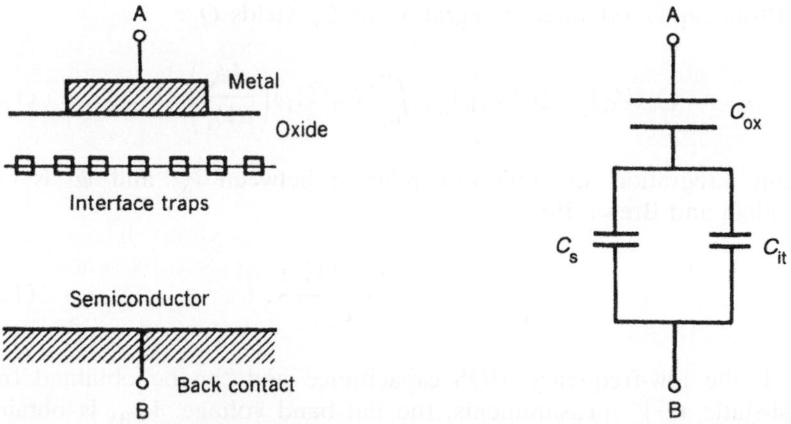

FIGURE 1.12 Equivalent circuit representation of a MOS capacitor with effect of interface traps included. (Reprinted by permission from John Wiley & Sons, Inc., from T. P. Ma and P. V. Dressendorfer, *Ionizing Radiation Effects in MOS Devices*, Wiley, New York (© 1989 John Wiley & Sons, Inc.).)

Note that Eq. (1.49) is obtained based on the following assumptions. First, D_{it} is restricted to the semiconductor band gap and is slow-varying over an energy range of a few kT. Second, only single-electron donor/acceptor states are considered. The interface trapped charge, Q_{it}, on the other hand, can be directly obtained from C_{it} without utilizing either of the two assumptions stated above [Yang *et al.*, 1991a]. Since Q_{it} is one of the key parameters quantifying the integrity of an insulator/semiconductor system, this technique should have a wide range of applications. Further, the information about both the amount and polarity of interface trapped charge, which cannot be obtained from the usual D_{it} calculations, provides a powerful means of delineating electron and hole trapping at the interface.

In obtaining Q_{it}, information about bulk oxide charge must be either available or assumed. Combined I–V and C–V techniques to separate oxide and interface charges have been reported [DiMaria, 1976; DiMaria and Stasiak, 1989; Nissan-Cohen *et al.*, 1985]. These techniques utilize the shifts in I–V curves to establish the extent of interface charge contribution. Their primary purpose was to determine the oxide charge component. Further, they examined the interface trapped charge only at a single Fermi energy, for example, at flat band. In comparison, the Q_{it} method [Yang *et al.*, 1991a] yields the amount of interface trapped charge over a range of Fermi energies covering accumulation, depletion, and inversion. Thus it provides further insight into interface trap characteristics.

From Eq. (1.48), direct integration of C_{it} yields Q_{it}:

$$Q_{it}(V_G) = Q_{it}(V_{FB}) - \int_{V_{FB}}^{V_G} C_{it}(V_G)\left(\frac{d\psi_s}{dV_G}\right) dV_G. \tag{1.50}$$

In this integration, the following relation between V_G and ψ_s is used [Nicollian and Brews, 1982].

$$\frac{d\psi_s}{dV_G} = 1 - \frac{C_{LF}(V_G)}{C_{ox}}. \tag{1.51}$$

C_{LF} is the low-frequency MOS capacitance and can be obtained from quasi-static C–V measurements, the flat-band voltage, V_{FB}, is obtained from high-frequency C–V measurements, and $Q_{it}(V_{FB})$ is related to V_{FB} by

$$Q_{it}(V_{FB}) + Q_0 = (\phi_{ms} - V_{FB})C_{ox}. \tag{1.52}$$

C_{it} can be conveniently obtained with the high–low frequency capacitance method [Nicollian, 1982].

Since it is not possible to distinguish between Q_0 and $Q_{it}(V_{FB})$ from C–V measurements alone, the function $Q_{it}(V_G)$ or $Q_{it}(\psi_s)$ is determined up to an unknown constant Q_0. For unstressed devices, Q_{it} and Q_0 are usually within the same order of magnitude (10^{10} q/cm^2). For stressed devices, change of Q_0 due to stressing can be extracted from positive-gate I–V measurements, based on the assumption that the tunneling current is affected not by the interface trapped charge but by the oxide field near the oxide–substrate interface. The change ΔQ_0 is simply $-\Delta V_G C_{ox}$, where ΔV_G is the corresponding change in gate voltage for the same current used in I–V measurements before and after stressing. The stressed $Q_{it}(\psi_s)$ function can be determined from Eqs. (1.50) and (1.52), with ΔQ_0 replacing Q_0. Thus the combination of C–V and I–V measurements for stressed devices yields both ΔQ_0 and $Q_{it}(\psi_s)$. The latter is accurate up to an unknown prestressed Q_0 value.

1.2.2.3 Nonuniform Substrate Doping

Our treatment thus far has been based on the assumption that the doping concentration in the silicon substrate is uniform. In real devices, because of various process steps involving the silicon surface region, even initially constant doping will become somewhat nonuniform (at least near the surface) after the entire device is fabricated. Thermal oxidation, for instance, can cause impurity redistribution. Several device performance enhancement techniques involve ion implantation into the channel region or the substrate below [see, for example, Sze 1981]. The detailed profiling of the substrate doping distribution can be obtained from high-frequency C–V measurements, with additional refinements to determine the profile for depths less than a Debye length from the surface [Ziegler *et al.*, 1975a; Toyabe *et al.*, 1989].

Regardless of the exact doping profile, the effect of nonuniform doping can be viewed as an additional source for shift in threshold voltage [Rideout *et al.*, 1975]. In particular, the doping profile $N(x)$ can be modeled by a step function, with a constant surface doping N_s and a constant bulk doping N_B. N_s is simply given by

$$(N_s - N_B)x_s = \int_0^\infty [N(x) - N_B]\,dx \equiv D_I, \tag{1.53}$$

where x_s is the step depth. If x_s is sufficiently deep (larger than W_m), then the entire surface region can be considered uniformly doped with N_s and the ideal threshold is simply given by Eqs. (1.15) and (1.16) with N_B replaced by N_s. For $x_s < W_m$, qD_I can be crudely viewed as an additional charge added to Q_B in the depletion region and the resulting threshold shift is essentially given by qD_I/C_{ox} [Sze, 1981]. In the limiting case where the surface charge could be modeled by a delta function at the SiO_2–Si interface, this result would be exact.

1.3 Principles of the MOSFET

Any study of hot-carrier effects in MOS systems requires a clear understanding of the MOSFET operation. The brief review presented here is intended to capture the essence of the basic overall physics without being overwhelmed by the algebraic details. The bulk charge model is given as the basis for this description. A summary of standard techniques for threshold determination is also presented, followed by the sections on subthreshold conduction and short-channel effects. Sections 1.3.1, 1.3.2, and 1.3.4 are extracted from an overview in Ma and Dressendorfer [1989].

1.3.1 Qualitative Description of MOSFET Operation

A schematic of an n-channel MOSFET is shown in Figure 1.13. It is a four-terminal device consisting of a gate, source, drain, and substrate. The gate structure, which controls the conduction, is simply a MOS diode.

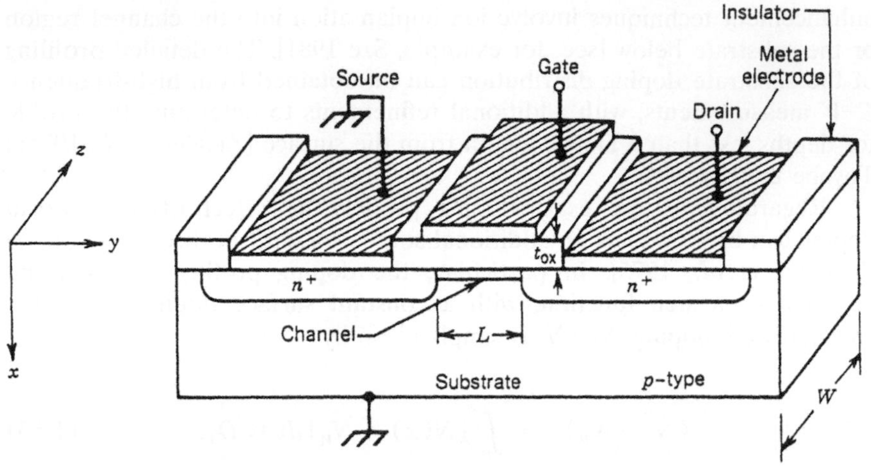

FIGURE 1.13 Schematical cross-sectional view of a simplified n-channel MOSFET showing its basic elements.

When the gate is biased such that the silicon surface under the gate is in accumulation or depletion ($V_G < V_{th}$), the channel region contains very few electrons, and no appreciable current can flow between the source and the drain. When V_G is biased so that inversion occurs ($V_G \geq V_{th}$), a significant number of mobile electrons are generated adjacent to the Si surface under the gate. This results in the formation of a conduction channel connecting the source and drain, as depicted in Figure 1.14a. Thus, the MOSFET structure allows one to modulate the source–drain conductance with the gate voltage.

After the inversion channel is formed, for a small V_D applied to the drain terminal, the conduction channel behaves just like a simple resistor, and the drain current is directly proportional to V_D. This is the linear I–V region and is shown as the line from the origin to point A in Figure 1.15. As V_D increase further, the potential drop along the channel due to the channel current starts to negate the inverting effect of the gate. Also, the surface potential is enhanced. As depicted in Figure 1.14b, the depletion region widens along the channel from source to drain, and the number of channel electrons decreases correspondingly. Thus, the channel conductance (dI/dV) decreases with increasing V_D, which is reflected as a decrease in the slope of the I–V characteristics, as shown in the A → B portion of the curve in Figure 1.15. The greatest increase in the depletion layer width, and therefore the greatest decrease in the number of channel electrons, occurs near the vicinity of the drain junction, and eventually for

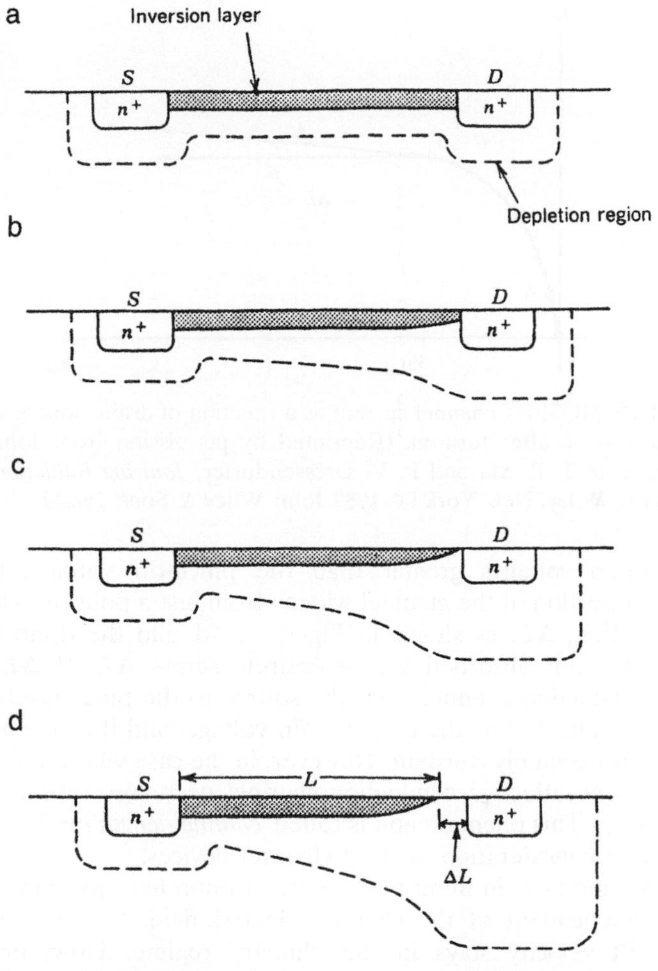

FIGURE 1.14 Schematic diagram of an n-channel MOSFET showing the conduction channel after the device is turned on: (a) $V_D = 0$; (b) $V_{D_{sat}} > V_D > 0$; (c) $V_D = V_{D_{sat}}$; (d) $V_D > V_{D_{sat}}$. (Reprinted by permission from John Wiley & Sons, Inc., from T. P. Ma and P. V. Dressendorfer, *Ionizing Radiation Effects in MOS Devices*, Wiley, New York (© 1989 John Wiley & Sons, Inc.).)

a sufficiently large drain voltage ($V_D \geq V_{Dsat}$), the inversion layer completely vanishes near the drain (Figure 1.14c). This condition is referred to as "pinch-off" because the normal conduction channel disappears adjacent to the drain. When the channel pinches off, the current saturation is observed; that is, the slope of the I–V characteristics becomes approximately zero, as shown beyond point B in Figure 1.15.

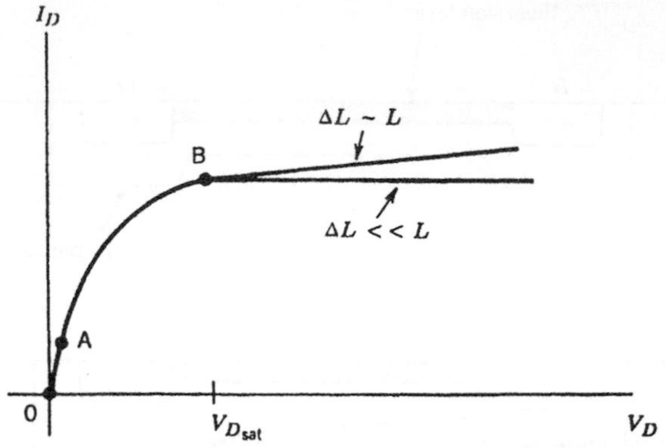

FIGURE 1.15 MOSFET channel current as a function of drain–source voltage at a given gate voltage after turn-on. (Reprinted by permission from John Wiley & Sons, Inc., from T. P. Ma and P. V. Dressendorfer, *Ionizing Radiation Effects in MOS Devices*, Wiley, New York (© 1989 John Wiley & Sons, Inc.).)

For drain voltages greater than the pinch-off voltage, V_{Dsat}, the pinched-off portion of the channel widens from just a point into a depleted channel section, ΔL, as shown in Figure 1.14d, and the drain voltage in excess of V_{Dsat} is absorbed almost entirely across ΔL. If $\Delta L \ll L$, the effective conduction channel from the source to the pinched-off region is essentially unaffected by the excess drain voltage, and the channel current remains approximately constant. However, in the case where ΔL is comparable to L, the effective conduction channel decreases substantially with increasing V_{D}. This phenomenon is called *channel-length modulation* and is an important consideration in short-channel devices.

One should bear in mind that the description here presumes a carrier mobility independent of the channel electric field. In other words, the carrier drift velocity stays in the "linear" regime. However, even in long-channel MOSFETs, carrier velocity saturation can occur at high drain voltages. Thus, strictly speaking, there is no pinch-off and the current saturation is simply due to carrier velocity saturation. A detailed discussion of this point can be found in a review article by Ko [Einspruch and Gildenblat, 1989].

The I–V curve shown in Figure 1.15 is for one gate voltage only. If the gate voltage is changed in a stepwise fashion, the familiar family of curves will be obtained as shown in Figure 1.16.

1.3.2 BULK CHARGE MODEL

To obtain the I_{D}–V_{D} behavior, we assume an n-channel enhancement mode device with a long channel in which the diffusion current is negligi-

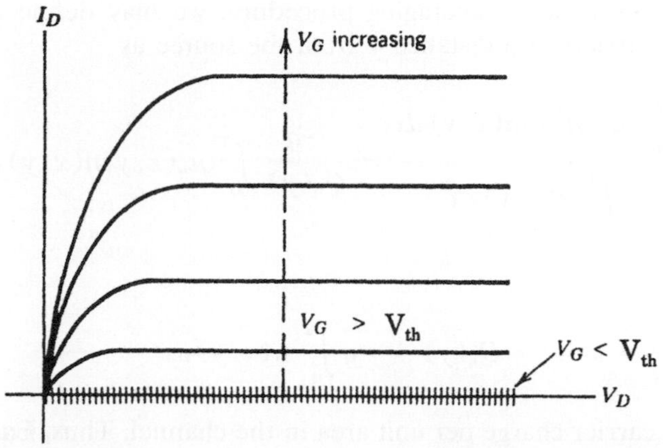

FIGURE 1.16 MOSFET channel current as a function of drain–source voltage for several different gate voltages. (Reprinted by permission from John Wiley & Sons, Inc., from T. P. Ma and P. V. Dressendorfer, *Ionizing Radiation Effects in MOS Devices*, Wiley, New York (© 1989 John Wiley & Sons, Inc.).)

ble compared with the drift current. When $V_G \geq V_{th}$, the channel is turned "on" and the current density in the channel is given by (the x, y and z coordinates are as defined in Figure 1.13)

$$J_n = -q\mu_n n\mathscr{E}_y = q\mu_n n \frac{dV}{dy}. \tag{1.54}$$

Integrating the current density over the cross-sectional area of the channel at an arbitrary point y gives

$$I_D = \int_0^{x_c}\int_0^W J_n \, dx\, dz = W\int_0^{x_c(y)} J_n \, dx = qW\frac{dV}{dy}\int_0^{x_c} \mu_n(x,y)n(x,y)\, dx, \tag{1.55}$$

where x_c is the channel depth.

Note that both the electron mobility, μ_n, and the electron concentration, n, are functions of x and y. The electron mobility in the channel is determined by various scattering mechanisms due to phonons, impurities, and surface roughness; the surface roughness scattering depends strongly on the electric field and the distance of the electron to the surface, causing μ_n to be a function of both x and y.

Using the standard averaging procedure, we may define an *average mobility* of carriers at a distance y from the source as

$$\overline{\mu}_n = \frac{\int_0^{x_c} \mu_n(n, y) n(x, y)\, dx}{\int_0^{x_c} n(x, y)\, dx} = \frac{-q}{Q_n(y)} \int_0^{x_c} \mu_n(x, y) n(x, y)\, dx, \quad (1.56)$$

where

$$Q_n(y) = -q \int_0^{x_c} n(x, y)\, dx \quad (1.57)$$

is the total carrier charge per unit area in the channel. Thus, Eq. (1.55) can be simplified as

$$I_D = -W \overline{\mu}_n Q_n \frac{dV}{dy}. \quad (1.58)$$

The current flowing in the channel, I_D, is independent of y due to the current continuity requirement. If we assume that $\overline{\mu}_n$ is constant throughout the channel, we may rearrange Eq. (1.58) and perform the integration to yield

$$\int_0^L I_D\, dy = I_D L = -W \overline{\mu}_n \int_0^{V_D} Q_n\, dV \quad (1.59)$$

or

$$I_D = -\frac{W}{L} \overline{\mu}_n \int_0^{V_D} Q_n\, dV. \quad (1.60)$$

1.3.2.1 The Square Law

To proceed further, we need to find the function $Q_n(V)$. For the special case $V_D = 0$, the carrier charge density per unit area in strong inversion can be obtained from analysis of a simple MOS capacitor:

$$Q_n = -C_{ox}(V_G - V_{th}). \quad (1.61)$$

For $V_D \neq 0$, the carrier density will be a function of y due to the potential distribution in the channel. At point y where the potential due to V_D is $V(y)$, the carrier charge density per unit area is approximately

$$Q_n(y) = -C_{ox}[V_G - V_{th} - V(y)]. \quad (1.62)$$

This is based on the key assumption that the depletion width does not vary along the channel. Substituting Eq. (1.62) into Eq. (1.60) and integrating, we obtain the I_D–V_D relationship

$$I_D = \frac{W}{L}\bar{\mu}_n C_{ox}\left[(V_G - V_{th})V_D - \frac{V_D^2}{2}\right] \quad \text{for } V_{Dsat} \geq V_D \geq 0. \tag{1.63}$$

The post-pinch-off portion of the behavior is approximately modeled by setting $I_D(V_D \geq V_{Dsat}) \equiv I_{Dsat}$, where V_{Dsat} corresponds to the drain voltage which causes I_D in Eq. (1.63) to be maximum. V_{Dsat} can be obtained by recognizing that at pinch-off, $Q_n(L) = 0$. Thus

$$V_{Dsat} \equiv V(L) = V_G - V_{th}. \tag{1.64}$$

From Eqs. (1.62) and (1.63)

$$I_{Dsat} = \frac{1}{2}\frac{W}{L}\bar{\mu}_n C_{ox}(V_G - V_{th})^2. \tag{1.65}$$

While Eqs. (1.63) and (1.64) describe the first-order behavior of the I–V characteristics of a long-channel enhancement mode MOSFET, some approximations used in the derivation may cause an overestimate of the current magnitude, especially for high-drain voltages. The major factor that contributes to the error is the assumption that the depletion depth for all channel points from the source to the drain remains fixed for $V_D \neq 0$. In reality, as shown in Figure 1.14b–d, the depletion depth increases from source to drain when $V_D > 0$. Therefore, Eq. (1.63) must be modified to take this into account. The more accurate I–V characteristics including this effect have been worked out and can be found in Sze [1981] and Muller and Kamins [1986].

1.3.2.2 Threshold Voltage

The threshold voltage, V_{th}, is one of the most important parameters from an operational point of view. As discussed previously, the threshold voltage is the gate voltage when $\psi_s = 2\phi_B$, or when strong inversion starts. For an ideal n-channel enhancement-mode MOSFET, Eq. (1.15) applies, and we have

$$V_{th} = 2\phi_B + \frac{2(qN_A\epsilon_s\phi_B)^{1/2}}{C_{ox}}. \tag{1.66}$$

Equation (1.66) suggests that V_{th} depends on the oxide thickness and the channel doping concentration. Since both terms in Equation (1.66) are positive quantities, V_{th} for an ideal n-channel device is positive, as expected intuitively.

The threshold voltage for a practical device must include the effects of work function difference, and on oxide and interface charges and may be expressed as

$$V_{th} = 2\phi_B + \frac{2(qN_A \epsilon_s \phi_B)^{1/2}}{C_{ox}} + \phi_{ms} - \frac{Q_0}{C_{ox}} - \frac{Q_{it}}{C_{ox}}. \tag{1.67}$$

In Eq. (1.67), the terms ϕ_{ms} and $-Q_0/C_{ox}$ are typically negative quantities, while the term $-Q_{it}/C_{ox}$ is typically positive for an n-channel device (due to charged acceptor-like interface traps above midgap, which make Q_{it} negative). Hence, depending on the balance of the positive and negative terms, it is possible to end up with a negative V_{th}; that is, the device is "on" even at $V_G = 0$, which is known as depletion mode.

In addition to the terms discussed above, the threshold voltage can be changed by biasing the substrate contact relative to the source. If a negative substrate bias, $V_{SUB} < 0$, is applied between the substrate terminal and the source, then the surface potential must reach $2\phi_B + |V_{SUB}|$, with an additional band bending of $q|V_{SUB}|$, before the surface inverts. Thus, the threshold voltage is modified from Eq. (1.67) to become

$$V_{th} = 2\phi_B + \frac{[2qN_A \epsilon_s(2\phi_B + |V_{SUB}|)]^{1/2}}{C_{ox}} + \phi_{mx} - \frac{Q_0}{C_{ox}} - \frac{Q_{it}}{C_{ox}}. \tag{1.68}$$

Note that V_{SUB} must be zero or negative to avoid forward biasing the source junction. The change of the threshold voltage resulting from the substrate bias is called the substrate bias effect, or the "body effect." Some circuits are designed to make use of this additional control to place the threshold voltage in a more desirable range. In some cases, however, the body effect may be introduced unintentionally due to the operation of adjacent devices, and it must be taken into account to ensure proper functioning of the circuit.

1.3.2.3 Transconductance

Another important parameter that characterizes a MOSFET is the transconductance, G_m, which is defined as

$$G_m \equiv \frac{\partial I_D}{\partial V_G}. \tag{1.69}$$

From Eqs. (1.63) and (1.69)

$$G_m = \bar{\mu}_n C_{ox} \frac{W}{L} V_D \quad \text{for } V_D < V_{Dsat}. \tag{1.70}$$

Assuming that the effective mobility of the carriers is independent of V_G, it is apparent from Eq. (1.70) that G_m increases linearly with V_D but is independent of V_G. For a given device geometry at a given V_D, G_m is proportional to $\bar{\mu}_n$. Since, among other things, $\bar{\mu}_n$ depends on carrier scattering due to the oxide and interface charges, one of the important consequences of hot carriers is the reduction of transconductance.

When $V_D \geq V_{Dsat}$, the transconductance can be obtained by combining Eqs. (1.65) and (1.69) with $V_D = V_{Dsat}$.

$$G_{msat} = \bar{\mu}_n C_{ox} \frac{W}{L} (V_G - V_{th}) \quad \text{for } V_D \geq V_{Dsat}. \tag{1.71}$$

Thus, G_{msat} is independent of V_D but linearly proportional to V_G.

In the above discussion, we assumed $\bar{\mu}_n$ to be independent of V_G. In reality, because the carrier scattering is a function of V_G, it may be necessary to modify Eqs. (1.70) and (1.71) to model more accurately the transconductance behavior. The physics involved in the various scattering processes for channel carriers has been extensively studied elsewhere [Sun and Plummer, 1980].

1.3.3 Threshold Voltage Determination

There are several ways to measure the threshold voltage of a MOS device. These techniques can be divided into two categories, one based on C–V and the other on I–V measurements.

From the high-frequency C–V behavior of the MOS diode (see Section 1.2), one can estimate the threshold voltage directly by locating the voltage at which depletion "curve" crosses C_{min}. A more accurate determination for low-interface-trapped-charge devices would be to compute the substrate doping from the measured C_{ox} and C_{min} [Eq. (1.27)], which then yields the ideal threshold voltage and flat-band capacitance [Eqs. (1.15) and (1.39), respectively]. The flat-band voltage can then be read off the C–V curve and the threshold voltage is found. This approach can be problematic when C_{it} is significant, giving rise to considerable stretch-out of the C–V curve in depletion. It is also unsuitable for devices with nonuniform substrate doping.

When the MOSFET is operating in the linear regime, i.e., $V_{\mathrm{Dsat}} > V_{\mathrm{D}} > 0$, for a fixed V_{D}, I_{D} varies linearly with V_{G} [Eq. (1.63)]. As we will see in the next section, the drain current does not vanish abruptly at threshold. The I_{D} vs. V_{G} (for a small fixed V_{D}) behavior in the linear regime can be extrapolated to zero I_{D} to yield V_{th}. Similarly, in the saturation regime, using the square law [Eq. (1.65), a $\sqrt{I_D}$ vs. V_{G} plot can also yield V_{th}.

It is quite apparent from the above discussion that for MOSFETs, I–V characteristics are generally used to determine threshold voltages. It is important to note that whichever technique one chooses to measure V_{th}, in order to examine threshold changes (as one does in studying hot-carrier degradation), the same technique should be employed throughout.

1.3.4 Subthreshold Conduction

In the discussion thus far it has been assumed that the channel does not contain any free carriers for $V_{\mathrm{G}} < V_{\mathrm{th}}$, and the current vanishes for all gate voltages below V_{th}. In reality, a small but finite drain current does exist at gate voltages below threshold. This current is called the *subthreshold current*, and it arises from the finite concentration of minority carriers in the channel that are present for $V_{\mathrm{G}} < V_{\mathrm{th}}$.

Referring to the band diagram for the ideal MOS capacitor shown in Figure 1.2, one realizes that, for the n-channel device in the regime $V_{\mathrm{th}} > V_{\mathrm{G}} > 0$, the majority carriers are depleted from the surface but the minority carriers are increasing with V_{G} so that the mass-action law $pn = n_{\mathrm{i}}^2$ is obeyed in quasi-equilibrium. When V_{G} is such that the Fermi level at the silicon surface still lies below the intrinsic level, $p_{\mathrm{s}} > n_{\mathrm{i}} > n_{\mathrm{s}}$, where n_{i} is negligibly small at room temperature. When V_{G} is increased such that the Fermi level at the surface lies above E_{i}, $n_{\mathrm{s}} > p_{\mathrm{s}}$, and we enter a regime called *weak inversion* before strong inversion is reached.

For an n-channel device in weak inversion, a positive voltage supplied at the drain creates an electron concentration gradient between source and drain, with the concentration at the source being higher. Because of the low carrier concentration in weak inversion the drain current is thus dominated by diffusion, and it can be shown [Sze, 1981] that the subthreshold I_{D}–V_{D} relationship for an n-channel device becomes

$$I_{\mathrm{D}} = \bar{\mu}_n \left(\frac{W}{L}\right) \frac{aC_{\mathrm{ox}}}{2} \left(\frac{n_{\mathrm{i}}}{\beta N_{\mathrm{A}}}\right)^2 \exp(\beta\psi_{\mathrm{s}})(\beta\psi_{\mathrm{s}})^{-1/2}[1 - \exp(-\beta V_{\mathrm{D}})], \tag{1.72}$$

where

$$a = \frac{\sqrt{2}(\epsilon_s/L_D)}{C_{ox}} \tag{1.73}$$

and

$$\psi_s = (V_G - V_{FB}) - \frac{a^2}{2}\left(\frac{1}{\beta}\right)\left\{\left[1 + \frac{4}{a^2}(\beta V_G - \beta V_{FB} - 1)\right]^{1/2} - 1\right\}. \tag{1.74}$$

Equations (1.72)–(1.74) indicate that in the subthreshold regime the drain current varies exponentially with V_G and becomes virtually independent of V_D for $V_D > 3$ kT/q. A typical set of subthreshold characteristics is shown in Figure 1.17 [Troutman, 1974]. As will be discussed in the next subsection, the subthreshold current could become a strong function of V_D for short-channel devices.

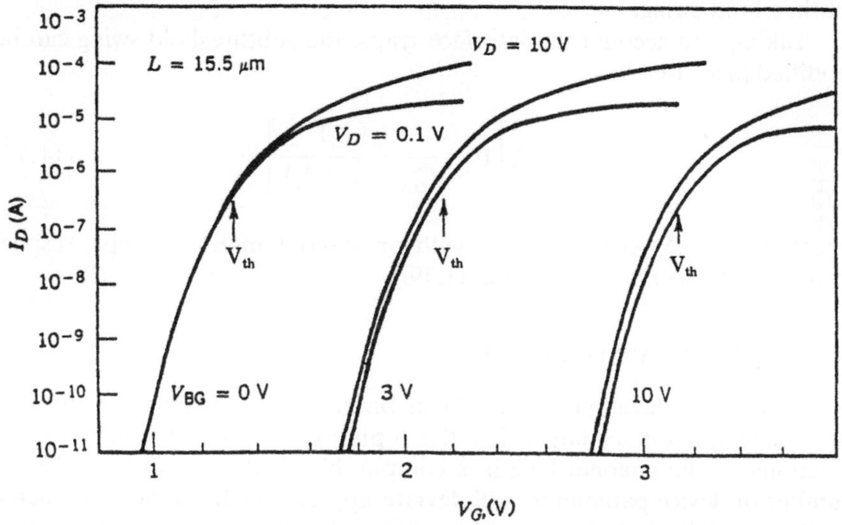

FIGURE 1.17 Experimental subthreshold characteristics at two drain voltages and three substrate biases. (Reprinted with permission from R. R. Troutman, "Subthreshold Design Considerations for IGFET," *J. Solid State Circ.* **9**, 55 (© 1974 IEEE).)

A useful parameter in the subthreshold regime is the gate voltage swing, S, defined by

$$S \equiv \log\left[\frac{dV_{\mathrm{G}}}{d(\ln I_{\mathrm{D}})}\right] = \frac{kT}{q}\log\left[1+\frac{C_{\mathrm{s}}(\psi_{\mathrm{s}})}{C_{\mathrm{ox}}}\right]\left\{1-\left(\frac{2}{a^2}\right)\left[\frac{C_{\mathrm{s}}(\psi_{\mathrm{s}})}{c_{\mathrm{ox}}}\right]^2\right\}^{-1}. \tag{1.75}$$

In the case $a \gg C_{\mathrm{s}}/C_{\mathrm{ox}}$, which is valid for weak inversion, Eq. (1.75) reduces to

$$S = \frac{kT}{q}\log\left(1+\frac{C_{\mathrm{s}}}{C_{\mathrm{ox}}}\right). \tag{1.76}$$

As might be expected, the parameters that cause the I–V characteristics above threshold to deviate from the ideal case also affect the subthreshold behavior. The effect of the work function difference and the oxide charge is to shift the I_{D}–V_{G} characteristic curve along the V_{G} axis by an amount equal to that in the post-threshold regime. The interface traps, however, will cause a change in the slope of the I_{D}–V_{G} curve and thus the subthreshold swing.

Taking into account the interface traps, the subthreshold swing can be modified [Sze, 1981] as

$$S_{\mathrm{it}} = S_0\left[1+\frac{C_{\mathrm{it}}(\psi_{\mathrm{s}})}{C_{\mathrm{ox}}+C_{\mathrm{s}}(\psi_{\mathrm{s}})}\right], \tag{1.77}$$

where S_{it} and S_0 correspond to with or without interface traps, respectively, and $C_{\mathrm{it}}(\psi_{\mathrm{s}})$ is given by Eq. (1.49).

1.3.5 Short-Channel Effects

So far, the development has been based on the assumption that the channel length is much larger than the depletion widths of the source–drain junctions. If the channel length is comparable to the depletion widths, a number of device parameters will deviate appreciably from those predicted by the long-channel theories. One major deviation is threshold voltage reduction. This is primarily due to the effect of the space charge from the source and drain regions extending into the channel region, causing a two-dimensional potential distribution, and charge sharing between the channel and source–drain regions. Hence, the amount of charge reflected to the gate for a given silicon surface potential is reduced, resulting in a decrease of the threshold voltage.

The subthreshold characteristics are also different for short-channel devices. In contrast to the long-channel device, it has been demonstrated both theoretically and empirically that the subthreshold current for a short-channel device is a strong function of V_D, due to the reduction of the effective channel length L_{eff} [Ng and Brews, 1990] as V_D increases. For a given V_D, the magnitude of the subthreshold current increases monotonically with decreasing channel length as in the long-channel case [see Eq. (1.72)].

In this subsection, several key short-channel effects are reviewed. Much of the material presented here is extracted from an article by Ko [Einspruch and Gildenblat, 1989]. The issues discussed center around carrier mobilities and electric field. The subject of hot carriers, which constitute a major short-channel effect, will be introduced separately in Section 1.4.

1.3.5.1 Mobility Degradation

It has been known for some time that the channel carrier mobility depends on the gate field. The scattering mechanisms responsible include those due to photons, ionized species, and surface roughness [Cheng, 1974]. For a good-quality Si–SiO_2 interface in strong inversion at room temperature, phonon scattering is generally believed to be the dominant mechanism. However, a satisfactory physics-based quantitative model is still lacking. A widely used empirical model for effective surface mobility is given by [Hoefflinger *et al.*, 1979]

$$\mu_{eff} = \mu_0/[1 + \theta(V_G - V_{th})], \tag{1.78}$$

which is based on the observation that the mobility often peaks around V_{th} and falls monotonically with increasing $V_G - V_{th}$. The parameter θ was found to be strongly process and substrate bias dependent. It is difficult to attach to it a coherent physical interpretation. In 1978, Sabnis and Clemens [1979] proposed a unified formulation based on the concept of an effective field $\mathscr{E}_{eff}$ given by

$$\mathscr{E}_{eff} = \frac{Q_B + (Q_n/2)}{\epsilon_s}. \tag{1.79}$$

The existence of a universal correlation between $\mathscr{E}_{eff}$ and the channel mobility of either carrier was subsequently verified by Sun and Plummer [1980]. Later, Liang *et al.* [1986] found that the correlation is still valid for gate oxide as thin as 4 nm, provided that the channel inversion charge is properly extracted and the Si–SiO_2 interface is of good quality, although there were earlier claims that surface roughness causes the mobility to become oxide thickness dependent for oxides thinner than 10 nm [Han *et al.*, 1985; Su *et al.*, 1985]. Eq. (1.79) can be readily derived if $\mathscr{E}_{eff}$ can be interpreted as the average transverse electric field, $\mathscr{E}_{av}$, experienced by carriers in the inversion layer:

$$\mathscr{E}_{eff} = \mathscr{E}_{av} \equiv \frac{\int_0^\infty n(x)\mathscr{E}_x(x)\,dx}{\int_0^\infty n(x)\,dx}. \tag{1.80}$$

Here $Q_n = -q\int_0^\infty n(x)\,dx$ is assumed.

It should be mentioned that Eq. (1.79) holds only if the mobility is phonon scattering limited. It fails at low temperature (e.g., 77 K) when Coulomb scattering becomes important [Hung, 1987]. Behaviors for the effective surface mobilities for electron and hole at room temperature as a function of $\mathscr{E}_{eff}$ are shown in Figure 1.18. The curves can be fitted to a simple empirical relation [Liang *et al.*, 1986]

$$\mu_{eff} = \frac{\mu_s}{1 + (\mathscr{E}_{eff}/\mathscr{E}_s)^{\nu}}. \tag{1.81}$$

Values for parameters μ_s, $\mathscr{E}_s$, and ν for surface electron and hole are shown in the first two columns of Table 1.1.

Equation (1.79) can be rewritten as a form that explicitly relates $\mathscr{E}_{eff}$ to the device parameters:

$$\mathscr{E}_{eff} = \frac{V_G - V_{th}}{6t_{ox}} + \frac{V_{th} + V_a}{3t_{ox}}. \tag{1.82}$$

Note that $Q_n = -C_{ox}(V_G - V_{th})$ is assumed. Since this assumption overestimates Q_n at low $V_G - V_{th}$ [Sodini *et al.*, 1984], $\mathscr{E}_{eff}$ according to Eq. (1.82) would be higher than its actual value. But, the error is small for $V_G - V_{th} > 1$ V. For an n^+ polysilicon gate CMOS technology with $V_{thn} = 0.7$ V and $V_{thp} = -0.7$ V, $V_a \simeq 0.5$ V was found to be applicable semiempirically [Hung, 1987]. The carrier effective mobilities for both NMOS and PMOS devices according to Eqs. (1.81), (1.82), and Table 1.1 are plotted against $V_G - V_{th}$ in Figure 1.19 for three different t_{ox} values.

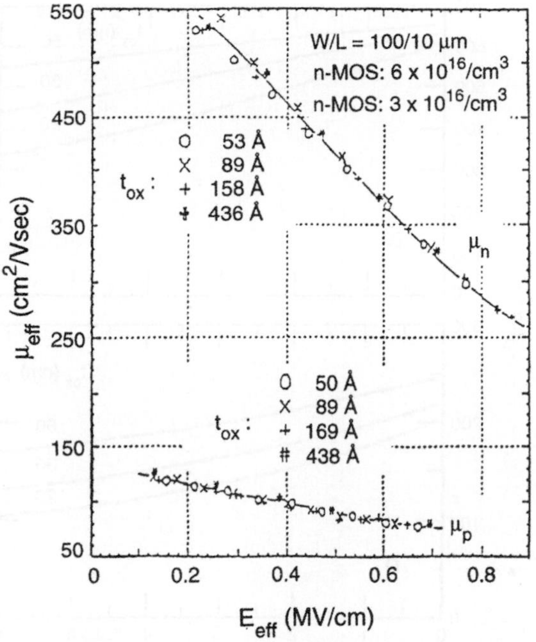

FIGURE 1.18 Measured universal μ_{eff} versus E_{eff} curves for electrons and holes in the inversion layer. (Reprinted with permission from Ko [1989].)

These curves are useful for rough estimation of device performance or as first-order calibrators in process monitor and control.

1.3.5.2 Carrier Velocity Saturation

Although it is widely accepted that the surface drift velocity of either electron or hole will saturate at high field, accurate experimental verification is extremely difficult and the results are often controversial. Based on the limited amount of data available from three more successful attempts

TABLE 1.1 Parameters for the Effective Mobility Models for Electrons and Holes

	Electron (surface)	Hole (surface)	Hole (buried-channel)
μ_s (cm^2/V)	670	160	290
$\mathscr{E}_s$ (MV/cm)	0.67	0.7	0.35
ν	1.6	1	1

Note. Reprinted with permission from Ko [1989].

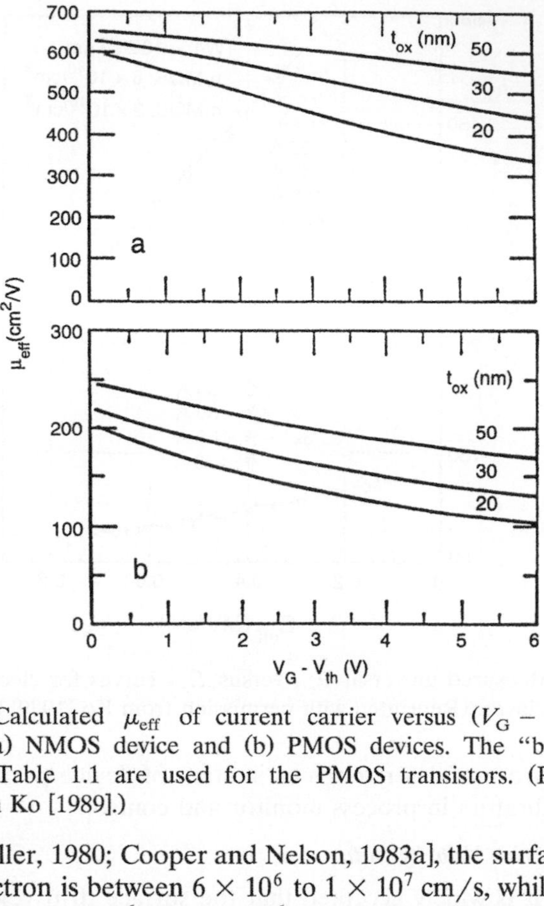

FIGURE 1.19 Calculated μ_{eff} of current carrier versus $(V_G - V_{\text{th}})$ for three different t_{ox}: (a) NMOS device and (b) PMOS devices. The "buried-channel" parameters in Table 1.1 are used for the PMOS transistors. (Reprinted with permission from Ko [1989].)

[Coen and Muller, 1980; Cooper and Nelson, 1983a], the surface saturation velocity of electron is between 6×10^6 to 1×10^7 cm/s, while that for the hole is between 4×10^6 to 8×10^6 cm/s. The dependences on t_{ox}, doping, and other processing conditions have not been established, although both velocities appear to be relatively independent of the gate field. Many empirical velocity-field models [Hofstein and Warfield, 1965b] have been published. Two most widely used models are as follows.

Model A:

$$v = \frac{\mu_{\text{eff}} \mathscr{E}}{\left[1 + (\mathscr{E}/\mathscr{E}_{\text{sat}})^{n_v}\right]^{1/n_v}}, \qquad n_v = 2 \quad \text{for electrons}, \qquad < 2 \quad \text{for holes}. \tag{1.83}$$

Model B:

$$v = \frac{\mu_{\text{eff}} \mathscr{E}}{1 + (\mathscr{E}/\mathscr{E}_{\text{sat}})}. \tag{1.84}$$

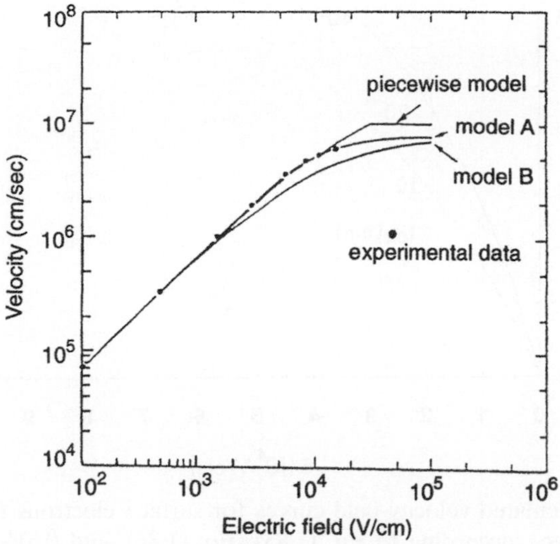

FIGURE 1.20 Comparison of velocity-field models: models A and B as given in Eqs. (1.83) and (1.84), respectively. The piecewise model is given in Eq. (1.85). $\mu_0 = 710\ \text{cm}^2/\text{V}$, $v_{sat} = 1 \times 10^7$. (Reprinted with permission from Ko [1989].)

Velocity saturation to model A occurs much more abruptly and at a lower $\mathscr{E}$. In Figure 1.20, we compare the two models for electrons. A piecewise model, which will be discussed below, is also included. Model A fits the experimental data quite well, while model B significantly underestimates the velocity at moderate field if the same saturation velocity is assumed. Unfortunately, using model A in the current transport equation, Eq. (1.58), yields a prohibitively complicated solution for I_D. We will adopt a two-region piecewise model that is a modified form of model B to circumvent the problem. The piecewise model is defined by

$$v = \frac{\mu_{eff}\mathscr{E}}{1 + (\mathscr{E}/\mathscr{E}_{sat})} \qquad \mathscr{E} < \mathscr{E}_{sat},$$
$$= v_{sat} \qquad \mathscr{E} \geq \mathscr{E}_{sat}. \tag{1.85}$$

In this model, the carrier velocity will saturate at $\mathscr{E}_{sat}$. $\mathscr{E}_{sat}$ can be expressed in terms of v_{sat} and μ_{eff} using Eq. (1.85) for $\mathscr{E} = \mathscr{E}_{sat}$.

$$\mathscr{E}_{sat} = 2v_{sat}/\mu_{eff}. \tag{1.86}$$

Comparison of the three models in Figure 1.20 [Sodini *et al.*, 1984] shows that this piecewise model more accurately models the experimental data over a wide range of electric field while slightly overestimating the carrier

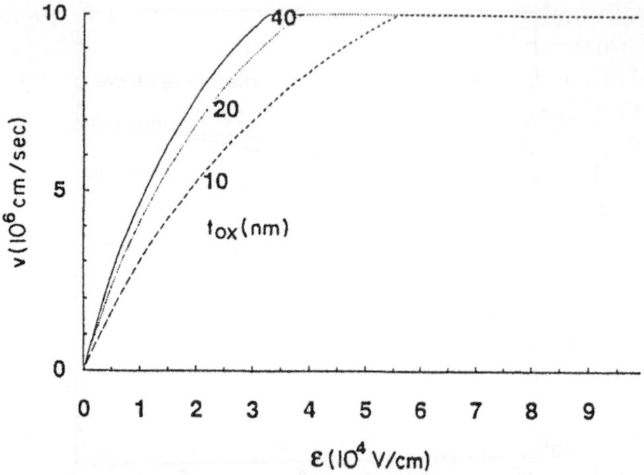

FIGURE 1.21 Calculated velocity-field curves for surface electrons for three gate-oxide thicknesses t_{ox} according to Eq. (1.85), Eq. (1.86), and $0.5V_{th} = 0.7$ V and $V_G - V_{th} = 2$ V. (Reprinted with permission from Ko [1989].)

velocity near $\mathcal{E}_{sat}$. It is important to realize that $\mathcal{E}_{sat}$ has been adjusted to a higher value than v_{sat}/μ_{eff} in order to achieve a better fit to the experimental data in the moderate field region. One should not interpret $\mathcal{E}_{sat}\,\mu_{eff}/2$ as the physical saturation velocity. Electron velocities calculated using Eqs. (1.85) and (1.86) are plotted against $\mathcal{E}$ for devices with three different gate oxide thicknesses in Figure 1.21. The devices are assumed to have a V_{th} of 0.7 V and the gate bias $V_G - V_{th}$ is 2 V. The v_{sat} used is 1×10^7 cm/s. Because v_{sat} remains the same, velocity saturation is more pronounced when the low-field mobility is high.

1.4 Survey of Device and Circuit Reliability Issues Related to Hot-Carrier Effects

Hot-carrier effects are generally manifested in threshold shifts and transconductance degradation. The physical origin consists of carrier trapping and interface state generation as a result of hot-carrier injection. Hot carriers can be "lucky carriers" and/or impact-ionized carriers [Ning, 1977; Takeda *et al.*, 1983e], which gain sufficient energy to surmount the Si–SiO_2 barrier without suffering an energy-losing collision in their paths. MOS device structures for many VLSI generations (from 2 to 0.3 μm) have been designed with such effects in mind. To reduce hot-carrier degradation in scaled MOS devices, two approaches have been taken so far: (1) hot-carrier-resistant device structures such as DDD (double diffused drain) [Takeda, *et al.*, 1982], LDD (lightly doped drain) [Ogura *et al.*, 1980], and GOLD (gate–drain overlapped device) [Izawa *et al.*, 1988]; (2) reduction of power supply voltage from 5 to 2.5 V [Takeda *et al.*, 1984]. The former 5-V power supply standard resulted in chips with high power consumption and process complexity. The low-voltage approach improves reliability, but must overcome low speed and low current density. A third approach recently being discussed is to make good use of AC effects accounting for the duty ratios of supply voltages. This calls for a deeper physical understanding of AC hot-carrier phenomena, followed by use of this understanding in the actual device/circuit design and supply voltage selection.

Hot-carrier injection and degradation mechanisms have been extensively investigated using DC voltage and current stress measurements. A model predicting DC hot-carrier lifetime, τ_{DC}, in which τ_{DC} is proportional to $\exp(\alpha/V_D)$ (α being constant) and $\tau_{DC} I_D$ is proportional to $(I_{SUB}/I_D)^{-3}$, was developed by Takeda and Suzuki [1983] and extended in more detail by Hu *et al.* [1985]. Many mechanisms of DC device degradation have been proposed but not yet completely clarified. This is mainly because it is still unclear which physical characteristics of the Si–SiO_2 interface are changed by hot-carrier injection. In addition, such phenomena are strongly dependent on gate oxide characteristics and fabrication processes. Among the proposed explanations, the hydrogen model developed by Fair and Sun [1981], which suggested that H_2 to H_2O plays a significant role in the generation of Si dangling bonds (traps and interface states), is still accepted as a possible scenario for hot-carrier degradation.

An AC hot-carrier degradation model which is based on DC models and adaptable to the actual circuits has been in great demand and investigated extensively. Bellens *et al.* [1990] reported on the significant influence (gate pulse–induced noise) of the wiring inductance in the measurement systems on AC hot-carrier effects. Other experiments [Takeda *et al.*, 1991] having complete precautions against noise revealed that AC hot-carrier degradation lifetime, τ_{AC}, can be directly estimated from DC lifetime using as $\tau_{AC} = \tau_{DC} R$, where R is the supply voltage duty ratio under normal circuit operation. However, a new hot-carrier phenomenon specific to AC operation was also reported by Tsuchiya [1989] and verified by Matsuzaki *et al.* [1991] and Yoshida *et al.* [1994] using ring oscillators and an electron beam tester for inspection of the circuit waveform. The enhanced degradation (V_{th} shift) is caused by electrons trapped in neutral trap centers produced by hot-hole injection and occurs under conditions greater than a certain drain voltage. In this case, the above lifetime relationship is not satisfied. Thus, at high drain voltage operation, this enhanced AC degradation must be taken into account.

In terms of hot carrier–resistant device structures, "drain engineering" methods, such as DDD and LDD proposed by Ogura *et al.* [1980], are being used to reduce the electric field near the drain. Moreover, "gate engineering" methods such as inverse-T [Huang *et al.*, 1986c] and GOLD [Izawa *et al.*, 1988] are becoming important in 0.8–0.3 μm regimes, because these device structures can provide not only high resistance against hot-carrier effects but also high transconductance. This is essentially due to a reduction of source resistance by the gate–source overlap. The gate engineering structure is being fabricated using a simple tilted implantation technique [Hori *et al.*, 1988].

Some hot-carrier degradation models are incorporated in an integrated circuit reliability simulator BERT (Berkeley Reliability Tool). Besides these hot-carrier models, BERT also incorporates an oxide time-dependent breakdown model and an electromigration model [Hu, 1993].

1.5 Summary

This introductory chapter begins with a brief historical account, followed by a comprehensive overview of the MOS diode and transistor. The levels of treatments for those two devices are such that the first-time learner, with some background in semiconductor physics, can grasp the essential concepts necessary for comprehending the materials to be covered in subsequent chapters. The chapter concludes with a qualitative survey of some important technological issues related to hot-carrier degradation. This survey serves as the bridge linking device physics with problems in technology.

CHAPTER

2

Hot-Carrier Injection Mechanisms

2.1 Introduction

In recent years, since advanced fine-line patterning technologies have made it possible for device miniaturization to approach its physical limits, interest in hot-electron injection and associated device degradation has increased. This is because hot-electron effects impose more severe constraints on very large-scale integration (VLSI) device design as device dimensions are reduced. Main interest has focused on channel hot-electron (CHE) injection. Substrate hot-electron (SHE) injection has also been used to investigate gate insulator qualities. Both of these injection mechanisms are schematically illustrated in Figure 2.1. Ning *et al.* [1977b] have demonstrated CHE effects extensively, since it is relatively easy to measure the injection gate current (I_G) with a 2- or 3-μm L_{eff} metal-oxide-semiconductor field-effect transistor (MOSFET).

However, when accounting for the causes of VLSI circuit degradation, it seems to be necessary for not only CHE but also other injection mechanisms to exist. For example, Fair and Sun [1981] have proposed a hot hole–induced degradation mechanism. In this chapter, not only the above two injections but also two newer hot-carrier injection mechanisms are discussed, both experimentally and theoretically. They are drain avalanche hot-carrier (DAHC) injection and secondarily generated hot-electron (SGHE) injection. Device degradation due to these injections is

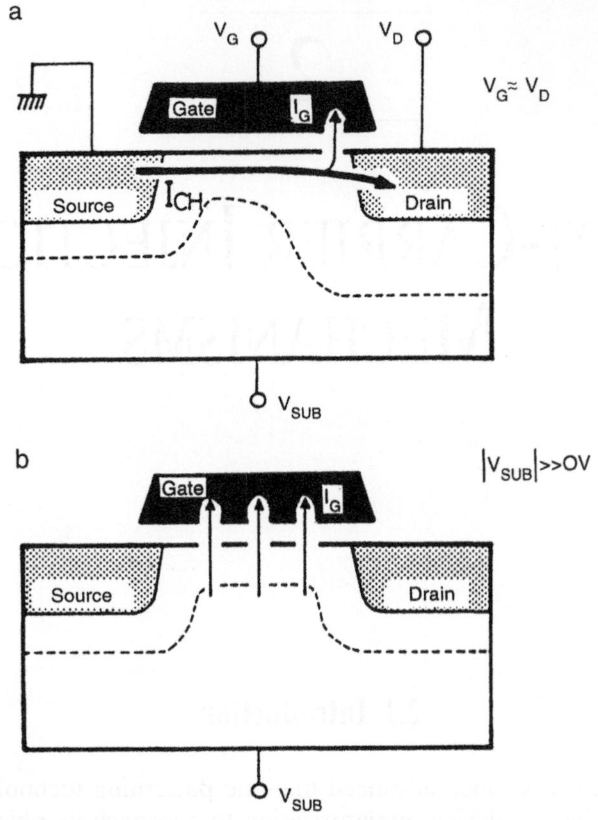

FIGURE 2.1 Schematic diagram of channel hot-electron (CHE) injection and substrate hot-electron (SHE) injection.

also clarified by comparing each with that due to CHE injection. Before moving on to hot-carrier injection mechanisms, the physics of hot carriers will be discussed briefly in the next section.

2.2 Avalanche Breakdown

Hot-carrier effects in Si VLSI circuits represent phenomena that are brought about by high-energy carriers created by the channel electric field. These fundamental physical mechanisms share some common characteristics with time-dependent dielectric breakdown (TDDB), radiation damage (RD) due to cosmic and α-rays, and electrostatic discharge (ESD), as

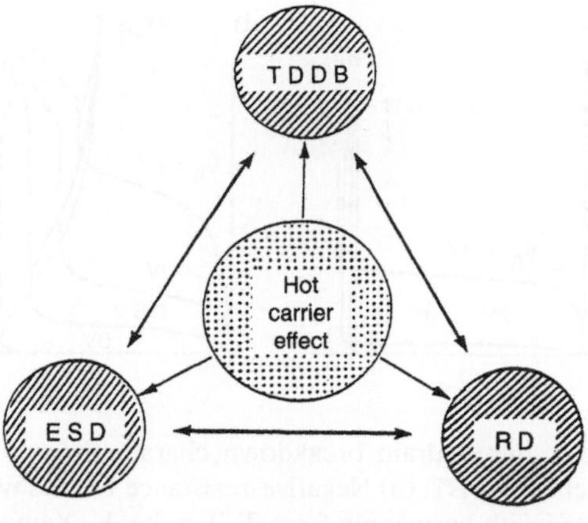

FIGURE 2.2 Problems related to hot-carrier effects.

shown schematically in Figure 2.2. The commonality lies in the phenomenon of avalanche multiplication. As a background on hot-carrier injection, the avalanche breakdown phenomenon is discussed below.

2.2.1 AVALANCHE MULTIPLICATION

The breakdown phenomenon determines the highest applicable voltage and limits the speed and power-handling capacity of discrete MOSFET devices or MOS integrated circuits. Therefore, breakdown voltage is as important as transconductance or threshold voltage in MOS device design. However, there have been few theories which account for the source–drain breakdown voltages of MOSFETs [Grove, 1967] due to the inability to estimate accurately the electric field distribution.

The avalanche breakdown characteristics in MOSFETs are divided into two types. One is normal breakdown observed in *p*-MOSFETs or long-channel *n*-MOSFETs. The breakdown voltage, BV_{DS}, decreases with an increase in gate voltage in normal breakdown. The other is negative-resistance breakdown, which is often observed in short-channel *n*-MOSFETs. The breakdown voltage of *p*-MOSFETs or long-channel *n*-MOSFETs increases with increase in gate voltage magnitude, as depicted in Figure 2.3a. However, BV_{DS} decreases with increase in V_G, and the sustain voltage, BV'_{DS}, apparently exists in the negative-resistance breakdown observed in short-channel *n*-MOSFETs as depicted in Figure 2.3b.

These avalanche breakdown characteristics are caused by the impact ionization of carriers in the high-field region, which develops around the reverse-biased drain-substrate junction. The avalanche breakdown in a

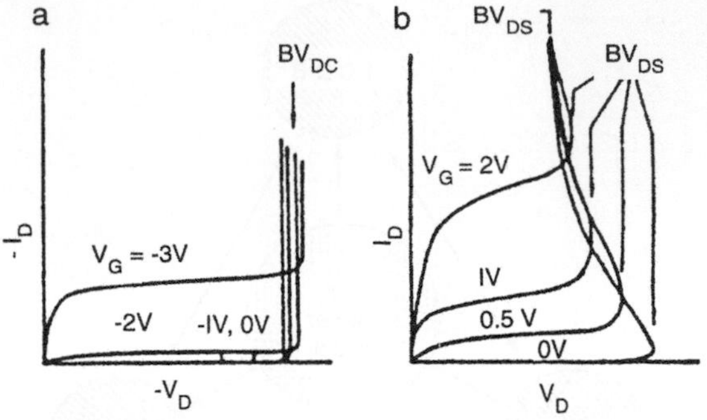

FIGURE 2.3 Typical source–drain breakdown characteristics. (a) Normal breakdown in a p-channel MOSFET. (b) Negative-resistance breakdown in an n-channel MOSFET. (Reprinted with permission from T. Toyabe, K. Yamaguchi, S. Asai, and M. S. Mock, "A Numerical Model of Avalanche Breakdown in MOSFETs," *IEEE Trans. Electron Dev.* **25**, 825–832 (© 1978 IEEE).)

planar p–n junction can be analyzed relative easily, since the electric field is known from an analytical solution of Poisson's equation [Sze, 1969]. The electric field distribution in the drain region of a MOSFET is more complicated because the field is intensified by the presence of the insulated gate.

The two-dimensional analysis reported by Toyabe *et al.* [1978] shows that carrier leaves the surface at the so-called pinch-off point and flows deep within the substrate before reaching the source, as depicted in Figure 2.4. The electric field is highest just beneath the surface. However, the seed current of avalanche multiplication in this region is the small reverse-biased saturation current, I_{D0}. In addition, the primary channel current, which is the source current, I_S, is multiplied by a relatively high electric field. Thus the drain current, I_D, can be written by

$$I_D = M^* I_S + M I_{D0}. \tag{2.1}$$

The multiplication factors, M and M^*, are given by

$$M = 1/(1 - I_{ion}) \tag{2.2}$$

and

$$M^* = 1/(1 - I_{ion}^*), \tag{2.3}$$

where I_{ion} and I_{ion}^* are ionization integrals. I_{ion} is obtained as

$$I_{ion} = \int \alpha_n \exp\left(-\int^x (\alpha_n - \alpha_p)\, dx' \right) dx, \tag{2.4}$$

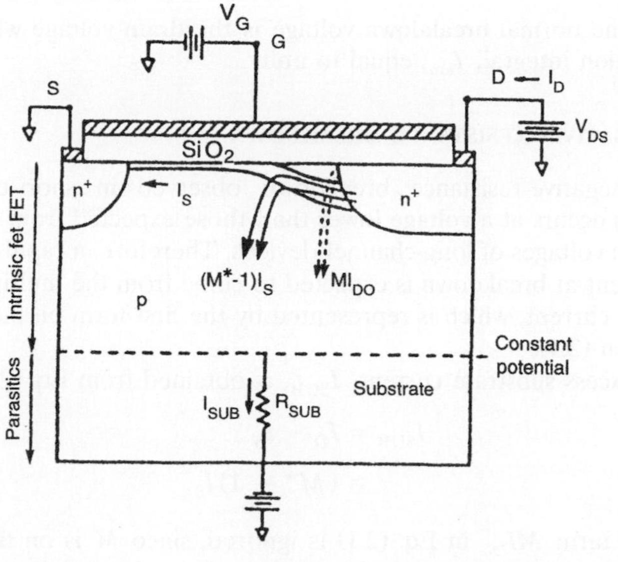

FIGURE 2.4 Cross section of an *n*-MOSFET showing the current components resulting from avalanche multiplication near the drain. (Reprinted with permission from T. Toyabe, K. Yamaguchi, S. Asai, and M. S. Mock, "A Numerical Model of Avalanche Breakdown in MOSFETs," *IEEE Trans. Electron Dev.* **25**, 825–832 (© 1978 IEEE).)

where α_n and α_p are the ionization rates of electrons and holes, respectively. The integration path is along the channel (in the x-direction). Similarly, I^*_{ion} is expressed as

$$I^*_{\text{ion}} = \int \alpha_n \exp\left(- \int^{\zeta} (\alpha_n - \alpha_p)\, d\zeta' \right) d\zeta, \tag{2.5}$$

where the integration is carried out along the channel current path represented by a curvilinear coordinate ζ. The channel current path is determined by locating the position of the maximum current density on each vertical grid line and connecting these positions. α_n and α_p are calculated as functions of the electric field using the expression given by Niehaus *et al.* [1973].

2.2.2 NORMAL BREAKDOWN

The normal breakdown as depicted in Figure 2.3a shows a current–voltage behavior similar to breakdown characteristics of a reverse-biased p–n junction. The second term on the right-hand side of Eq. (2.1) is responsible for this breakdown. The device is then expected to have normal breakdown when the multiplication factor M becomes infinity. Thus the normal breakdown voltage is the drain voltage which makes the ionization integral, I_{ion}, equal to unity.

2.2.3 NEGATIVE-RESISTANCE BREAKDOWN

The negative-resistance breakdown observed in short-channel *n*-MOSFETs occurs at a voltage lower than those expected from the normal breakdown voltages of long-channel devices. Therefore, a rapid increase in drain current at breakdown is expected to come from the multiplication of the source current, which is represented by the first term on the right side of Equation (2.1).

The excess substrate current, I_{SUB}, is obtained from Eq. (2.1) as

$$\begin{aligned} I_{SUB} &= I_D - I_S \\ &\cong (M^* - 1)I_S, \end{aligned} \tag{2.6}$$

where the term MI_{D0} in Eq. (2.1) is ignored, since M is on the order of unity and I_{D0} is much smaller than I_S. A voltage drop of $R_{sub}I_{SUB}$ across the substrate resistance acts as a back bias in the forward direction. This positive back-biasing increases the source current, I_S, and, subsequently, I_{SUB}. This increase continues until the substrate–source junction is turned on. Then, a smaller voltage is sufficient to sustain a higher current, resulting in negative resistance.

The potential at the lower boundary of the intrinsic FET is assumed to be constant as shown in Figure 2.4. Thus the back bias (substrate-to-source voltage), V_{SUB}, is given as

$$V_{SUB} = -I_{SUB}R_{sub} + V_{SUB0}, \tag{2.7}$$

where V_{SUB0} is the external back-bias voltage. From Eqs. (2.6) and (2.7), the relation

$$(M^* - 1)I_S = (V_{SUB0} - V_{SUB})/R_{sub} \tag{2.8}$$

is obtained. The negative-resistance breakdown ("snap-back") in MOSFETs can be analyzed based on Eqs. (2.7) and (2.8).

Besides hot-carrier generation due to impact ionization shown above, there is another hot-electron generation due to the escape of "lucky" electrons from the channel, which obtain sufficient energy to surmount the Si–SiO_2 barrier without suffering an energy-losing collision. These electrons are called "channel hot electrons," forming the main part of gate current, as in the write mode in FAMOS [Frohman-Bentchkowsky, 1974]. To formulate the gate current, there are two approaches, as will be discussed later in this chapter: effective hot-electron temperature and lucky electron, where electron temperature (T_e) and hot-electron mean free path (λ) are key parameters, respectively. Both types of hot-carrier generation mechanisms are important in our discussions.

2.3 Hot-Carrier Injection Mechanisms and Gate Currents

Six types of injection modes exist:

1. Channel hot-electron (CHE) injection [Ning, 1977],
2. Drain avalanche hot-carrier (DAHC) injection [Takeda *et al.*, 1982c; Nakagome *et al.*, 1983],
3. Secondary generated hot-electron (SGHE) injection [Nakagome *et al.*, 1983; Toriumi *et al.*, 1985b; Tam *et al.*, 1982a],
4. Substrate hot-electron (SHE) injection [Ning *et al.*, 1977b],
5. Fowler–Nordheim (F–N) tunneling injection, and
6. Direct tunneling (DT) injection [Eitan and Frohman-Bentchkowsky 1981].

CHE injection is due to the escape of "lucky" electrons from the channel, causing a significant degradation of the oxide and the Si–SiO_2 interface, especially at low temperature (77 K). This degradation process involves a sizable gate current. On the other hand, DAHC injection results in both electron and hole gate currents due to impact ionization, giving rise to the most severe degradation around room temperature. This is because hot holes, as well as hot electrons, take part in this process. SGHE injection is due to minority carriers from secondary impact ionization, or more likely bremsstrahlung radiation, and becomes a problem in ultra-small MOS devices. F–N tunneling injection is used in the "write" mode in EEPROM (electrical erasable programmable read-only memory) and is a significant cause of time-dependent dielectric breakdown (TDDB) in thin insulators. Finally, DT injection is used in MNOS-type EEPROMs. Transconductance degradation for $V_D < 3$ V might be partially due to this mode. Thus, for deep submicrometer devices, it is important to attempt to account for the effects resulting from combinations of some or all of these six injection processes. This can be done only if the operating conditions in each VLSI circuit are known and well understood.

2.3.1 CHANNEL HOT-ELECTRON (CHE) INJECTION

The gate current is usually several orders of magnitude smaller than the substrate current due to avalanche multiplication. Gate current with its peak at $V_G \cong V_D$ generally results from channel hot-electron injection. Ning [1977] reported that, if an *n*-channel MOSFET is operating at $V_G = V_D$, the conditions would be optimum for CHE injection of "lucky electrons" [Cottrell *et al.*, 1979]. Such electrons gain sufficient energy to surmount the Si–SiO_2 barrier without suffering an energy-losing collision in the channel. Typical gate currents due to CHE injection are shown as a function of V_G with V_D being a parameter (Figure 2.5). The device used had a 20-nm gate-oxide thickness (t_{ox}) and a 0.8-μm effective channel length (L_{eff}). In many cases, this gate current is responsible for device degradation as a result of carrier trapping. No gate current can be measured for $V_G < V_D$, since CHE injection is retarded. However, if V_D is large, reduction of V_G intensifies the electric field at the drain to the point where avalanche multiplication due to impact ionization may substantially increase the supply of both hot electrons and hot holes. Thus, it is conceivable that avalanche hot electrons and hot holes are injected into the gate in the same way as CHE. This DAHC injection mechanism is schematically illustrated in Figure 2.6 and described below.

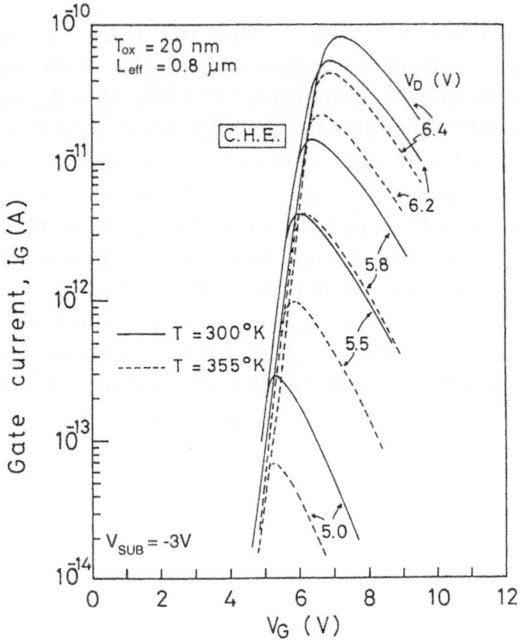

FIGURE 2.5 Typical gate currents due to CHE injection. t_{ox} = 20 nm; L_{eff} = 0.8 μm.

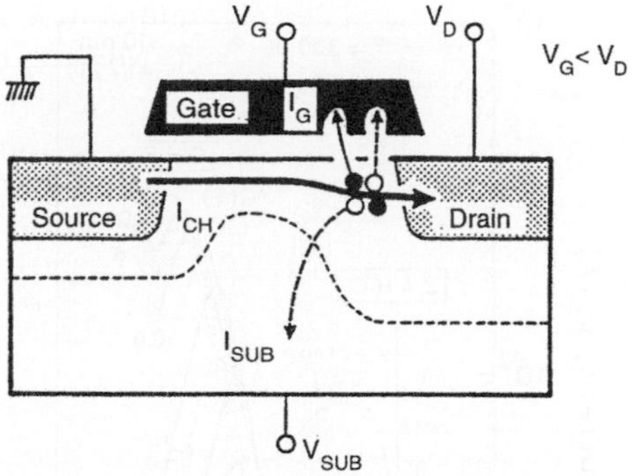

FIGURE 2.6 Drain avalanche hot-carrier injection mechanism.

2.3.2 DRAIN AVALANCHE HOT-CARRIER (DAHC) INJECTION

Gate currents directly measured down to 10^{-14} A with various V_G and V_D values are shown in Figure 2.7 for a device with $t_{ox} = 10$ nm and $L_{eff} = 0.8$ μm. There are two additional peaks in the $V_G < V_D$ region in addition to the peak due to CHE injection. These peaks correspond to hole and electron injections resulting from avalanche multiplication at the drain. This implies that it is necessary to take DAHC, as well as CHE, into consideration when one examines the device for possible hot-carrier effects. Analyzing drain avalanche hot-carrier behavior is difficult because both hot holes and hot electrons are injected simultaneously into the oxide and across the drain junction just below the substrate surface. In the following section, DAHC injection will be shown to have significant influence on device degradation even for devices having $t_{ox} \geq 20$ nm.

2.3.3 SECONDARILY GENERATED HOT-ELECTRON (SGHE) INJECTION

For a MOS device with $L_{eff} = 1.0$ μm and $t_{ox} = 6.8$ nm, gate currents are plotted as a function of V_G with V_D as a parameter (Figure 2.8). It should be noted that gate currents different from those due to DAHC injection become prominent in the $V_G \leq 5$ V region. They are also strongly dependent on the back-bias voltage, V_{SUB}. Furthermore, these gate currents have a strong correlation with the substrate current, I_{SUB}. This empirical result is sufficient evidence to determine that the mechanism is SGHE injection [Takeda, 1984]. For a device with $t_{ox} = 7$ nm, especially with a large back bias such as -5 V, the gate current behavior shows a strong correlation with I_{SUB} as demonstrated in Figure 2.9. Thus this

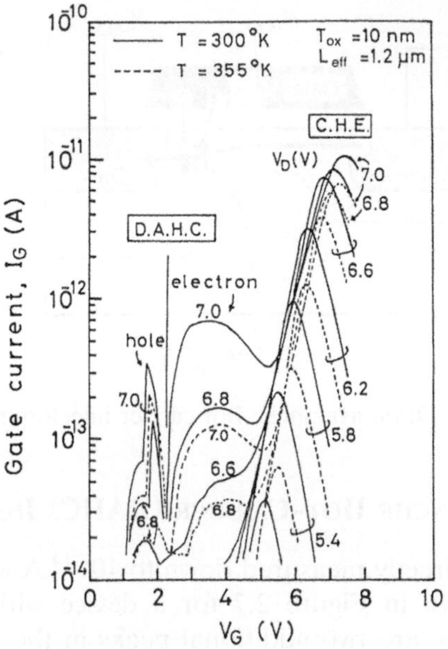

FIGURE 2.7 Gate current directly measured down to a 10^{-14} A under varied bias conditions for V_G and V_D for a device with L_{eff} = 1.2 μm and t_{ox} = 10 nm. (——) T = 300 K; (- - -) T = 355 K.

injection can be considered to be due to secondarily generated hot electrons.

Secondary impact ionization by hot holes (as shown in Figure 2.10) and photoinduced generation processes [Tam *et al.*, 1983] have been reported as "secondary" minority carrier generation mechanisms. Figure 2.11 compares the temperature dependence of I_{SUB} and that of electron diffusion current, I_d, for a device with t_{ox} = 7 nm and l_{eff} = 2.0 μm, measured using the setup shown in the inset [Takeda, 1984]. It should be noted that although these two currents show excellent correlation, as expected, I_d exhibits much less temperature dependence than I_{SUB}. If this minority carrier generation is due to secondary impact ionization, I_d should have a stronger temperature dependence, as demonstrated below. Assuming that secondary impact ionization exists, resulting in a mild avalanche condition, the expression for I_d can be shown to be [Takeda, 1984]

$$I_d \propto I_{ion}^{s*} I_{SUB}, \tag{2.9}$$

where I_{ion}^{s*} is the ionization integral and is given by $I_{ion}^{s*} = \int \alpha_s^* \, dx$, where α_s^* is the effective ionization rate.

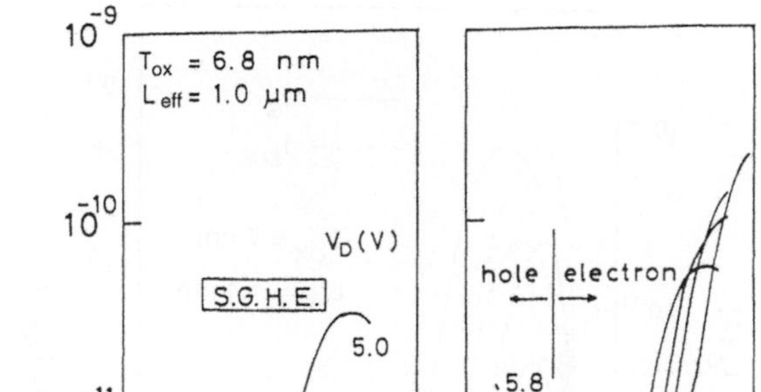

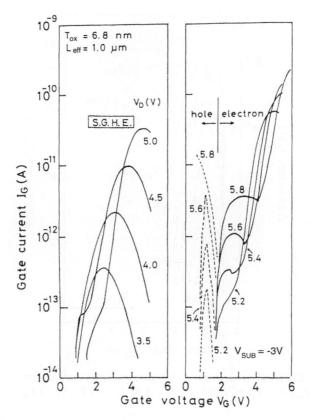

FIGURE 2.8 Gate currents as a function of V_G with V_D as a parameter for a device with $L_{eff} = 1.2$ μm and $t_{ox} = 6.8$ nm.

From Eq. (2.9), the temperature dependence of I_d and I_{SUB} can be compared as follows.

$$\left|\frac{\Delta \log I_d}{\Delta T}\right| = \left|\frac{\Delta \log I_{ion}^{s*}}{\Delta T}\right| + \left|\frac{\Delta \log I_{SUB}}{\Delta T}\right| > \left|\frac{\Delta \log I_{SUB}}{\Delta T}\right|. \tag{2.10}$$

This fact implies that a photoinduced generation process, believed to be bremsstrahlung radiation [Shewchun and Wei, 1965], rather than secondary impact ionization, is more likely to be the origin of SGHE. Furthermore, the existence of a considerably greater electron diffusion

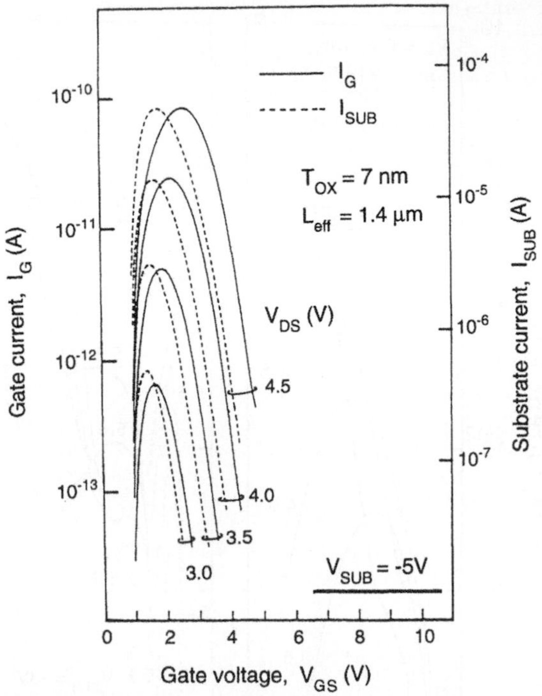

FIGURE 2.9 Correlation between SGHE gate currents and substrate current I_{SUB}. t_{ox} = 7 nm, L_{eff} = 1.4 μm, and V_{SUB} = −5 V. (——) I_G; (- - -) I_{SUB}.

current than would be expected from secondary impact ionization demonstrates the presence of this photoinduced generation mechanism. These mechanisms are summarized in Figure 2.12. It is essential that all three mechanisms are taken into account when considering hot-carrier effects in VLSI [Toriumi, 1987].

Toriumi [1989] also measured photon energy emitted from near the drain and investigated a relation between hot-carrier effects and electron temperature. Figure 2.13a shows a photoemission spectrum. It is found that energy distribution function $f(E)$ of photon intensity can be given as

$$f(E) = C \exp\left(-\frac{E}{kT_e}\right), \tag{2.11}$$

which implies bremsstrahlung radiation. Electron temperature, T_e, can be deduced using Eq. (2.11) and Figure 2.13a. Figure 2.13b shows the relationship between T_e and the maximum electric field at the drain edge.

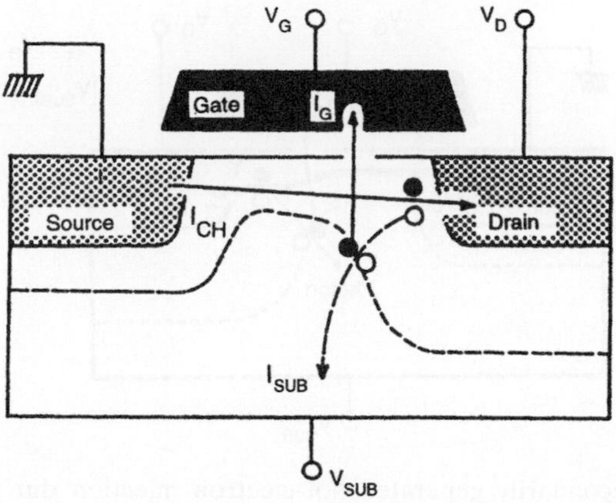

FIGURE 2.10 Substrate current induced hot-electron injection mechanism. $V_G < V_D$; $|V_{SUB}| > 0$.

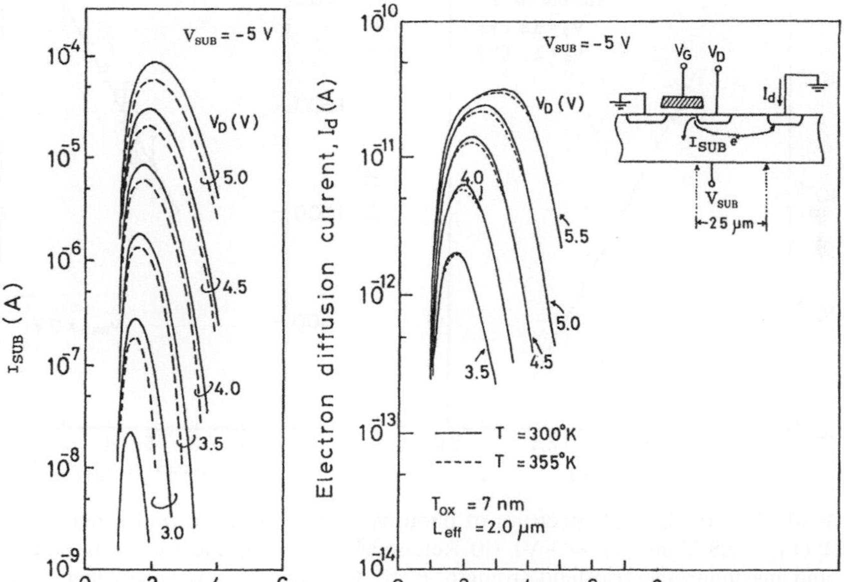

FIGURE 2.11 Temperature dependence of I_{SUB} and electron diffusion I_d for a device with $L_{eff} = 2\ \mu m$ and $t_{ox} = 7$ nm under the condition of $V_{SUB} = -5$ V.

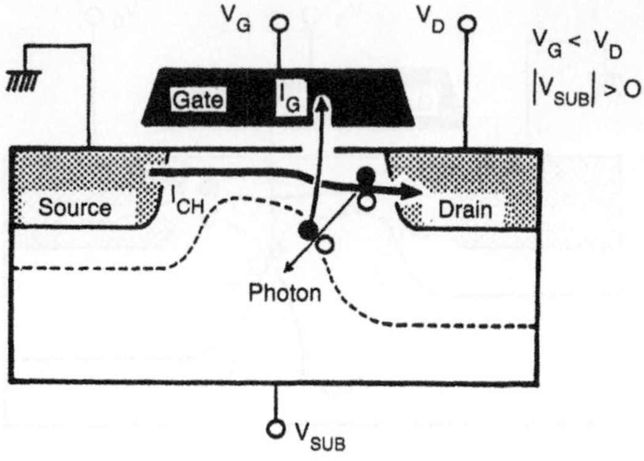

FIGURE 2.12 Secondarily generated hot-electron injection due to photon emission.

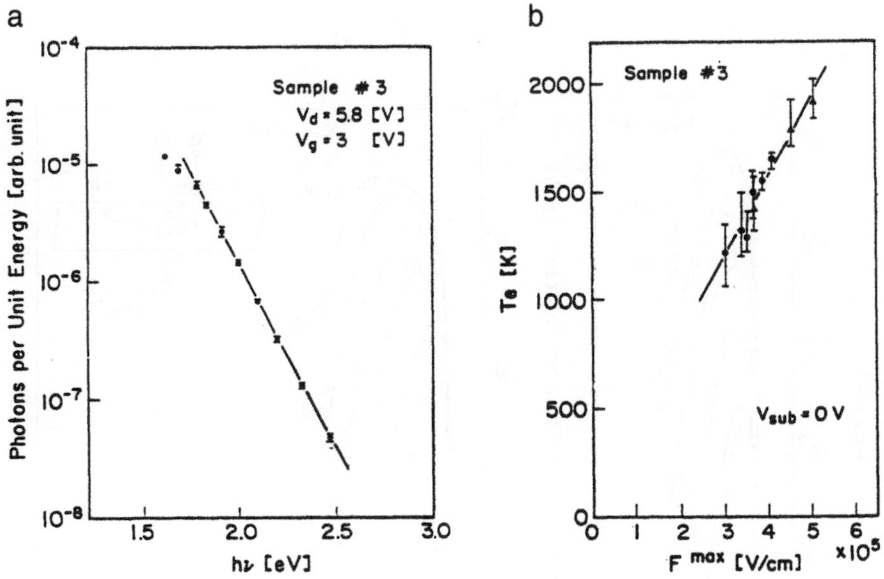

FIGURE 2.13 (a) Energy spectrum of photons emitted from an n-channel MOSFET ($V_D = 5.8$ V and $V_G = 3$ V). (b) Relationship between electron temperature T_e and maximum electric field strength F^{max}. Closed circles were obained from drain bias dependence and triangles were from gate bias dependence. (Reprinted from *Solid State Electronics*, **32**, A. Toriumi, "Experimental study of hot carriers in small size Si-MOSFETs," 519–1525, Copyright 1989, with kind permission from Elsevier Science Ltd., The Boulevard, Langford Lane, Kidlington OX5 1GB, UK.)

2.3.4 SUBSTRATE HOT-ELECTRON (SHE) INJECTION

The advantage of using this injection mode is that the transport of hot electrons within the silicon substrate is one-dimensional. Thus analysis of electron and hole trapping in SiO_2 is simpler than that of drain avalanche injection. SHE injection has also been used to investigate gate insulator qualities. In this case, hot electrons are thermally and/or radiatively generated or injected from the forward-biased *p–n* junction into the substrate high-field region. This injection, although not present in most circuits other than bootstrap circuits, is used to evaluate dielectrics quality (Figure 2.14a).

2.3.5 FOWLER–NORDHEIM (F–N) TUNNEL INJECTION

This injection mechanism is used in EEPROM devices and also is a significant cause of dielectric breakdown in thin insulators [I.-C. Chen,

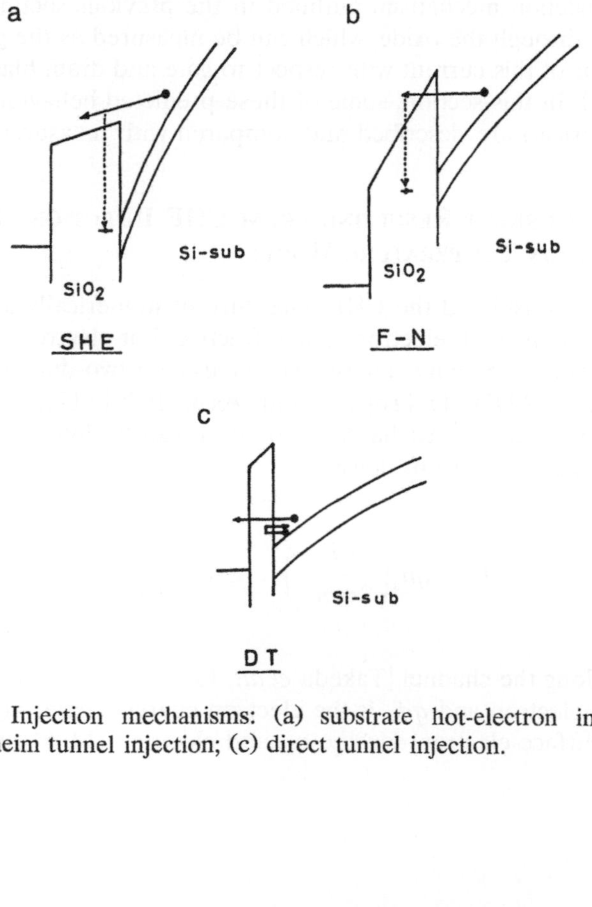

FIGURE 2.14 Injection mechanisms: (a) substrate hot-electron injection; (b) Fowler–Nordheim tunnel injection; (c) direct tunnel injection.

1985]. Hot electrons are injected from the channel inversion layer into the high-field gate dielectrics [Crook, 1979] (Figure 2.14b).

2.3.6 DIRECT TUNNEL INJECTION

This mode is used in MNOS-type EEPROM devices. Small-geometry MOS devices having thin gate insulators (less than 5 nm) may suffer from degradation due to this injection. Transconductance degradation under the condition $V_D < 3$ V might be due partly to this mode as well [Takeda *et al.*, 1984]. Thus, for deep-submicrometer devices ($< 0.5\ \mu$m), it will be important to include this mechanism in hot carrier–induced degradation studies (Figure 2.14c).

2.4 Gate Current Modeling

Each injection mechanism outlined in the previous section can result in a current through the oxide, which can be measured as the gate current. The behavior of this current with respect to gate and drain biases can then be predicted. In this section, some of these predicted behaviors (analytical or semiempirical) are described and compared with measurement results.

2.4.1 GATE CURRENT RESULTING FROM CHE INJECTION (EFFECTIVE ELECTRON TEMPERATURE MODEL)

We have considered the CHE gate current numerically as thermionic emission from heated electron gas (effective hot-electron temperature approach) over the Si–SiO_2 energy barrier using a two-dimensional analysis program (CADDET) [Toyabe and Asai, 1979]. Unlike the "lucky electron" approach, the Richardson equation can be integrated to obtain the thermionic gate current density

$$I_G = qn_s\left(\frac{kT_e}{2\pi m^*}\right)^{1/2} \exp\left(\frac{-q\phi_e}{kT_e}\right) \tag{2.12}$$

at a point along the channel [Takeda *et al.*, 1982]. Here, m^* is the effective mass of the electron and $q\phi_e$ is the electron energy barrier at the Si–SiO_2 interface. Surface electron density, n_s, and electric field, $\mathscr{E}$, are calculated numerically.

The electron temperature, T_e, can be obtained from the energy conservation equation [Toyabe and Kodera, 1974]

$$qn\mathscr{E}_x v - \frac{d}{dx}\left(\frac{5nkT_e v}{2}\right) - \frac{3nkT_e}{2\tau_E} = 0. \tag{2.13}$$

Combined with Eq. (2.12), one obtains

$$T_e(x) = \frac{2q}{5k}\int_0^\infty \mathscr{E}_x(x-u)\exp\left(\frac{-3u}{5\tau_E v_{sat}}\right) du. \tag{2.14}$$

$\mathscr{E}_x$ is the component of the electric field along the channel and τ_E is the energy relaxation time [Toyabe and Kodera, 1974].

Comparison between theory and experiment for an As-drain device with varying gate voltage is shown in Figure 2.15a. The agreement is

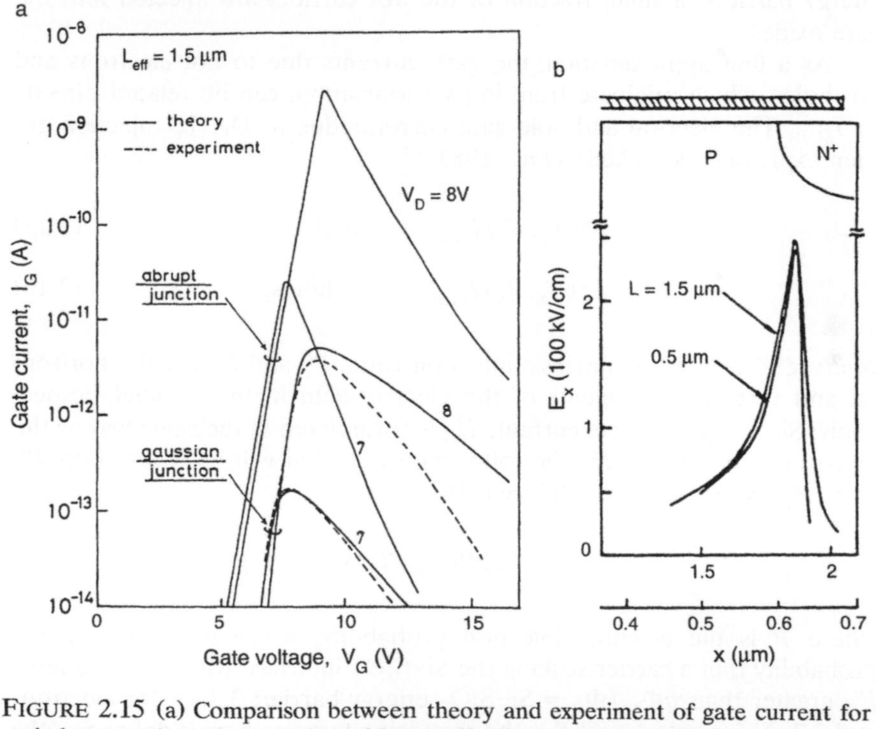

FIGURE 2.15 (a) Comparison between theory and experiment of gate current for varied gate voltage in a device with an effective channel length of 1.5 μm. (b) Electric field profile for a 1.5-μm MOSFET (V_D = 8 V, V_G = 8.5 V) and a 0.5-μm MOSFET (V_D = 8/3 V, V_G = 3.2 V) near the drain junction.

remarkable when the energy relaxation time used for estimating τ_E is 8×10^{-14} s. The characteristic length of nonlocality ($= 5\tau_E v_{sat}/3$) is equal to 16 nm. Figure 2.15b shows a strongly peaked electric field distribution for devices with effective channel lengths of 0.5 μm and 1.5 μm. The position of the peak roughly coincides with the location of the most vigorous injection. The localized nature of the hot-electron injection indicates that it is closely related to structural parameters which determine the electric field distribution in a MOSFET. Thus this approach, making use of a two-dimensional device simulator which accepts arbitrary impurity profiles, is a powerful tool in minimizing gate current [Takeda *et al.*, 1981].

2.4.2 Gate Current Resulting from DAHC Injection

Hot-electron and hot-hole injections by channel impact ionization have been applied to EEPROM. If the carriers produced by impact ionization near the drain junction are hot enough to surmount the Si–SiO_2 energy barriers, a small fraction of the hot carriers are injected into the gate oxide.

As a first approximation, the gate currents due to hot electrons and hot holes, which originate from impact ionization, can be related directly to I_{SUB}. The electron and hole gate currents due to DAHC injection are then expressed as [Takeda *et al.*, 1983e]

$$I_G^e = \xi^e(\mathscr{E}_x, \mathscr{E}_y) I_{SUB} \quad \text{for electrons,} \tag{2.15}$$

$$I_G^h = \xi^h(\mathscr{E}_x, \mathscr{E}_y) I_{SUB} \quad \text{for holes,} \tag{2.16}$$

where $\xi(\mathscr{E}_x, \mathscr{E}_y)$ is the intrinsic injection ratio. $\mathscr{E}_x$ and $\mathscr{E}_y$ are the horizontal and vertical components of the electric field in the channel, respectively. Since the hole gate current, I_G^h, is formulated in the same way as the electron gate current, I_G^e, the following discussion will be focused on I_G^e. $\xi(\mathscr{E}_x, \mathscr{E}_y)$ is expressed as [Bulucea, 1974]

$$\xi(\mathscr{E}_x, \mathscr{E}_y) = \gamma_A P(E_B, T_e) S(E_B, T_e), \tag{2.17}$$

where P is the essential injection probability, which is defined as the probability that a carrier striking the Si–SiO_2 interface will have an energy E_B greater than $q\Phi_B$ (Φ_B = Si–SiO_2 energy barrier: 3.2 eV for electrons and 3.7 eV for holes), and S is the scattering factor, which is defined as the probability that a carrier striking the Si–SiO_2 interface with an energy E_B greater than the barrier will be able to escape over it, avoiding the barrier relaxation associated with its transverse momentum. The injection ratio, γ_A, has been determined experimentally and has a simple dependence on V_G as shown later [Hu *et al.*, 1985]. T_e is obtained as a function of $\mathscr{E}$ using the following approximation [Ning and Yu, 1974]:

$$kT_e = \frac{E_0}{1/2 + [1/4 + (E_0/rE_r)]^{1/2}}, \tag{2.18}$$

where

$$E_0 = \frac{(q\mathscr{E}l_r)^2}{rE_r} \tag{2.19}$$

and l_r is the mean free path ($\sim$ 6 nm), E_r is the Raman optical phonon energy (= 0.063 eV), and r is the ratio ($\sim$ 3.2 for Si) of l_r to the mean free path for ionization interaction, l_i. Bartelink's theory [Bartelink, 1981] can be expected to be valid in near-avalanche fields shown here. $\mathscr{E}$ has been calculated using CADDET, taking its localized nature into consideration, and is shown in Figure 2.16. From Eqs. (2.18) and (2.19), T_e can be obtained.

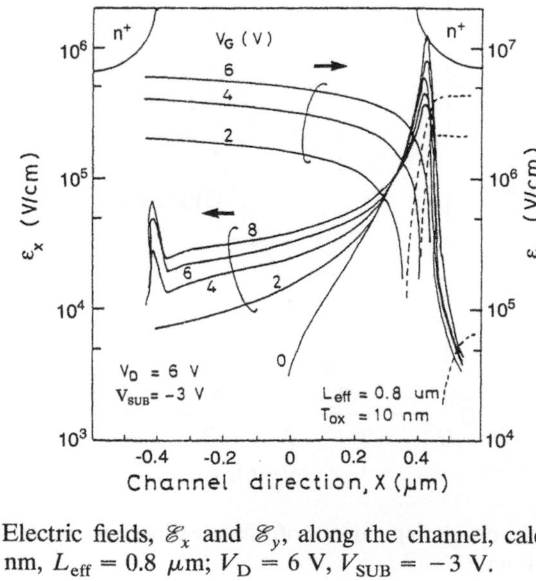

FIGURE 2.16 Electric fields, $\mathscr{E}_x$ and $\mathscr{E}_y$, along the channel, calculated by CADDET. t_{ox} = 10 nm, L_{eff} = 0.8 μm; V_D = 6 V, V_{SUB} = -3 V.

The essential injection probability, P, is empirically shown to be [Bulucea, 1974]

$$P = A\exp(-B\epsilon_B) \qquad \text{for} \quad \epsilon_B > 4, \tag{2.20}$$

where

$$\epsilon_B \equiv \frac{E_B}{kT_e}$$

with $A = 1.874$ and $B = 0.926$.

The scattering factor S is given by [Bulucea, 1974]

$$S \simeq \frac{1 + (f_1/f_0)\cos\theta}{1 + (f_1/f_0)}, \tag{2.21}$$

where θ is the angle between the electric field and the normal to the $Si-SiO_2$ interface. f_0 and f_1 are the zeroth and first-order harmonics of the distribution function in momentum space. Thus, the gate currents (electrons and holes) could be calculated using Eqs. (2.15), (2.17), (2.20), and (2.21). The gate voltage dependence of the injection ratio, γ_A, is demonstrated in Figure 2.17. γ_A is an exponential function of V_G and

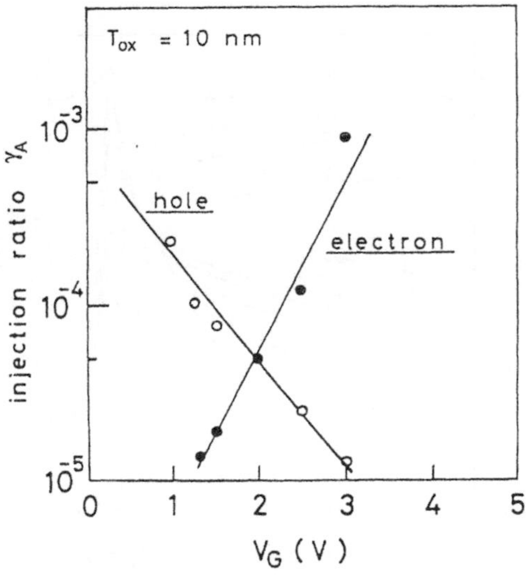

FIGURE 2.17 Gate voltage dependence of the injection ratio γ_A for both electrons and holes. $t_{ox} = 10$ nm.

exhibits different behaviors for electrons and holes. Carriers in the drain depletion region are physically below the surface and at potential energies lower than the surface potential. Thus the barrier height and the probability of collisions in this region are considerable. These effects are clearly manifested in the γ_A versus V_G plots. Comparison between measured gate currents and simulation results is shown in Figure 2.18. Agreement is remarkable in spite of the large number of variables involved in this injection process.

Noting the condition for the maximum avalanche, gate current fitting was carried out using the parameter A in Eq. (2.20) with the energy relaxation time assumed to be 8×10^{-14} s. There is a report on the energy relaxation time obtained experimentally (6×10^{-14} s) using the measured

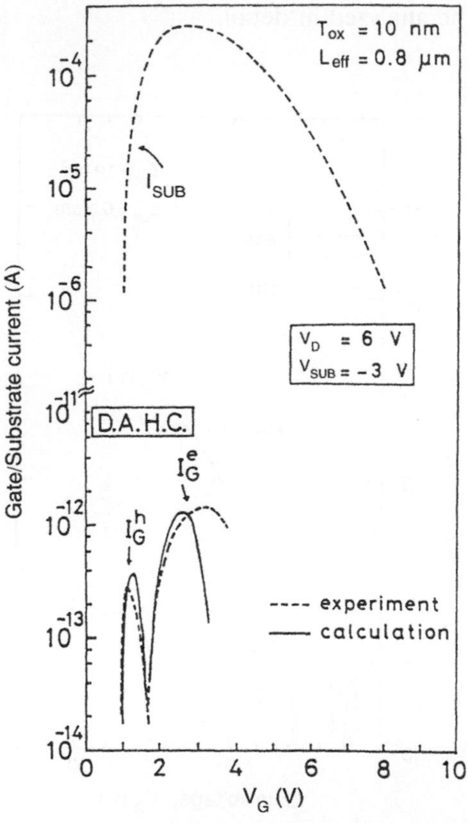

FIGURE 2.18 Comparison between measured gate currents and simulation results. $t_{ox} = 10$ nm; $L_{eff} = 0.8$ μm; $V_D = 6$ V; $V_{SUB} = -3$ V.

value of the electron temperature [Toriumi, 1987], which is comparable to previously reported values (8×10^{-14} and 5×10^{-13} s) [Takeda *et al.*, 1982c; Fukuma and Uebbing, 1984]. The experimental and simulated results of the gate current due to the avalanche hot holes where $t_{ox} = 10.2$ nm and $L_{eff} = 0.8\ \mu$m are shown in Figure 2.19 [Hamada *et al.*, 1988].

While experimental and calculated results seem to be quantitatively in good agreement, further understanding of avalanche hot holes and hot electrons is needed. This agreement indicates that the effective electron temperature approximation is valid not only for the channel hot electrons but also for avalanche hot carriers. Further investigation of the dependence on device parameter would also be of interest. The discrepancies between experiment and simulation are probably due to a lack of sufficient understanding of the dependence of parameters such as A and τ_E on the gate voltage. To solve this problem fully, energy as well as momentum conservation must be analyzed in detail.

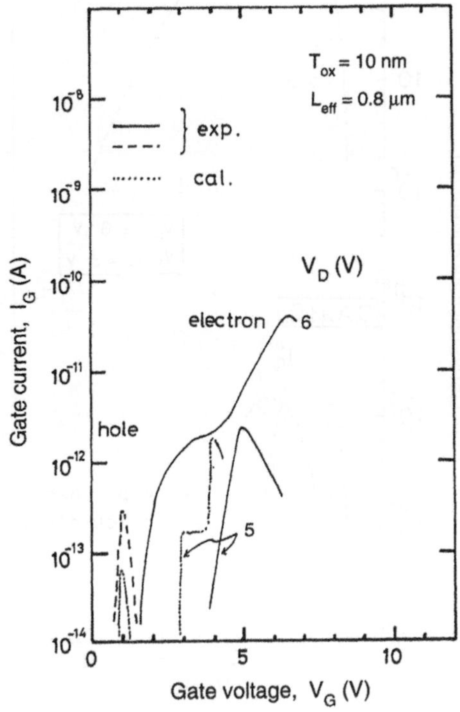

FIGURE 2.19 Comparison of experiment and simulation. Results for avalanche hot-hole injection.

2.4.3 EFFECTIVE ELECTRON TEMPERATURE VS LUCKY ELECTRON MODEL

Besides the effect electron temperature approach, the lucky electron approach can also be used in order to formulate gate currents [Tam *et al.*, 1984]. In order for channel hot electrons to reach the gate, the hot electrons have to gain sufficient kinetic energy from the channel field and have their momenta redirected toward the $Si-SiO_2$ interface to surmount the $Si-SiO_2$ potential barrier. To quantify the probability that these electrons could eventually be collected by the gate, several scattering mechanisms have to be considered. Here, the scattering mean free path becomes important and should depend on both the optical-phonon scattering mean free path and the impact-ionization mean free path. We can then formulate the probability that an electron will acquire kinetic energy greater than the $Si-SiO_2$ potential barrier. Thus, in the lucky electron approach, CHE gate current can be expressed as

$$I_G \propto \exp\left(\frac{-\phi_e}{\mathscr{E}_x \lambda}\right), \tag{2.22}$$

where $\mathscr{E}_x$ is the channel electric field at the drain end and λ is the scattering mean free path of hot electrons. The probability that a channel hot electron will travel a distance d $(= \phi_e/\mathscr{E}_x)$ or more without suffering any collision is then $\exp(-d/\lambda)$. From Eqs. (2.12), (2.14), and (2.22), a correlation of $kT_e/q = \mathscr{E}_x \lambda$ between the hot-electron temperature and lucky electron approaches can be obtained.

2.5 Summary

Hot-carrier injection mechanisms in scaled MOSFETs are presented and briefly discussed. Some modeled behaviors are compared with recent experimental results. Modeling of gate currents using the effective-electron-temperature model and a two-dimensional device simulator provides improved understanding of hot-carrier injection. This combined approach serves as the basis for discussions of device degradation mechanisms in Chapter 3.

CHAPTER

3

HOT-CARRIER DEVICE DEGRADATION

3.1 Introduction

Current ultra large-scale integration (ULSI) microfabrication technologies are being pushed to the extremes by the dynamic random access memory (DRAM) community, thereby expediting research on 64-Mbit DRAMs using a 0.3-μm design rule. Since process and device technologies in such a deep submicrometer region are approaching practical limits, reliability problems as well as improvement of performance are becoming increasingly significant. In particular, the scaled MOSFET (ULSI building block) is suffering from hot-carrier degradation and the downsliding of scaling merit due to high-field effects. Among ULSI reliability problems, the hot-carrier effect is an important factor in determining MOS device structure and power supply voltage. Figure 3.1 shows device structures and hot-carrier breakdown voltage for each DRAM generation. It can be seen that several kinds of hot-carrier-resistance device structures such as double diffused and lightly doped drains (DDD and LDD) have been proposed so far, but a 5-V supply voltage cannot be used even in a LDD with L_{eff} less than 0.8 μm. Therefore, in the regime of 0.5–0.3 μm, two approaches will be taken according to the type and purpose of ULSI circuits. These approaches involve (1) new hot-carrier-resistant devices usable with a 5-V supply and (2) reduction of power supply voltage to the 3–1.5 V range. In the former approach, a deeper physical understanding of high-field effects

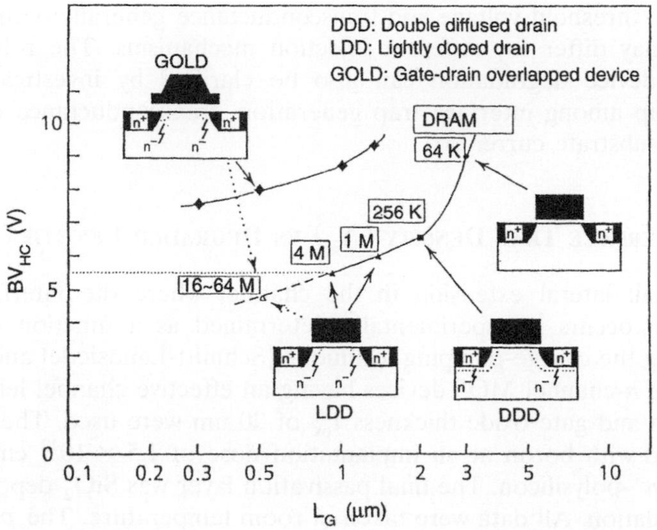

FIGURE 3.1 Device structures and hot-carrier breakdown voltage for each DRAM generation.

such as hot carriers and short channels is needed and the power consumption problem becomes significant. The second approach must also overcome a low-speed problem because subthreshold voltage is difficult to scale, resulting in the low channel current. Optimal device operation for any ULSI device can be achieved by choosing the best hot-carrier-resistant device structure. Also, since hot-carrier effects are phenomena caused by carriers with high energy, the fundamental physical mechanisms have something in common with the time-dependent dielectric breakdown (TDDB), radiation damage (RD) due to the cosmic and x-rays, and electrostatic discharge (ESD), as discussed in Chapter 2. Such a trend requires deeper physical insight into hot-carrier effects in the deep submicrometer MOSFETs and feedback of this knowledge to device structure and circuit design. The extent to which the device operation is affected constitutes the subject of this chapter.

3.2 Device Degradation Due to Various Hot-Carrier Injections

The manifestation of device degradation due to hot carriers is invariably an increase in interface trap density ($N_{it} \equiv Q_{it}/q$). Corresponding changes in threshold voltage and transconductance generally occur. These changes may differ for different injection mechanisms. The role of hot holes in device degradation can also be clarified by investigating the relationship among interface trap generation, transconductance degradation, and substrate current.

3.2.1 INTERFACE TRAP DENSITY (N_{it}) VS DEGRADED LENGTH (δL)

A small lateral extension in the channel where the interface trap generation occurs is experimentally determined as a function of stress time, using the charge-pumping technique [Schmitt-Landsiedel and Dorda, 1981]. The n-channel MOS devices having an effective channel length L_{eff} of 0.8 μm and gate oxide thickness t_{ox} of 20 nm were used. The channel was doped with boron at an implantation dose of 1.5×10^{12} cm^{-2}. The gate was n^+-polysilicon. The final passivation layer was SiO_2 deposited via silane oxidation. All data were taken at room temperature. The p-channel MOS devices with $L_{eff} = 1.0$ μm and $t_{ox} = 35$ nm were also used in order to distinguish the role of hot-hole injection from that of hot-electron injection.

Measurement of the charge pumping current [Elliot, 1976; Gesch *et al.*, 1982; Takeda *et al.*, 1983f], I_{cp}, was carried out with a setup as shown in Figure 3.2. The repetition frequency and width of the gate pulse were 200 kHz and 3 μs, respectively. A pulse train was applied to the gate and the resulting DC substrate current was recorded. The technique is based on the fact that trapped electrons at the interface recombine with holes from the p-substrate after the channel is turned off. This recombination current is the charge pumping current and is proportional to the signal frequency, degraded channel area, and interface trap density.

The fractional change in interface trap density (cm^{-3}), $\Delta N_{it}/N_{it0}$, can be expressed as follows:

$$\Delta I_{cp}/I_{cp0} = \delta L/L_{eff} \cdot \Delta N_{it}'/N_{it0} \equiv \Delta N_{it}/N_{it0}, \tag{3.1}$$

where δL is the lateral length of the degraded channel region. In this equation, ΔN_{it} has been defined as the increase in interface trap density, which is normalized for the entire channel, while $\Delta N_{it}'$ is the increase in interface trap density in a very small degraded region. Since the charge pumping current decreases linearly with extension of the drain depletion width (W_d) due to screening of the degraded channel region by the

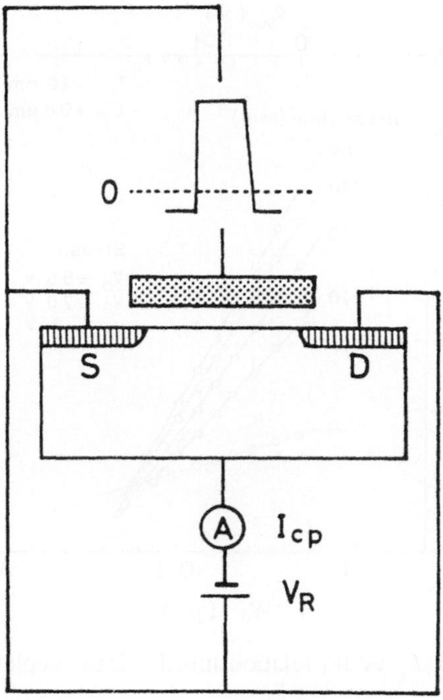

FIGURE 3.2 Charge-pumping bias setup.

depletion region, δL can be evaluated from the I_{cp} versus W_d relation using the expression for the depletion width

$$W_d(V_{SUB}) = [2\epsilon_s(|V_{SUB}| + V_{bi})/qN_A]^{1/2}. \tag{3.2}$$

δL is defined as $[W_d(V_0) - W_d(0)]$, where V_0 is the substrate bias corresponding to equal I_{cp} under stressed and unstressed conditions.

Figure 3.3 shows an I_{cp} versus W_d relationship.

Figure 3.4 shows the stress time dependence of δL with V_G as a parameter. δL is found to stretch slowly toward the source region with the time dependence as [Schmitt-Landsiedel and Dorda, 1981; Lombardi *et al.*, 1982]

$$\delta L \propto t^m, \qquad m \cong 0.1. \tag{3.3}$$

This stress time dependence of δL can become an important issue with the reduction of L_{eff} for submicrometer devices. It is clear from Figure 3.4 that the time dependence of δL due to channel hot electron (CHE)

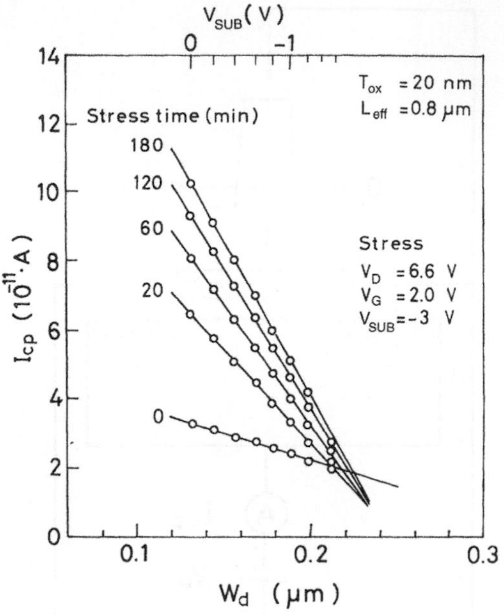

FIGURE 3.3 I_{cp} vs. W_d relationship: W_d is the depletion width.

injection ($V_G = V_D = 6.6$ V) is different from that due to drain avalanche hot carrier (DAHC) injection.

3.2.2 $\Delta G_m / G_{m0}$ VS $\Delta N_{it} / N_{it}$ RELATION

The observed time variation of tranconductance degradation $\Delta G_m/G_{m0}$ and the normalized change in interface traps $\Delta N_{it}/N_{it0}$ are shown in Figure 3.5 for a device with $L_{eff} = 0.8\ \mu$m and $t_{ox} = 20$ nm for both the DAHC injection ($V_G < V_D$) and the CHE injection ($V_G \cong V_D$) modes. It is found that the DAHC injection causes more severe G_m degradation and larger N_{it} increase than CHE. It should also be noted that the change in G_m follows the same time dependence as the N_{it} increase, and the following relation is well satisfied:

$$\Delta G_m/G_{m0} \propto \Delta N_{it}/N_{it0}. \tag{3.4}$$

This implies that the G_m degradation is caused by the interface trap generation, rather than the fixed charges of trapped electrons in SiO_2.

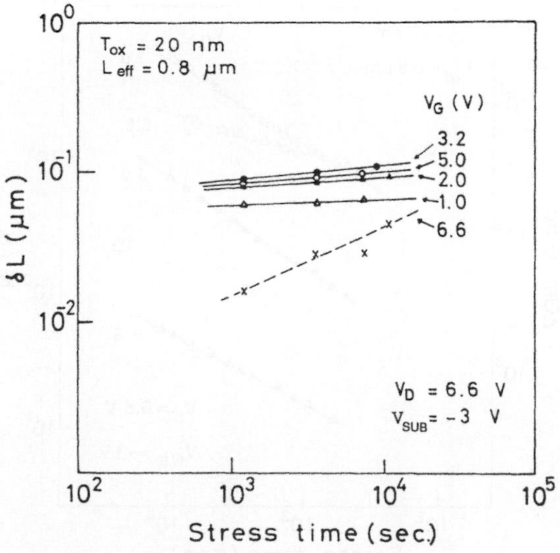

FIGURE 3.4 Time dependence of the localized lateral length of the degraded channel region with V_G as a parameter.

Further, there is a difference of more than one order of magnitude in the increase in N_{it} between the CHE and the DAHC injections, despite the fact that the CHE injection provides larger gate currents.

As with device degradation due to hot-carrier injection, both the G_m degradation and V_{th} shift are, of course, important. However, in the case of the conventional silox passivation layer, G_m degradation, rather than V_{th} shift measured at channel currents below 1 μA, is more noticeable. Figure 3.6 shows the stress time variation of G_m and V_{th} shift measured at $I_D = 10$ nA for a device with $L_{eff} = 0.8$ μm and $t_{ox} = 20$ nm. It should be noted that the V_{th} shift for a low channel current of 10 nA is less pronounced than the G_m degradation. V_{th} shift at the high channel current of 1 mA appears to be quite noticeable. This shift implies G_m degradation. This is because the tailing coefficient [$\equiv dV_G/d(\log I_D)$] increases, which means that the slope of I_D versus V_G decreases as G_m degrades, and the V_{th} shift of high channel current emerges even if V_{th} measured at low channel current hardly shifts. On the other hand, with the use of a plasma–nitride passivation layer and gate nitride, V_{th} shift at low channel current and G_m degradation are both noticeable. Thus G_m degradation is the most important criterion when investigating hot-carrier effects.

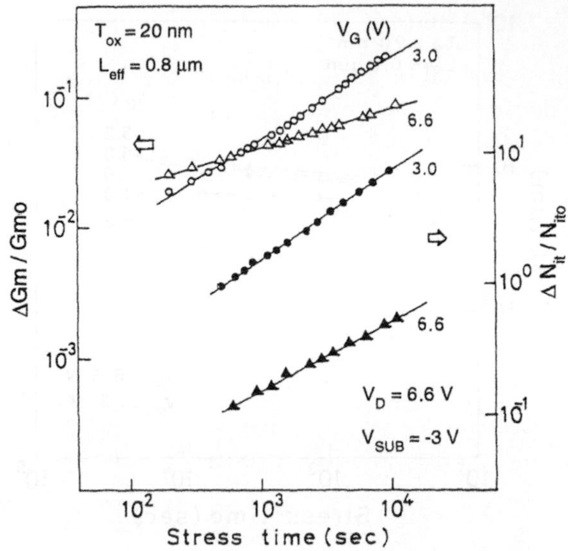

FIGURE 3.5 Time dependence of transconductance degradation and relative change in the generated interface traps for a device having L_{eff} = 0.8 μm and t_{ox} = 20 nm.

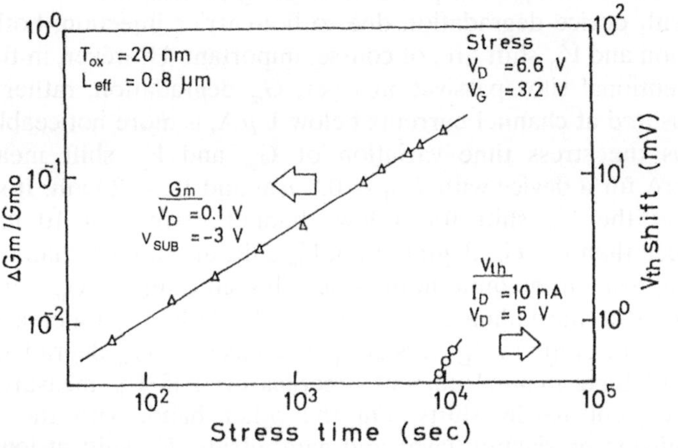

FIGURE 3.6 Stress time variation of G_m degradation and V_{th} shift at I_D = 10 nA for a device with L_{eff} = 0.8 μm and t_{ox} = 20 nm. Stress: V_D = 6.6 V; V_G = 3.2 V.

3.2.3 ROLE OF HOT HOLES

Figure 3.7 shows I_{SUB}, G_m degradation, and fractional change in interface trap density as functions of V_G for a device with $L_{eff} = 0.8$ μm and $t_{ox} = 20$ nm. It is found that there is a remarkable correlation among these three characteristics. The gate voltages at which the maximums for all three quantities occur are the same. This fact suggests that the N_{it} increase, and thereby G_m degradation, results from hot-hole injection, rather than CHE injection, because hot holes with larger effective mass may cross the $Si-SiO_2$ interface and create more interface traps or trapping centers than hot electrons do.

The discrepancy in behavior between G_m degradation and N_{it} increase in the $V_G \cong V_D$ region is attributed to the generation of donor states resulting from hot-hole injection, while the interface traps which are observed and characterized by a change in charge pumping current are mainly accepter type.

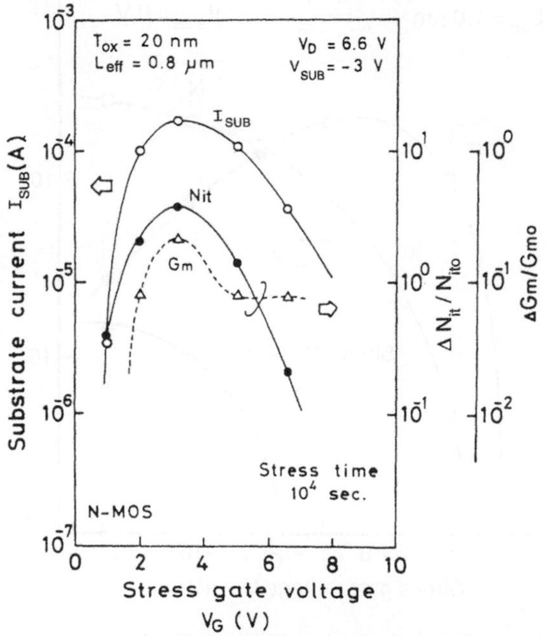

FIGURE 3.7 Correlation between substrate current, G_m degradation, and fractional change in interface traps for an n-channel device with $L_{eff} = 0.8$ μm and $t_{ox} = 20$ nm.

A similar effect of hot holes is more clearly demonstrated by investigating *p*-channel MOS devices. Figure 3.8 shows the N_{it} increase and substrate current as a function of V_G for a *p*-MOSFET having $L_{eff} = 1.0$ μm and $t_{ox} = 35$ nm. It is found that there is no correlation between the N_{it} increase and the substrate current. The bias condition which causes maximum N_{it} increase is about $V_G = V_D$. This condition corresponds to that which causes the maximum channel hot-hole injection in the *p*-channel device. Drain avalanche hot-electron injection at $|V_G| < |V_D|$, which mainly produces the gate current for *p*-channel devices [Takeda *et al.*, 1983b], is found to generate fewer interface traps. Thus regardless of the injection mode, hot holes play a significant role in device degradation and/or interface trap generation for both *n*- and *p*-channel MOS devices.

It is obvious that avalanche hot-carrier injection under bias conditions of $V_G < V_D$ imposes much more severe limitations on device design. It should also be noted that the As–P drain structure puts up a stronger

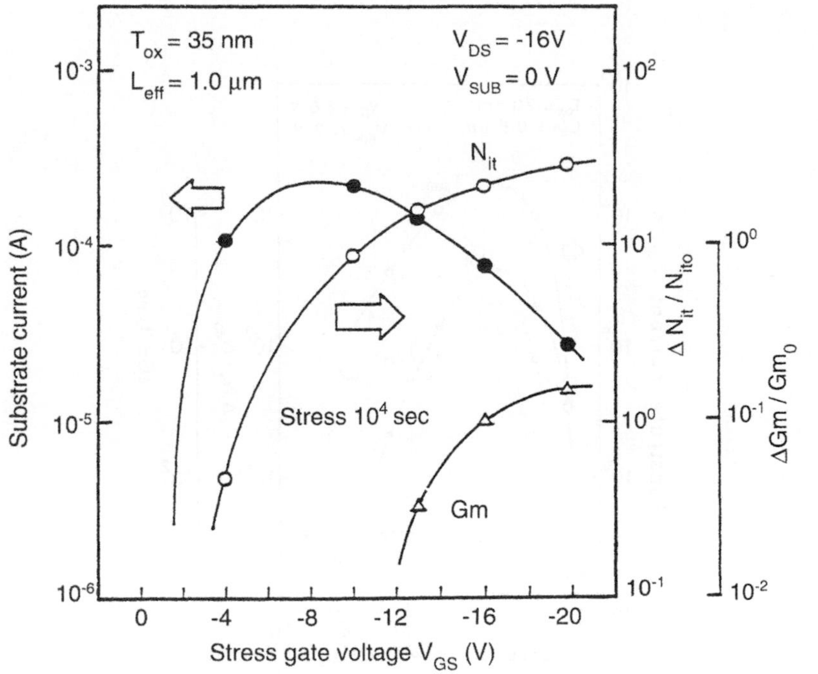

FIGURE 3.8 Correlation between substrate current, G_m degradation, and fractional change in interface trap density for a *p*-channel device with $L_{eff} = 1.0$ μm and $t_{ox} = 35$ nm.

resistance (by one or two orders of magnitude) to V_{th} shift degradation when compared with the As drain. The reason for avalanche hot-carrier injection imposing more stringent constraints is probably that hot holes with a larger effective mass than that of electrons, which may create interface traps or trapping centers, are injected into the gate oxide at the same time as the hot electrons. This speculation can be proved by the experimental results shown in Figure 3.9.

Figure 3.9 shows the time dependence of the V_{th} shift under bias conditions of $V_G < 3$ V that supply a larger number of hot holes, for an As drain having an effective channel length of 1.0 μm. It can be seen that during the short stress time, negative threshold-voltage shifts occur due to trapped holes. As stress time increases, the V_{th} shifts change to the positive direction. This implies that in the beginning, hot holes are mainly trapped in the gate oxide, and later, as the injected hot holes begin to create interface traps or trapping centers, hot electrons which are injected with hot holes are more easily trapped in the generated and original trapping centers. Therefore, with a long stress time, a V_{th} shift in the positive direction for *n*-channel MOS devices occurs in spite of the injection of hot holes [Takeda *et al.*, 1982c]. The fact that avalanche hot-carrier injection causes more severe degradation than channel hot-electron injection is very important for VLSI circuit operation.

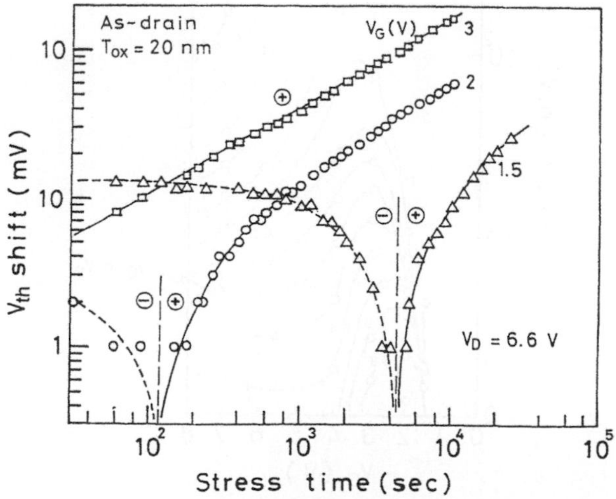

FIGURE 3.9 Time dependence of the V_{th} shift for $V_G < 3$ V.

3.2.4 DAHC VS CHE INJECTIONS

In Figure 3.10 V_{th} shifts for a device with L_{eff} = 1.2 μm and t_{ox} = 20 nm are shown as a function of V_G with V_D as a parameter. Stress time was 500 s. It is important to note that the V_{th} shifts behave similarly to substrate current. This clearly shows that device degradation due to DAHC injection is directly related to substrate current, which in turn reflects the number of excess carriers generated by impact ionization at the drain.

The effects of scaling down MOSFET dimensions are characterized by measuring the electron gate current and hole substrate current. Figure 3.11 shows examples of gate current and substrate current as a function of V_G with V_D as a parameter for a device with an effective channel length of 1.1 μm. The difference in voltage dependence between the gate current and the substrate current clearly demonstrates a difference in the physical mechanisms underlying these currents. The gate current peaks for $V_D \approx$

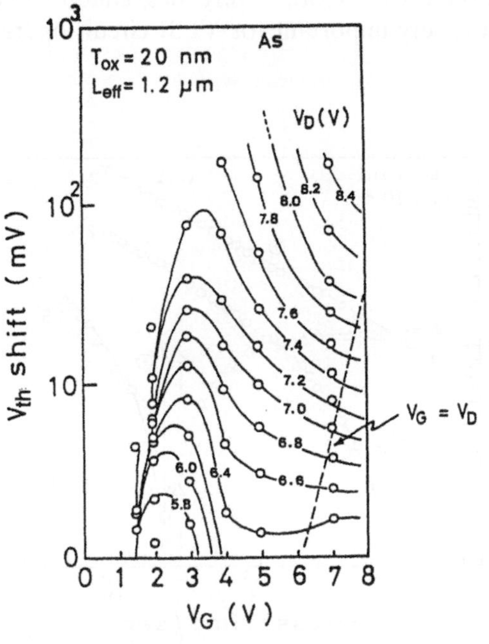

FIGURE 3.10 V_{th} shifts as a function of V_G with V_D as a parameter for a device wtih L_{eff} = 1.2 μm and t_{ox} = 20 nm.

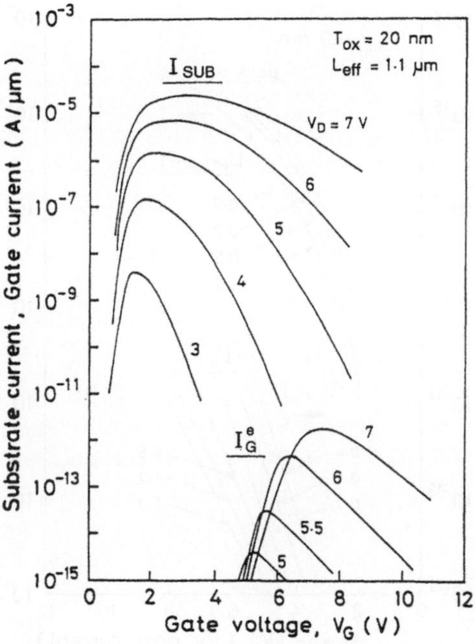

FIGURE 3.11 An example of gate currents and substrate currents as a function of V_G with V_D as a parameter, for a device with an effective channel length of 1.1 μm.

V_G. On the other hand, the substrate current reaches its maximum under the biasing conditions of $V_D > V_G > V_{th}$.

The effect of effective channel length on gate and substrate currents is illustrated in Figure 3.12. The magnitudes of both currents show strong dependences on the effective channel length. Scaling down the effective channel length to submicrometer size brings about a drastic lowering of the highest applicable voltage, BV_{DC}, and the drain sustaining voltage, BV_{DS}.

The time dependence of transconductance degradation and V_{th} shift due to DAHC injection for a MOS device having $L_{eff} = 0.8$ μm and $t_{ox} = 20$ nm is shown in Figure 3.13, compared with those due to CHE injection. The $V_G = 3$ V bias condition for DAHC injection was chosen so that degradation would be most severe. It also corresponds to the gate voltage at which the substrate current is maximum for the applied drain bias of 6.6 V. This comparison shows that DAHC injection imposes more stringent constraints on short-channel MOS device designs in terms of both G_m degradation and V_{th} shift. It should also be pointed out that

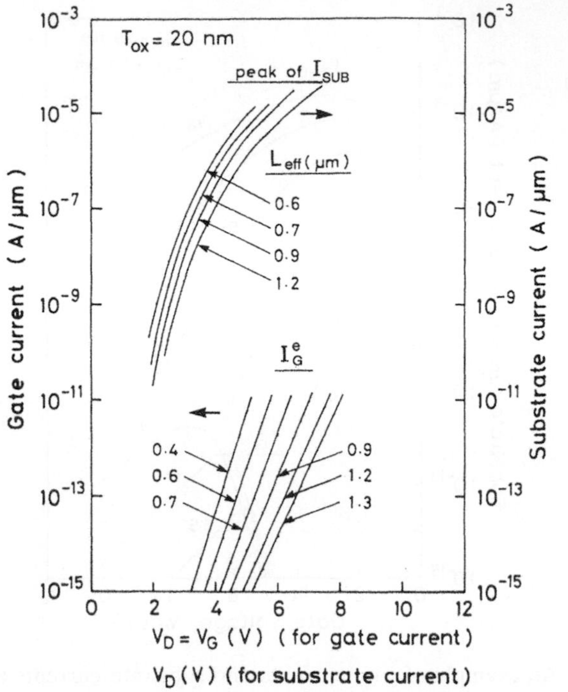

FIGURE 3.12 Gate current and substrate current due to hot-carrier generation for conventional As-drain structure.

DAHC injection appears to affect, in particular, G_m degradation quite strongly in spite of the fact that gate currents due to CHE injection are larger than those due to DAHC injection. Small gate current due to DAHC injection can, indeed, be measured in the range down to 10^{-15} A, as shown in Figure 3.11. This is attributed to hot holes, with a larger effective mass, being injected into the gate oxide at the same time as the hot electrons, creating interface traps.

3.2.5 SGHE vs DAHC Injections

The time dependence of G_m degradation and V_{th} shift due to SGHE injection for a device having $L_{eff} = 1.0$ μm and $t_{ox} = 6.8$ nm is shown in Figure 3.14, compared with that due to DAHC injection. It is obvious that second-order injection such as secondarily generated hot electron (SGHE) injection also causes considerable V_{th} shift for devices with gate oxides thinner than 10 nm. This suggests that SGHE injection will impose

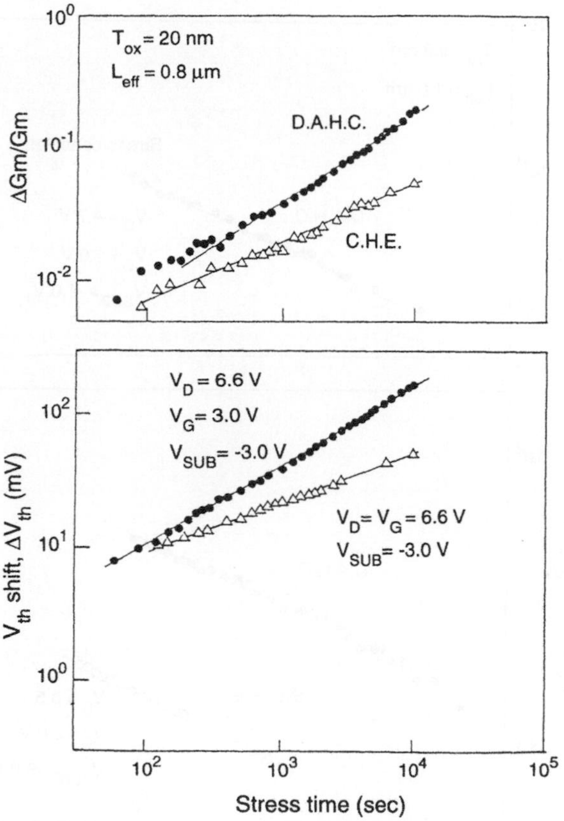

FIGURE 3.13 Time dependence of transconductance degradation and V_{th} shift due to DAHC injection and CHE injection for a MOS device with $L_{eff} = 0.8\ \mu$m and $t_{ox} = 20$ nm.

limitations on VLSI circuits designed with, for example, half-micrometer MOS devices even at $V_D = 3$ V. However, from Figures 3.13 and 3.14, one can conclude that this injection mechanism provides less G_m degradation than either DAHC or CHE injection. Such a tendency is understandable for SGHE injection, which is uniform due to emission of electrons from the Si substrate. Thus, the SGHE injection mechanism is probably substantially different from DAHC or CHE. The very localized nature of hot-carrier injections in the latter two cases may be weakened in SGHE injection.

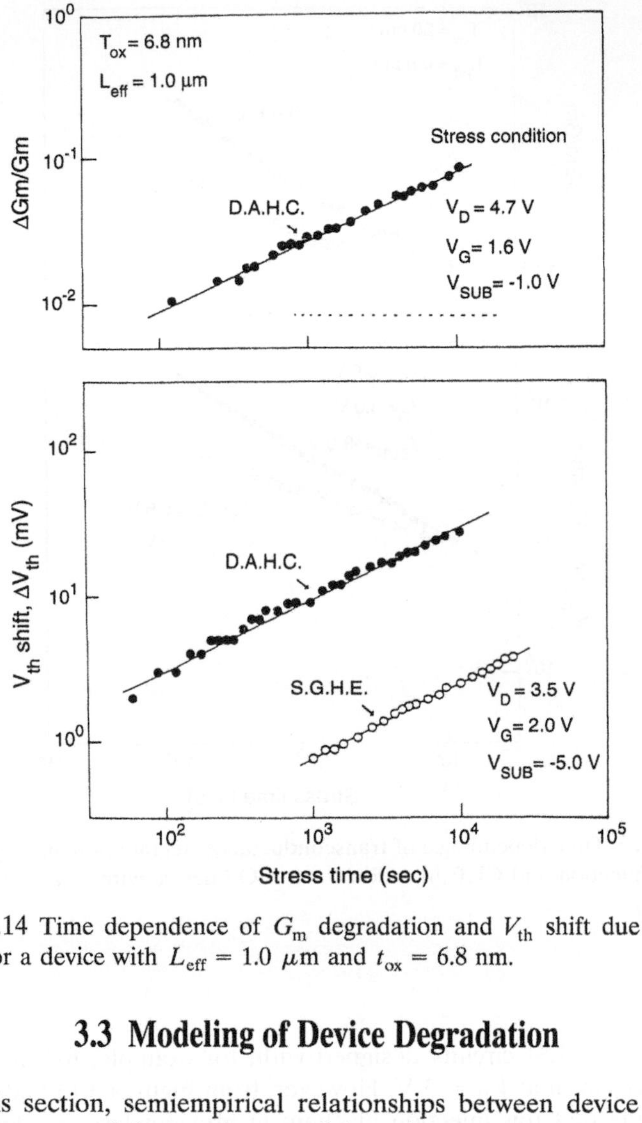

FIGURE 3.14 Time dependence of G_m degradation and V_{th} shift due to SGHC injection for a device with $L_{eff} = 1.0\ \mu m$ and $t_{ox} = 6.8$ nm.

3.3 Modeling of Device Degradation

In this section, semiempirical relationships between device degradation parameters and drain voltage and/or substrate current are present in an attempt to elucidate experimental findings [Takeda and Suzuki 1983].

3.3.1 SUBSTRATE CURRENT MODELING

Figure 3.15 is a typical plot of the peak substrate current versus $1/V_D$ with effective channel length as a parameter [Takeda *et al.*, 1982c]. The

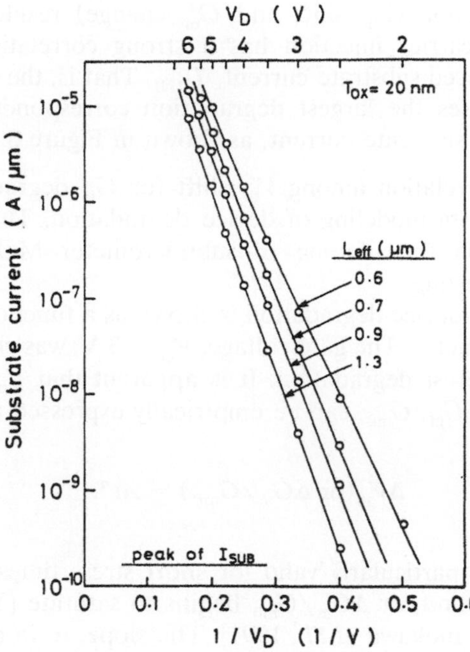

FIGURE 3.15 A typical plot of the peak of substrate current for a given V_D vs. V_D^{-1} with effective channel length as a parameter.

plot reveals the behavior, $I_{SUB} \propto \exp(-b/V_D)$. This relation is explained analytically in the next subsection.

The scaling down of other structural parameters, such as gate-oxide thickness, junction depth, inverse channel doping, and inverse substrate doping, also intensifies the electric field near the drain if the drain voltage is not decreased, resulting in an increase in hot-carrier generation [Eitan and Frohman-Bentchkowsky, 1981; Troutman, 1976; Toyabe and Asai, 1979].

3.3.2 Device Degradation Modeling—Hot-Carrier Lifetime

In this subsection, an empirical model for device degradation due to hot-carrier injection under the severest bias condition is presented. This model is based on the following assumptions:

1. Avalanche hot-carrier injection due to impact ionization at the drain, rather than channel hot-electron injection composed of "lucky electrons," imposes the severest constraints on device design.

2. Device degradation (V_{th} shift and G_m change) resulting from drain avalanche hot-carrier injection has a strong correlation with impact ionization–induced substrate current, I_{SUB}. That is, the gate bias condition which causes the largest degradation corresponds to that which yields the peak substrate current, as shown in Figure 3.16.

Significant correlation among V_{th} shift (or G_m degradation), V_D and I_{SUB} can result from modeling of device degradation. Thus, it should be possible to predict the lifetime of submicrometer MOSFETs without long-term stress testing.

In Figure 3.17 device degradation is shown as a function of stress time with V_D as a parameter. The gate voltage, $V_G = 3$ V, was chosen because it produced the severest degradation. It is apparent that V_{th} shift, ΔV_{th}, or G_m degradation, $\Delta G_m/G_{m0}$, can be empirically expressed as

$$\Delta V_{th}(\text{or } \Delta G_m/G_{m0}) = At^n. \tag{3.5}$$

This expression is particularly valid for short stress times, while for long stress times, ΔV_{th} and/or $\Delta G_m/G_{m0}$ begins to saturate [Yao *et al.*, 1987; Yang *et al.*, 1988; Inokawa *et al.*, 1991]. The slope, n, in a log–log plot is strongly dependent on V_G but has little dependence on V_D. This suggests that n changes according to hot-carrier injection mechanisms. In the case of drain avalanche hot-carrier injection, which causes the most degradation, n was 0.5–0.6 over the range of $L_{eff} = 0.35$–2 μm and $t_{ox} = 6.8$–20 nm.

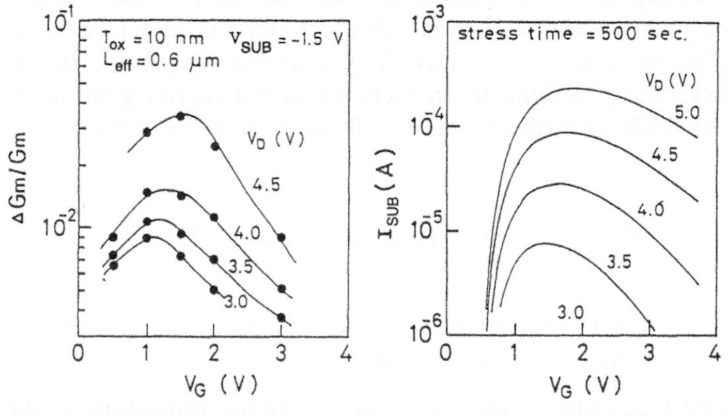

FIGURE 3.16 Relationship between G_m degradation and substrate currents as a function of V_G for a device having $L_{eff} = 0.6$ μm and $t_{ox} = 10$ nm.

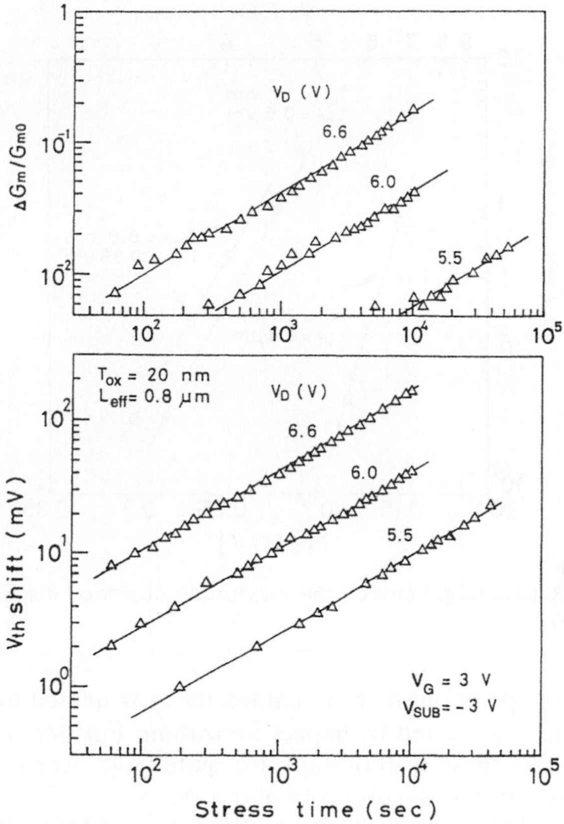

FIGURE 3.17 Device degradation in V_{th} and G_m as a function of stress time with V_D as a parameter for a MOS device having $L_{eff} = 0.8\ \mu m$ and $t_{ox} = 20$ nm.

On the other hand, the magnitude of degradation, A, is strongly dependent on V_D and has little dependence on V_G. In particular,

$$A \propto \exp(-\alpha/V_D). \tag{3.6}$$

This was verified empirically, as shown in Figure 3.18. The peak substrate current, I_{SUB}^{m}, which is indicative of the number of electron–hole pairs generated by impact ionization at the drain, also has the same dependence as shown in Figure 3.19.

$$I_{SUB}^{m} \propto \exp(-\beta/V_D) \propto A^{\beta/\alpha}. \tag{3.7}$$

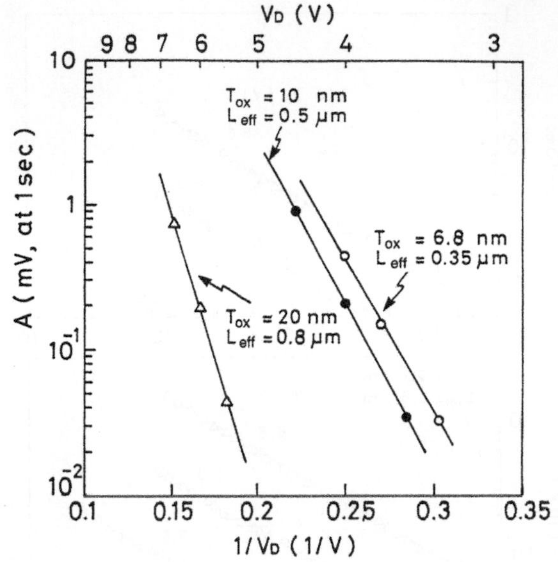

FIGURE 3.18 Relationship between the magnitude of device degradation A and drain voltage, V_D.

It is important to point out that the parameter A is related to the number of excess carriers generated by impact ionization. Furthermore, it should also be noted that these relationships are quite valid over wide ranges of L_{eff} and t_{ox}, as shown in Figures 3.18 and 3.19.

Using Eqs. (3.5), (3.6), and (3.7), the lifetime τ of MOS devices can be expressed as

$$\tau \propto \exp(b/V_D) \tag{3.8}$$

where

$$b = \alpha/n. \tag{3.9}$$

Here, τ is defined as the aging time in which the threshold voltage is shifted by 10 mV or the transconductance is reduced by 10%. Also, in the same way, τ can be obtained as a function of I_{SUB}^{m} as follows [Hu *et al.*, 1985; Takeda, 1985]:

$$\tau \propto (I_{SUB}^{m})^{-l} \tag{3.10}$$

where

$$l = (\alpha/\beta)/n. \tag{3.11}$$

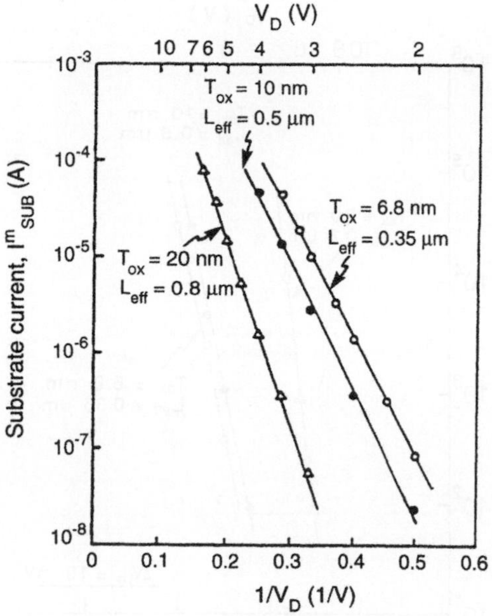

FIGURE 3.19 Relationship between the peak substrate current $I_{\mathrm{SUB}}^{\mathrm{m}}$ and V_{D}.

The range of l is 3.2–3.4 over a wide range of device dimensions. The relationships between τ and V_{D} and between τ and $I_{\mathrm{SUB}}^{\mathrm{m}}$ are empirically verified in Figures 3.20 and 3.21. The agreement between model and experiment is quite remarkable.

Thus, it is quite possible to predict the lifetime of MOS devices using Eqs. (3.8) and (3.10) and some experimental data. In particular, Eq. (3.10) can be useful for designing "hot-carrier-resistant" device structures and/or developing higher-quality gate insulators, because the substrate current, which can easily be measured, relates directly to the lifetime of a device subject to hot-carrier injection.

DAHC injection, which originates from impact ionization near the drain, determines the lifetime of MOS devices. From the impact ionization rate, $\gamma \propto \exp(-c_0/\mathscr{E})$, it can be shown that τ has the following relationships with V_{D}, I_{SUB}, and L_{eff} [Poorter and Zoestbergen, 1984; Takeda, 1984]:

$$\tau \propto \exp(\alpha/V_{\mathrm{D}}), \tag{3.12}$$

$$\tau \propto (I_{\mathrm{SUB}})^{-\rho}, \qquad \rho = 3.0, \tag{3.13}$$

$$\tau \propto \exp(-\beta/L_{\mathrm{eff}}). \tag{3.14}$$

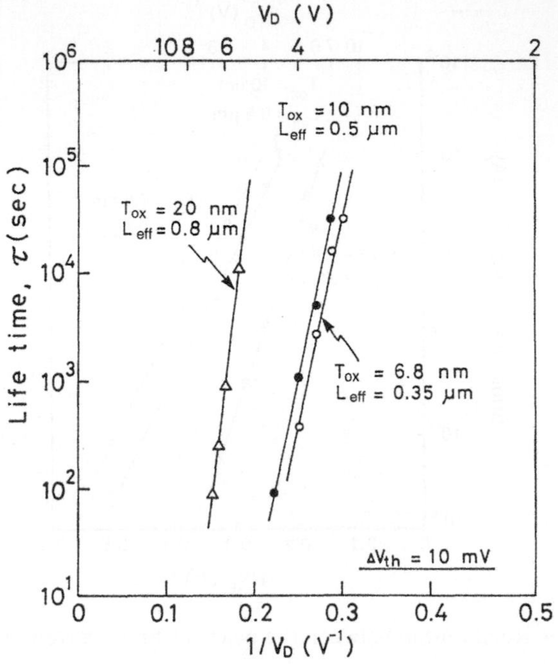

FIGURE 3.20 Dependence of the lifetime of MOS devices on $1/V_D$.

From Eqs. (3.12) and (3.14), the hot-carrier breakdown voltage, BV_{HC}, at $\Delta G_m/G_{m0} = 0.1$ is expressed as

$$BV_{HC} = C_1 L_{eff}/(L_{eff} + C_2). \tag{3.15}$$

Equation (3.15) is verified experimentally as shown in Figure 3.22 [Toyabe *et al.*, 1978].

Time-dependent dielectric breakdown (TDDB) lifetime in thin oxides has been reported [I.-C. Chen *et al.*, 1985], assuming that localized field enhancement is caused by hole trapping due to impact ionization in SiO_2. This lifetime can be expressed as

$$\tau \propto \exp(C_3/\mathscr{E}_p). \tag{3.16}$$

Where $\mathscr{E}_p$ is the lateral electric field at the drain edge.

Equation (3.16) is valid mainly for intrinsic breakdown. However, extrinsic breakdown, which is strongly dependent on the process parameters, is more important in actual DRAM failure.

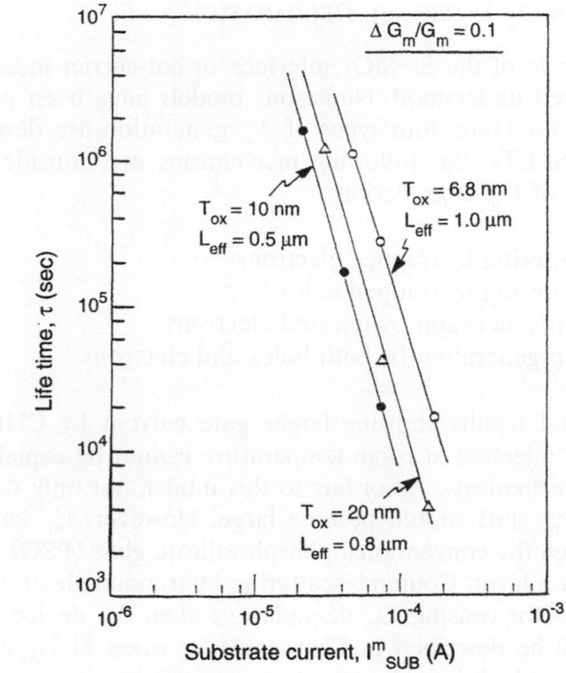

FIGURE 3.21 Dependence of the lifetime of MOS devices on I_{sub}^{m}.

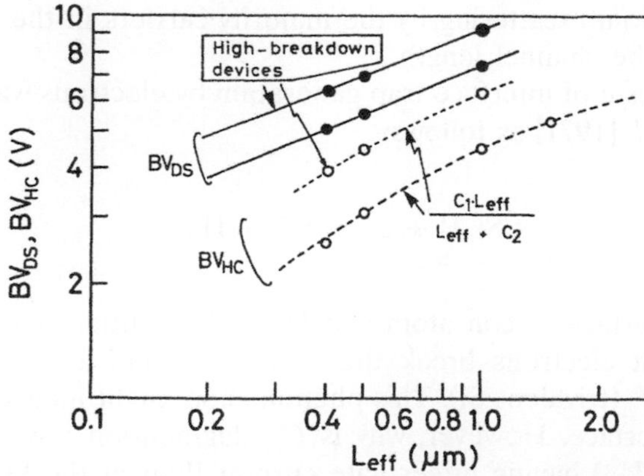

FIGURE 3.22 Hot-carrier breakdown voltage (BV_{HC}) and drain sustaining voltage (BV_{DS}) vs. effective channel length (L_{eff}). Relationships: $BV_{HC} = C_1 L_{eff}/(L_{eff} + C_2)$ and $BV_{DS} \propto L_{eff}^{\gamma}$, are satisfied.

3.3.3 THE Si–SiO$_2$ INTERFACE DEGRADATION

The exact role of the Si–SiO$_2$ interface in hot-carrier-induced degradation is not well understood. Numerous models have been proposed to elucidate this role. Here, four types of N_{it} generation are described. For n-channel MOSFETs, the following mechanisms are considered as the physical origins of G_m degradation:

1. Coulomb scattering by trapped electrons
2. Coulomb scattering by trapped holes
3. Interface trap generation by injected electrons
4. Interface trap generation by both holes and electrons.

Experimental results showing larger gate current by CHE injection than by DAHC injection at room temperature cannot be explained by the first of these mechanisms. According to this model, not only G_m degradation but also V_{th} shift should become large. However, V_{th} shift is small (Figure 3.6) when the conventional phosphosilicate glass (PSG) film is used as the passivation layer. Coulomb scattering by trapped electrons becomes the dominant factor causing G_m degradation when the device is operated at 77 K (as will be described in Chapter 4), or when Si_3N_4 is used as a passivation layer which increases the trapping efficiency.

Scattering by trapped holes, however, play a key role in increasing G_m for n-channel MOSFETs. Similarly, for p-channel MOSFETs, G_m increases as a result of scattering by trapped electrons. If hot carriers with opposite polarity to channel carriers are injected, the trapped carriers tend to suppress Coulomb scattering by the majority carriers in the channel, in effect reducing the channel length.

The mechanism of interface trap generation by electrons was proposed by Nicollian *et al.* [1971] as follows:

$$\equiv \mathrm{Si_sH} + e^- \rightarrow \mathrm{Si}^* + \mathrm{H_i}, \qquad (3.17)$$

where $\mathrm{Si_s}$ is a surface silicon atom and $\mathrm{H_i}$ an interstitial hydrogen atom. The injected hot electrons break the $\equiv \mathrm{Si_s}$–H bond and generate the interface trap Si* (trivalent Si). This phenomenon might indeed happen at the Si–SiO$_2$ interface. However, why is G_m degradation less in the CHE injection ($V_G \geq V_D$) having larger gate current than in the DAHC injection ($V_G < V_D$)? This experimental fact is very significant in discussing degradation mechanisms.

The following model was proposed to illustrate interface trap generation by both holes and electrons [Fair and Sun, 1981]. Usually, H_2 anneal occurs at 400° C in order to saturate the unbonded Si with H at the Si–SiO_2 interface. The following reaction is likely to occur.

$$\text{hot e}^- + \text{hot h}^+ + H_2 \rightarrow 2H_i. \qquad (3.18)$$

Energy from the injected electrons and holes breaks the H_2 molecular bonds at the Si–SiO_2 interface. Then the created H_i reacts with $\equiv Si_sH$ in the following way:

$$\equiv Si_sH + H_i \rightarrow Si^* + H_2. \qquad (3.19)$$

A model based on hole trapping results in the same reaction [Svensson, 1978]. We believe this model of interface trap generation can be used to explain experimental results of G_m degradation. According to this model, to break the H–H bond with a bond energy of 4.5 eV, both the hot electron and hot hole are required to have a minimum energy of 2.25 eV each. This model suggests that the hot-carrier effect can be important even at low-voltage operation ($V_D < 3$V), as will be discussed in Chapter 5.

It is also possible that only hot holes break the $\equiv Si_s$–H bond. Tsuchiya showed experimentally that scattering by trapped electrons is dominant in the initial stage of G_m degradation and then interface trap generation becomes more dominant in long-time stressing [Tsuchiya, 1987, 1987b]. Figure 3.23 helps to separate G_m degradations caused by trapped carriers

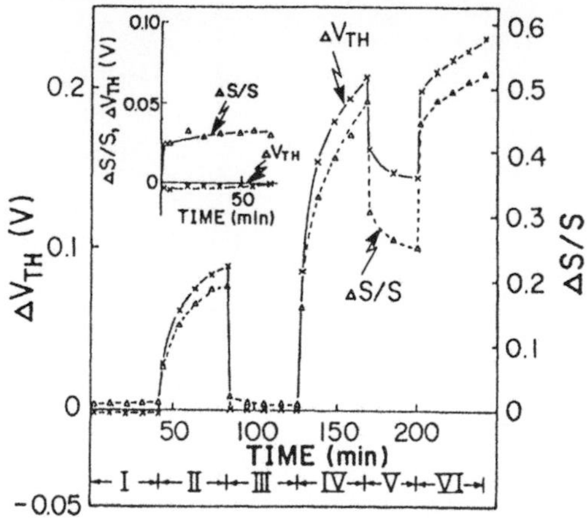

FIGURE 3.23 Changes in ΔV_{TH} and $\Delta S/S$ as a function of time. $L_{eff} = 0.7\ \mu$m. Terms I and III: hot-hole injection ($V_{AG} = 0.6$ V, $V_{AD} = 5.5$ V); terms II, IV, and VI: hot-electron injection ($V_{AG} = V_{AD} = 5.5$ V); term V: trapping of electrons ($V_{AG} = -3$ V, $V_{AD} = 0$ V). The insert of the figure is for only hot-electron injection. (Reprinted with permission from T. Tsuchiya, T. Kobayashi, and S. Nakajima, "Hot-Carrier-Injected Oxide Region and Hot-Electron Trapping as the Main Cause in Si n-MOSFET Degradation," *IEEE Trans. Electron Dev.* **34**, 386–391 (© 1987 IEEE).)

and interface trap generation. The effect of interface trap generation becomes increasingly dominant as G_m degradation continues.

3.4 Summary

Hot-carrier device degradation in ULSI devices has been discussed. In some cases, the true physical origin of the hot-carrier phenomenon is still controversial. In particular, the degradation mechanism related to interaction between hot carriers and Si atoms at the $Si-SiO_2$ interface requires further elucidation in order to realize more sophisticated ULSI devices and enhanced reliability for future applications.

CHAPTER

4

AC AND PROCESS-INDUCED HOT-CARRIER EFFECTS

4.1 Introduction

The basic device degradation mechanisms due to hot carriers have been discussed in depth in Chapter 3. The discussion was based on DC electrical stress measurement results. In practice, the operating device and circuit are subject to AC signals. Thus a study of dynamic or AC stress effects on device degradation is essential for an improved understanding of the hot-carrier phenomena. Further, other factors such as fabrication process and changes in material properties can have a significant influence on the observed device degradation. These topics are discussed at some length in this chapter.

4.2 Dynamic (AC) Stress Effects

To reduce hot-carrier degradation in scaled metal-oxide semiconductor (MOS) devices, two approaches have been taken so far: (1) hot-carrier-resistant device structures such as lightly doped drain (LDD) and gate–drain overlapped device (GOLD) and (2) reduction of power supply voltage to 1.5–3 V [Aoki *et al.*, 1989]. The former with a 5-V power supply suffers from power consumption problems and process complexity. The second approach must overcome the problem of reduced speed. A third approach is to make good use of AC effects including the duty ratio. This calls for a deeper physical understanding of AC hot-carrier phenomena and the subsequent use of this understanding in device/circuit design and supply voltage selection. Despite many experiments and analyses on AC hot-carrier effects [Choi *et al.*, 1987b], noise caused by the wiring *inductance* in ultra large-scale integrated (ULSI) circuits and in measurement systems [Bellens *et al.*, 1990] has contributed to the controversy in data interpretations. The effect of this noise is most significant near the falling and rising edges of a gate pulse. In this section, a universal guideline on AC hot-carrier effects is proposed from the viewpoints of (1) gate pulse fall/rise time (Δt), (2) AC frequency, and (3) device structures.

4.2.1 Gate Pulse-Induced Noises

The results summarized here are based on recent work by Takeda *et al.* [1991]. Figure 4.1 shows AC fluctuations of drain and source voltages and substrate current with no precautions and with precautions taken against noises. The devices used in this study are n-MOSFETs with gate length, L, of 1.4 μm and channel width of 10 or 100 μm. The gate pulse has a frequency of 500 kHz, a duty ratio of 0.5, and a fall/rise time of 10 ns. Drain voltage is DC. During AC stress, an external 50-Ω resistor is automatically connected to the gate electrode to alternate the imposition of AC pulses and measurement of DC device characteristics. For precautions against gate pulse-induced noise, an experimental setup as shown in Figure 4.2 was used. If no precautions are taken, source and drain voltage noises and thereby substrate current noise are generated as shown in Figure 4.1.

4.2.2 AC Hot-Carrier Degradation Due to Noises

a. Degradation Mechanism As schematically illustrated in Figure 4.3, it can be seen that the gate pulse-induced I_{SUB} noise is generated by the wiring-inductance-induced electromotive force at the falling edge of the gate pulse. A-mode is the electron injection current from source due to negative bias (ΔV_S). B-mode is the extra hole current generated by impact ionization of injected electrons at the drain edge. Thus, due to ΔV_s, a number of extra hot carriers are created spontaneously, resulting in an enhancement of hot-carrier degradation.

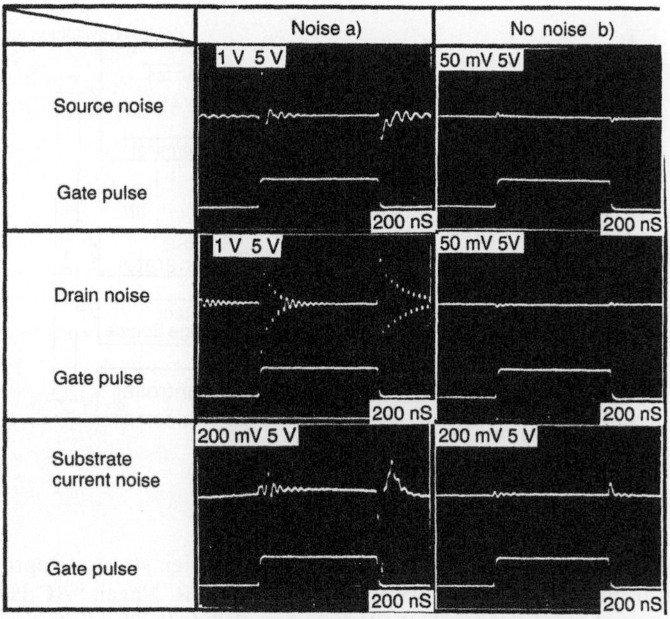

FIGURE 4.1 AC fluctuations of drain and source voltages and substrate current with no precautions (a) and precautions taken against noises (b). (Reprinted with permission from E. Takeda, R. Izawa, K. Umeda, and R. Nagai, "AC Hot-Carrier Effects in Scaled MOS Devices," *IEEE 1991 Int. Reliability Physics Symp.*, 118–122 (© 1991 IEEE).)

b. Noise Formulation Figure 4.4 shows the source noise (ΔV_S) as a function of the wiring impedance (Z) for two different channel widths (W). It is found that ΔV_S is a logarithmic function of Z and is negligible when Z is smaller than 250 mΩ, which is an important fact for circuit design. Figure 4.5 shows the relation between ΔV_S and gate voltage V_G with Z as a parameter. From these experimental results and taking the wiring-inductance-induced electromotive force into account, the source noise can be formulated as

$$(\Delta V_S)^2 = 2q\,\Delta f(\Delta I_S)Z^2 \simeq 2q\left(\frac{1}{\Delta t}\right)\frac{W}{L}V_G e^{q\,\Delta V_S/kT}Z, \tag{4.1}$$

where Δf, ΔI_S and Δt are the AC frequency spread, injection current from source, and gate pulse fall time, respectively. Empirically, the source

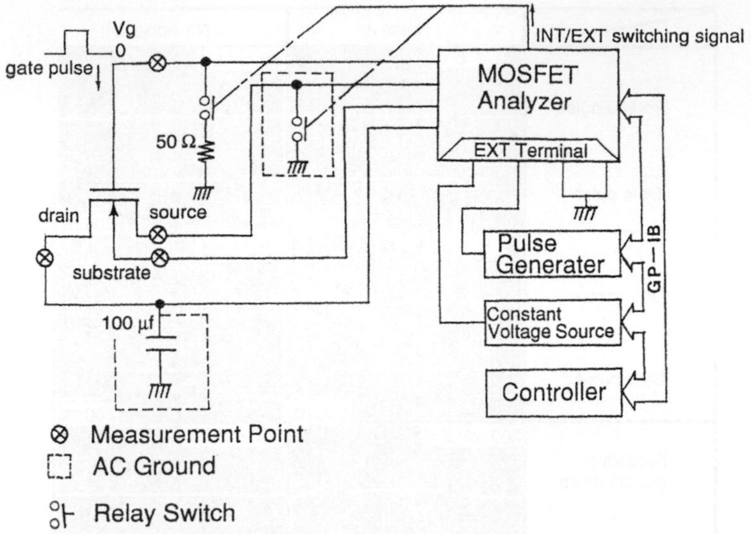

FIGURE 4.2 A measurement system for AC hot-carrier stress. (Reprinted with permission from E. Takeda, R. Izawa, K. Umeda, and R. Nagai, "AC Hot-Carrier Effects in Scaled MOS Devices," *IEEE 1991 Int. Reliability Physics Symp.*, 118–122 (© 1991 IEEE).)

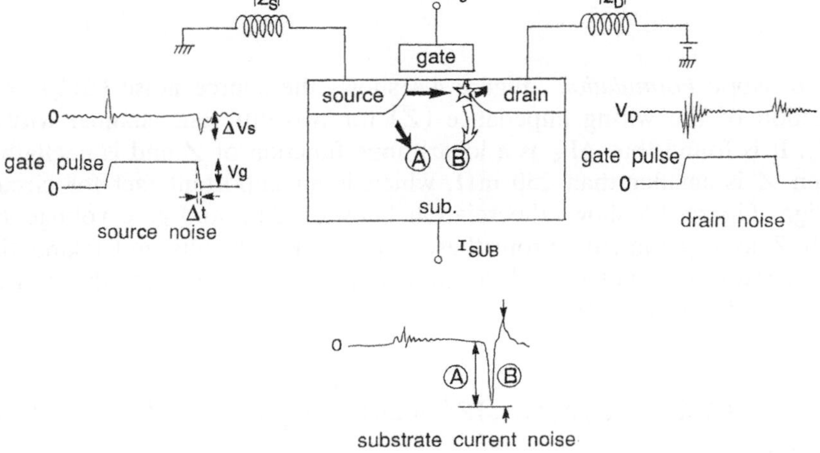

FIGURE 4.3 Gate pulse-induced noise model. (Reprinted with permission from E. Takeda, R. Izawa, K. Umeda, and R. Nagai, "AC Hot-Carrier Effects in Scaled MOS Devices," *IEEE 1991 Int. Reliability Physics Symp.*, 118–122 (© 1991 IEEE).)

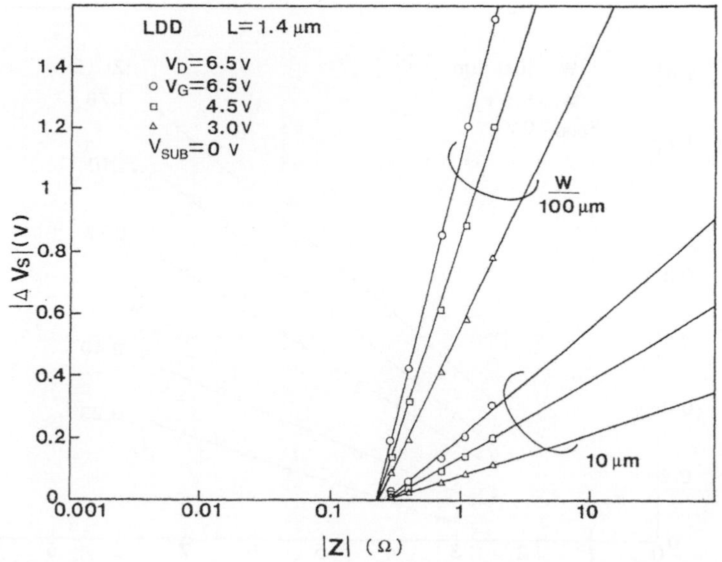

FIGURE 4.4 Source noise as a function of the wiring impedance. (Reprinted with permission from E. Takeda, R. Izawa, K. Umeda, and R. Nagai, "AC Hot-Carrier Effects in Scaled MOS Devices," *IEEE 1991 Int. Reliability Physics Symp.*, 118–122 (© 1991 IEEE).)

noise varies with the wiring impedance approximately as follows:

$$\Delta V_s \simeq \frac{W}{L} V_G (\Delta t)^X \log Z, \tag{4.2}$$

where X is somewhat dependent on Δt. The physical connection between Eqs. (4.1) and (4.2) is still controversial, although the latter fits well the experimental data shown in Figure 4.6.

c. Device Degradation Figure 4.7 shows the time dependence of G_m degradation with Z as a parameter. First, it can be seen that G_m degradation increases with Z increase and consequently with ΔV_S increase. On the other hand, in the case of complete precautions against noise, AC hot-carrier degradation is almost the same as DC degradation (at least in LDD structures) in terms of the effective stress time, which takes the duty ratio into account.

Figure 4.8 shows G_m degradation as a function of substrate current noise I_{SUB} (B-mode), where stress time is 10,000 s. It is found that with I_{SUB} increase, the degradation is enhanced as expected from the above

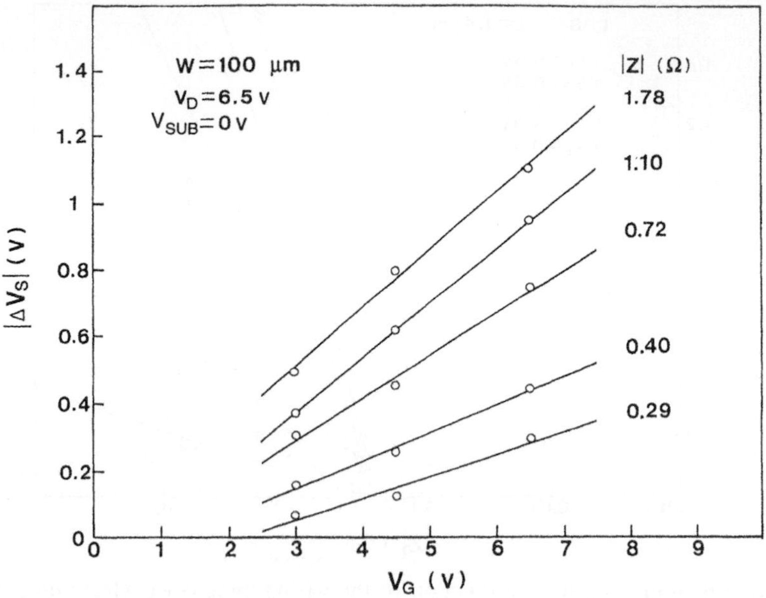

FIGURE 4.5 Source noise vs. gate voltage relationship with inductance as a parameter. (Reprinted with permission from E. Takeda, R. Izawa, K. Umeda, and R. Nagai, "AC Hot-Carrier Effects in Scaled MOS Devices," *IEEE 1991 Int. Reliability Physics Symp.*, 118–122 (© 1991 IEEE).)

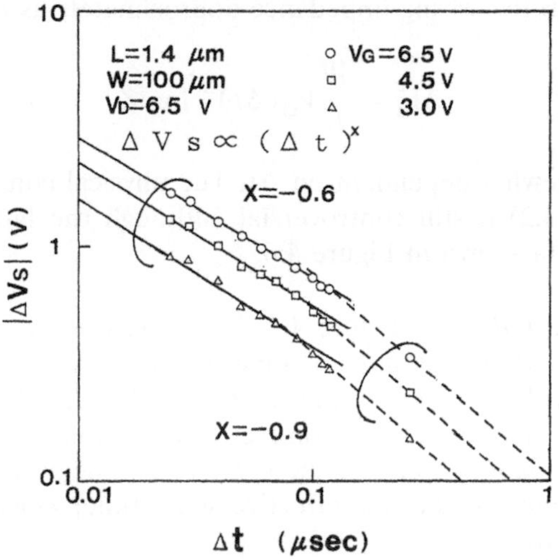

FIGURE 4.6 Source noise vs. gate pulse fall/rise time relationship. (Reprinted with permission from E. Takeda, R. Izawa, K. Umeda, and R. Nagai, "AC Hot-Carrier Effects in Scaled MOS Devices," *IEEE 1991 Int. Reliability Physics Symp.*, 118–122 (© 1991 IEEE).)

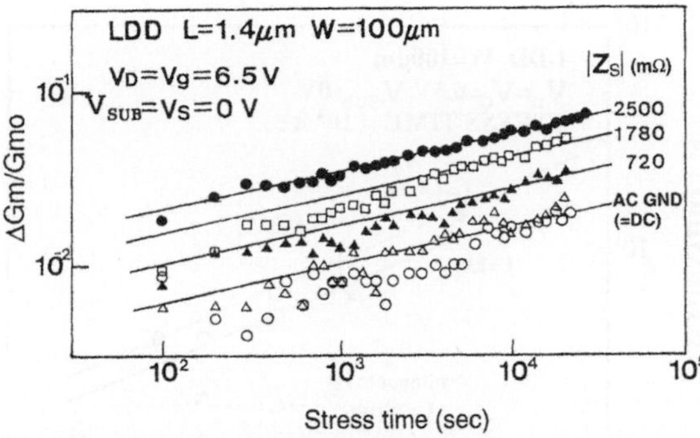

FIGURE 4.7 Time-dependent G_m degradation with the impedance as parameter. (Reprinted with permission from E. Takeda, R. Izawa, K. Umeda, and R. Nagai, "AC Hot-Carrier Effects in Scaled MOS Devices," *IEEE 1991 Int. Reliability Physics Symp.*, 118–122 (© 1991 IEEE).)

model. Also, in the same way as DC stress, device degradation ($\Delta G_m/G_{m0}$, ΔV_{th}) has the behavior (I_{SUB} noise)^n, where $n \approx 0.5$.

4.2.3 AC Hot-Carrier Effects without Noises

a. Gate Pulse Fall / Rise Time (Δt) Dependence Figure 4.9 illustrates the dependence of G_m degradation on Δt without noise, as compared to

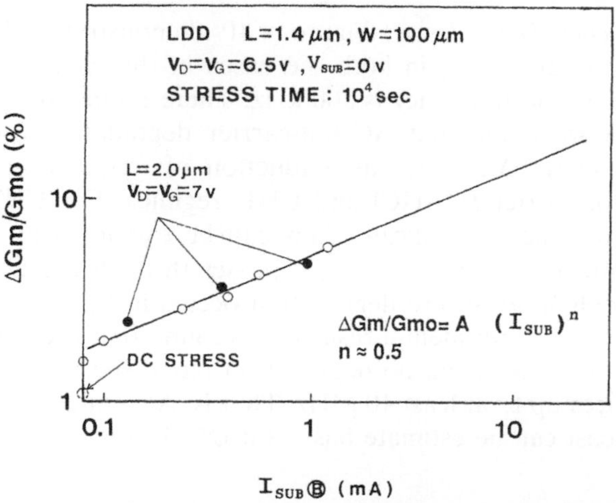

FIGURE 4.8 G_m degradation vs. I_{SUB} noise relationship. (Reprinted with permission from E. Takeda, R. Izawa, K. Umeda, and R. Nagai, "AC Hot-Carrier Effects in Scaled MOS Devices," *IEEE 1991 Int. Reliability Physics Symp.*, 118–122 (© 1991 IEEE).)

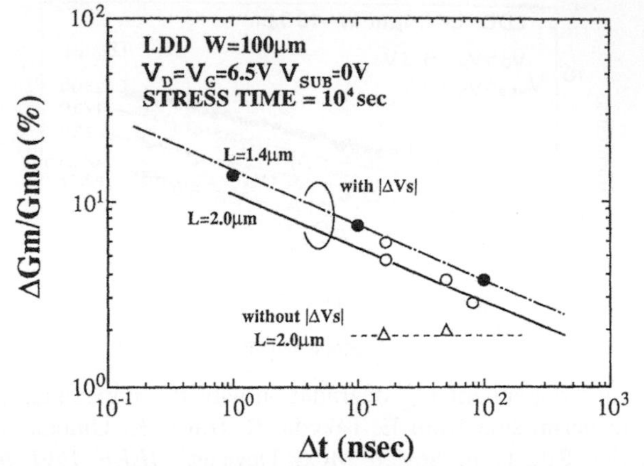

FIGURE 4.9 G_m degradation vs. Δt relationship. (Reprinted with permission from E. Takeda, R. Izawa, K. Umeda, and R. Nagai, "AC Hot-Carrier Effects in Scaled MOS Devices," *IEEE 1991 Int. Reliability Physics Symp.*, 118–122 (© 1991 IEEE).)

that with noise. It should be noted that G_m degradation in case of no noise does not depend on the gate falling time. These experimental results suggest that AC hot-carrier effects in LDD can be evaluated on the basis of DC degradation at least in the range of 1–10 ns fall time if the circuit's duty ratio is taken into account.

b. Frequency Dependence Figure 4.10 demonstrates the frequency dependence of $\Delta G_m/G_{m0}$ in both the cases in the channel hot electron (CHE) regime. The frequency is 500 kHz. There seems to be a significant difference between DC and AC hot-carrier degradation without noises. Figure 4.11 shows $\Delta G_m/G_{m0}$ as a function of frequency in both drain avalanche hot carrier (DAHC) and CHE regimes. For CHE, frequency dependence is indeed significant. This can be explained by the fact that when the gate voltage rises or falls, it passes through the DAHC regime, where relatively more severe degradation occurs in LDD and single drain (SD). From these experimental results, it is found that if complete precaution against noise is taken, no degradation mode specific to AC exists in LDD structures up to at least 10 MHz. That is, AC hot-carrier degradation in LDD at least can be estimate based on DC degradation results.

4.2.4 DEVICE STRUCTURE DEPENDENCE OF AC HOT-CARRIER EFFECTS

a. Device Structure Dependence Figure 4.12 shows G_m degradation as a function of stress time under AC and DC conditions, for SD, LDD, and GOLD structures. No difference between AC and DC stresses was ob-

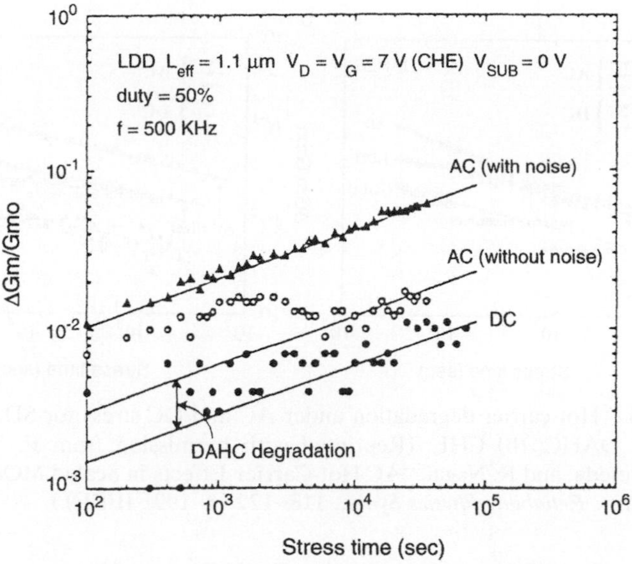

FIGURE 4.10 Comparison of DC and AC degradation in both no noise and noise cases. (Reprinted with permission from E. Takeda, R. Izawa, K. Umeda, and R. Nagai, "AC Hot-Carrier Effects in Scaled MOS Devices," *IEEE 1991 Int. Reliability Physics Symp.*, 118–122 (© 1991 IEEE).)

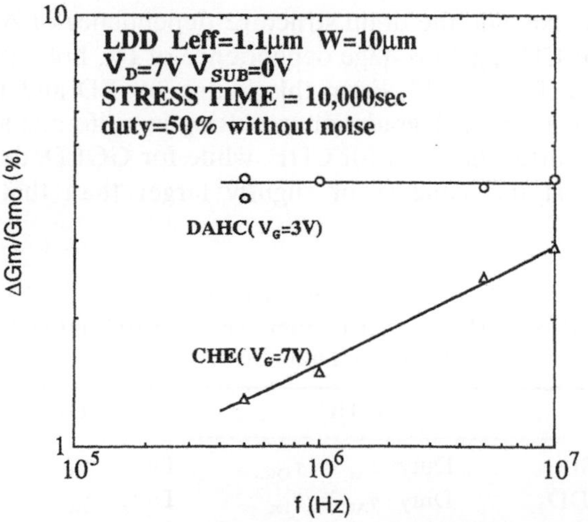

FIGURE 4.11 Frequency dependence of G_m degradation in both the DAHC and CHE regimes. (Reprinted with permission from E. Takeda, R. Izawa, K. Umeda, and R. Nagai, "AC Hot-Carrier Effects in Scaled MOS Devices," *IEEE 1991 Int. Reliability Physics Symp.*, 118–122 (© 1991 IEEE).)

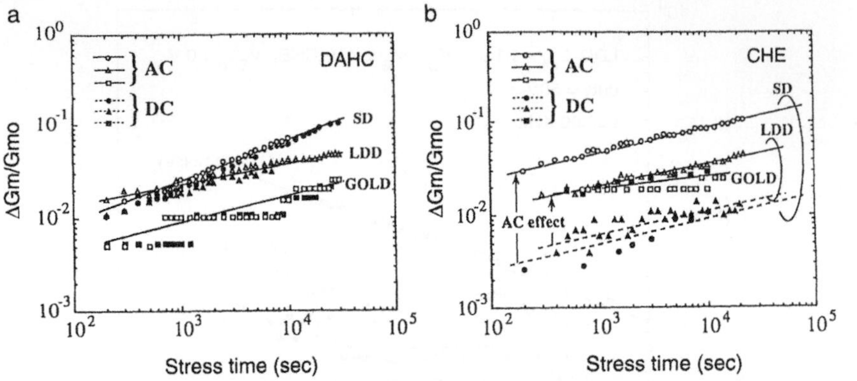

FIGURE 4.12 Hot-carrier degradation under AC and DC stress for SD, LDD, and GOLD. (a) DAHC; (b) CHE. (Reprinted with permission from E. Takeda, R. Izawa, K. Umeda, and R. Nagai, "AC Hot-Carrier Effects in Scaled MOS Devices," *IEEE 1991 Int. Reliability Physics Symp.*, 118–122 (© 1991 IEEE).)

served for all drain structures in the DAHC case. However, for CHE, enhanced AC hot-carrier degradation was seen for both SD and LDD as mentioned above, while no enhancement was noticed for GOLD. The above results are summarized in Table 4.1 in terms of lifetime, which has the usual definition of the time stress required to reach 10% G_m degradation.

In order to examine the drain structure dependence of AC hot-carrier degradation for CHE, gate voltage dependence of DC hot-carrier degradation is shown in Figure 4.13. From this figure, for SD and LDD, DAHC provides the most severe degradation, resulting in a lifetime several orders of magnitude shorter than that for CHE, while for GOLD, degradation for CHE was almost the same as or slightly larger than that for DAHC.

TABLE 4.1
Comparison of Hot-Carrier Lifetime between AC and DC Stress for Three Drain Structures

	DAHC	CHE
SD	$\text{Duty} \cdot \tau_{AC} \doteq \tau_{DC}$	$\text{Duty} \cdot \tau_{AC} \ll \tau_{DC}$
LDD	$\text{Duty} \cdot \tau_{AC} \doteq \tau_{DC}$	$\text{Duty} \cdot \tau_{AC} < \tau_{DC}$
GOLD	$\text{Duty} \cdot \tau_{AC} \doteq \tau_{DC}$	$\text{Duty} \cdot \tau_{AC} \doteq \tau_{DC}$

Note. Reprinted with permission from E. Takeda, R. Izawa, K. Umeda, and R. Nagai, "AC Hot-Carrier Effects in Scaled MOS Devices," *IEEE 1991 Int. Reliability Physics Symp.*, 118–122 (© 1991 IEEE).

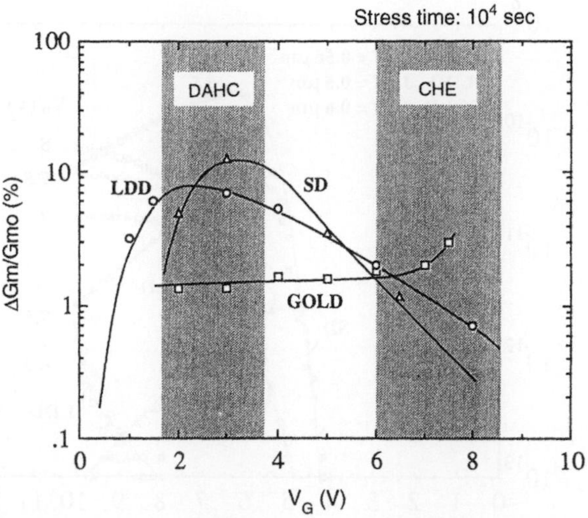

Figure 4.13 Gate voltage dependence of G_m degradation under DC stress. (Reprinted with permission from E. Takeda, R. Izawa, K. Umeda, and R. Nagai, "AC Hot-Carrier Effects in Scaled MOS Devices," *IEEE 1991 Int. Reliability Physics Symp.*, 118–122 (© 1991 IEEE).)

During the rise and fall intervals in the AC CHE stress condition, devices were undergoing DAHC. This DAHC stress, which is much stronger than CHE for SD and LDD, is thought to be the main cause of the accelerated AC degradation.

Figure 4.14 shows the I_G characteristics for the three drain structures. I_G for GOLD is considerably larger than that for LDD, because of lower channel electric field in the latter. GOLD's large I_G, probably due to large oxide field at the gate drain overlapped region giving rise to electron injection, correlates with the relatively high degradation rate in CHE stress, as seen in Figure 4.13.

Figure 4.15 shows the correlation between lifetime under AC CHE stress and stress frequency, together with lifetime calculated from DC degradation. Lifetime is seen to decrease with increasing frequency, as was expected. The model is in good agreement with the LDD stress data and thus confirms that the DAHC stress during gate pulse transients is the main cause of enhanced degradation.

b. New AC Degradation Mechanism In several devices, measured SD lifetime under AC stress was much smaller than that predicted by the model. This suggests the presence of another new acceleration factor inherent to AC stress. To clarify this new AC degradation mechanism in SD structure, the following experiment was carried out.

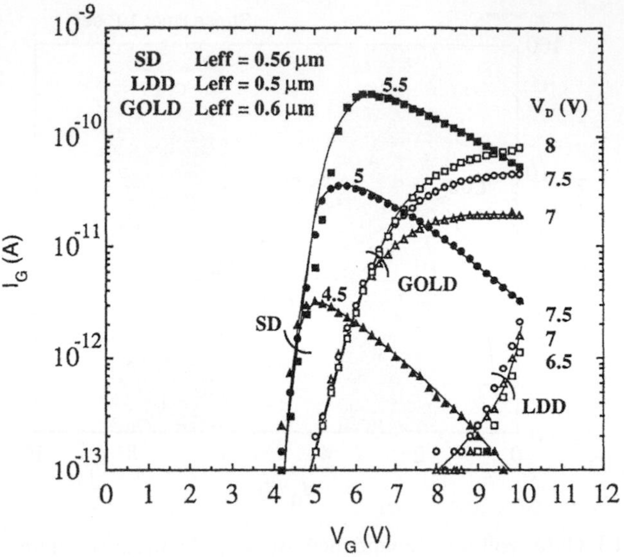

FIGURE 4.14 Gate current characteristics. (Reprinted with permission from E. Takeda, R. Izawa, K. Umeda, and R. Nagai, "AC Hot-Carrier Effects in Scaled MOS Devices," *IEEE 1991 Int. Reliability Physics Symp.*, 118–122 (© 1991 IEEE).)

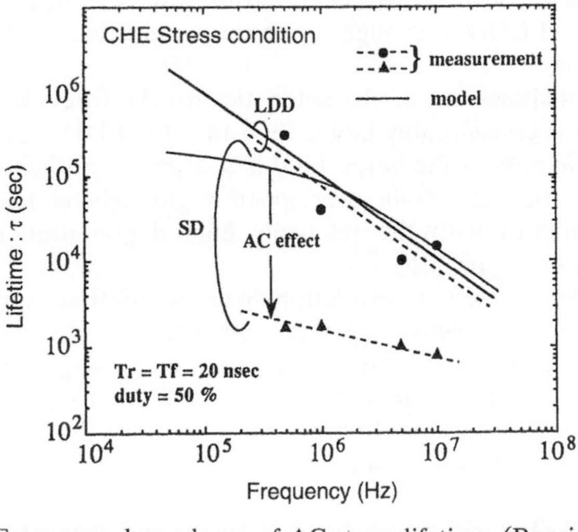

FIGURE 4.15 Frequency dependence of AC stress lifetime. (Reprinted with permission from E. Takeda, R. Izawa, K. Umeda, and R. Nagai, "AC Hot-Carrier Effects in Scaled MOS Devices," *IEEE 1991 Int. Reliability Physics Symp.*, 118–122 (© 1991 IEEE).)

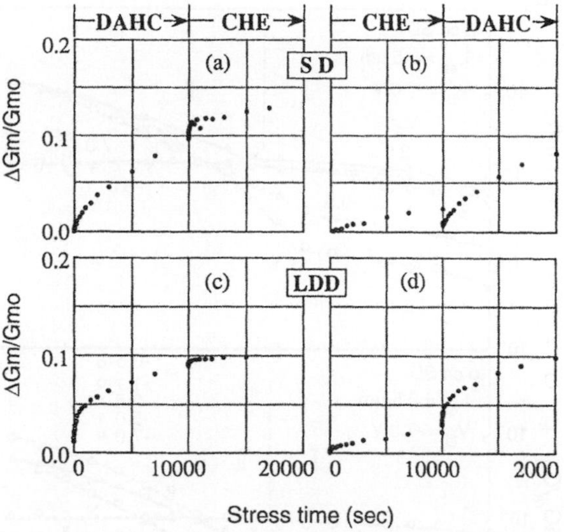

FIGURE 4.16 Hot-carrier degradation, CHE following DAHC (a and c) and DAHC following CHE (b and d), for SD and LDD. (Reprinted with permission from E. Takeda, R. Izawa, K. Umeda, and R. Nagai, "AC Hot-Carrier Effects in Scaled MOS Devices," *IEEE 1991 Int. Reliability Physics Symp.*, 118–122 (© 1991 IEEE).)

Figure 4.16 shows G_m degradation with CHE stress following DAHC stress and vice versa, for both SD and LDD MOSFETs. For SD, rapid degradation was observed under CHE immediately after DAHC stress (Figure 4.16a), which is thought to be caused by trapping of injected channel hot electrons (high gate voltage) at neutral traps generated during the DAHC (low gate voltage) stress period [Doyle *et al.*, 1990b]. Such enhanced degradation was not seen in LDD. This mechanism is thought to be the possible acceleration factor for AC stress in SD, while for LDD, low I_G and the fact that spacer-induced degradation is dominant render this mechanism unimportant.

4.2.5 Initial Stage Degradation

In Figures 4.17 and 4.18, V_{th} shift and G_m degradation in both *p*-channel SD and LDD devices from 1 μs to 1000 s are shown [Igura and Takeda, 1987]. It can be seen that the degradation has two phases. Mechanisms of these phenomena will be discussed below.

In the first phase (1 $\mu s < t <$ 10 ms), the ΔV_{th} behavior saturates quite rapidly and keeps increasing very slowly up to about 10 ms. In the

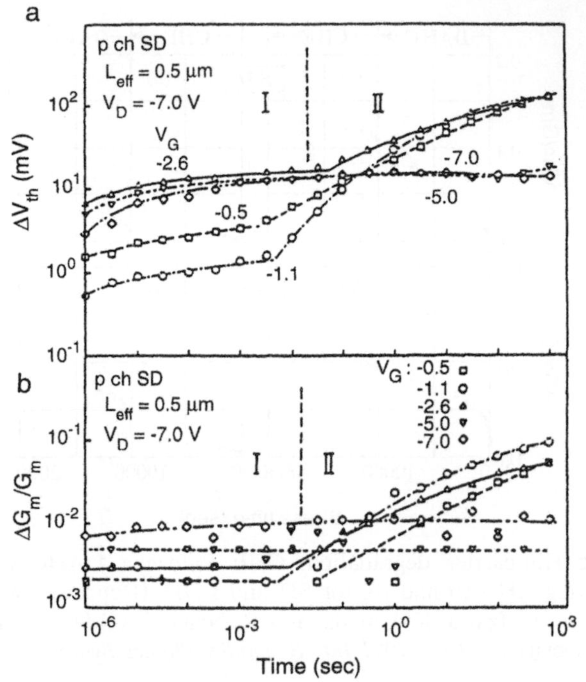

FIGURE 4.17 Degradation vs. stress time in *p*-channel SD devices. (a) V_{th} shift; (b) fractional G_m change (increase).

second phase ($t > 10$ ms), a new degradation mode appears and dominates the degradation. In Figure 4.19, the time evolution of the V_{th} shift is shown as a function of stress gate voltage. The corresponding I_{SUB} and I_G are also shown.

In phase I, the V_{th} shift is the severest under the bias condition which yields the maximum substrate current. Under this condition, many holes and electrons exist near the Si–SiO_2 interface, resulting in interface trap generation and electron trapping by these traps. Since this phenomenon occurs in the region close to the interface, it might have a rather short time constant.

In phase II, on the other hand, an I_G peak condition gives the largest V_{th} shift. Since the number of injected electrons in this case is at a maximum, electron trapping in the oxide traps might be the main cause of the degradation. This process might occur in the entire oxide layer and might have a longer time constant.

The G_m variation (in fact, it is a G_m increase) seems to show almost the same characteristics as the V_{th} shift (Figure 4.17). The G_m degradation is most likely due to electron trapping by the interface or oxide traps.

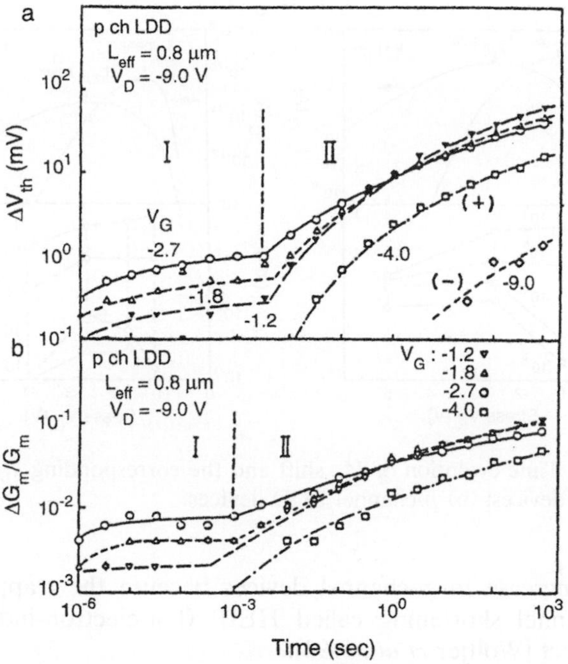

FIGURE 4.18 Degradation vs. stress time in *p*-channel LDDs. (a) V_{th} shift; (b) fractional G_m change (increase).

In Figure 4.20, V_{th} shift and G_m degradation in *n*-channel LDD devices are shown. Here, the drain voltage is 7 V. The V_{th} shift shows characteristics similar to those of *p*-channel devices. In phase I ($t < 1$ s), the I_{SUB} peak condition gives the largest V_{th} shift. In phase II ($t > 1$ s), on the other hand, the I_G peak condition yields the most degradation. This can be attributed to the fact that electron trapping is responsible for the V_{th} shift. Phase I reflects the electron trapping by the interface traps, and phase II corresponds to electron trapping by the oxide traps and/or the interface traps generated by hot carriers.

G_m degradation in *n*-channel devices is considered to be caused by interface trap creation. The degradation in phase I corresponds to interface trap generation by injected holes. Thus the lower V_G condition ($V_G = 2$ V) gives the largest degradation. In phase II, on the other hand, the degradation is caused by holes as well as electrons [Fair and Sun, 1981]. So the I_{SUB} peak condition gives the largest degradation.

As has already been seen, this study clarifies the roles which electrons and holes play in the degradation process. Electrons are trapped by the interface traps or oxide traps, and they cause the V_{th} shift in both *n*-channel and *p*-channel devices. Electron trapping is also the main cause

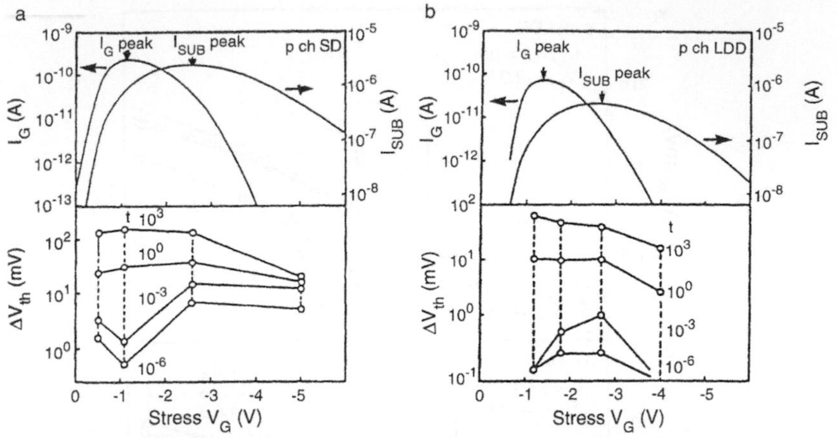

FIGURE 4.19 Time evolution of V_{th} shift and the corresponding I_{SUB} and I_G. (a) *p*-channel SD devices; (b) *p*-channel LDD devices.

of the G_m increase in *p*-channel devices because the trapped electrons cause a channel shortening called HEIP (hot-electron-induced punch-through) effect [Woltjer *et al.*, 1993].

Injected holes, on the other hand, play an essential role in the interface trap creation, and this results in the large G_m degradation. Figure 4.21 shows the initial stage degradation with the same V_G but with three different values of V_D in *p*-channel SD devices. All are under the I_{SUB} peak condition. The saturated value of V_{th} shift in phase I increases with V_D. This fact implies that the degradation in phase I is not simply electron trapping by interface traps which exists from the beginning. Rather, it is a complex process of interface trap generation due to injected holes and electron trapping by these created traps. In the case of *n*-channel devices, hole injection without electrons ($V_G = 2$ V) creates interface traps rapidly (G_m degradation in phase I). The injection of both holes and electrons causes the severest G_m degradation. The existence of both types of carriers might be responsible for the creation of traps in the oxide, and this process may have a rather long time constant. This corresponds to the usual long-time stress experiments.

4.3 Process Effects on Hot-Carrier Degradation

Fine-line patterning technologies such as electron-beam lithography and reactive ion etching are radiative processes. These process steps are

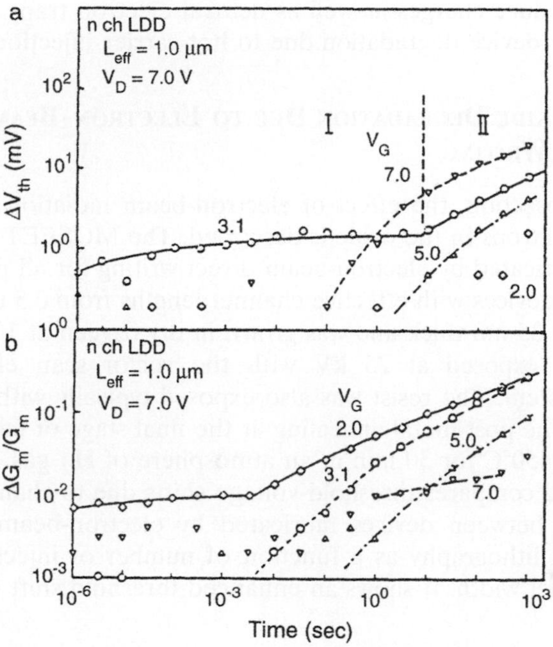

FIGURE 4.20 Degradation vs. stress time in n-channel LDDs. (a) V_{th} shift; (b) G_m degradation.

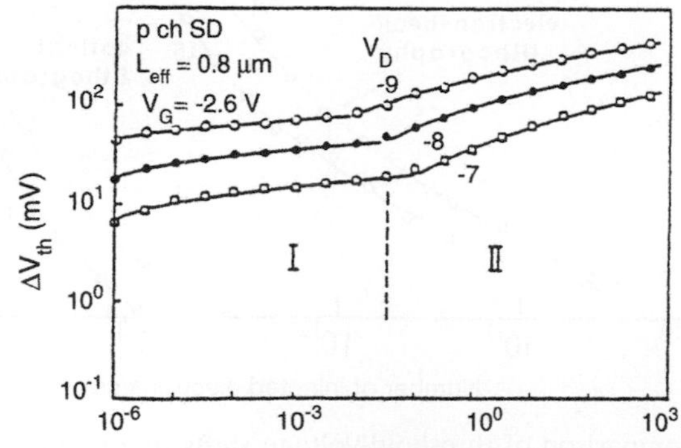

FIGURE 4.21 Initial stage degradation of p-channel devices with three different values of V_D.

known to introduce charges as well as neutral electron traps in the oxide, which enhance device degradation due to hot-carrier injection.

4.3.1 Gate Oxide Degradation Due to Electron-Beam Direct Writing

In this subsection, the effect of electron-beam radiation on the trapping of hot electrons in the oxide is described. The MOSFETs used in this study were fabricated by electron-beam direct writing for all process steps, which include devices with effective channel lengths from 0.5 to 6 μm. The gate oxide was 35 nm thick and was grown in dry oxygen at 1000°C. Resist patterns were exposed at 25 kV with the vector scan electron-beam lithography system. The resist was also exposed typically with a dosage of 20 $\mu C/cm^2$. The post-metal annealing at the final stage of fabrication was carried out at 450°C for 30 min in an atmosphere of H_2 gas.

Figure 4.22 compares threshold-voltage shifts due to channel hot-electron injection between devices fabricated by electron-beam lithography and by optical lithography as a function of number of injected electrons per unit channel width. It shows an enhanced threshold shift by about two

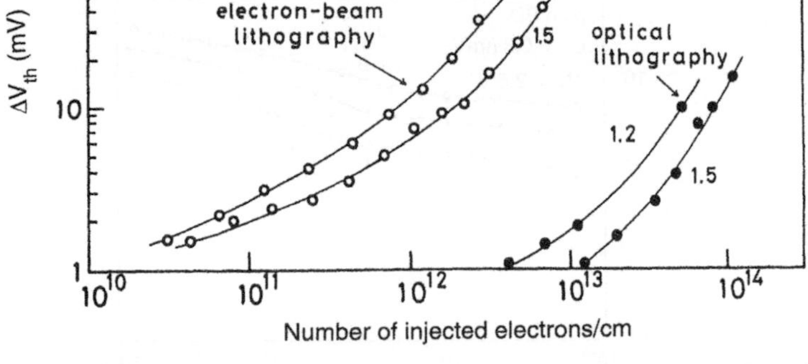

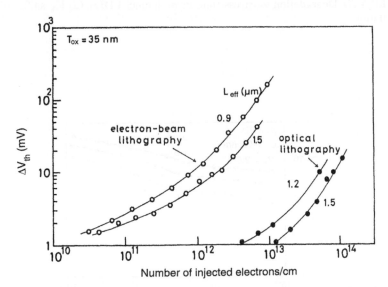

FIGURE 4.22 Comparison of threshold-voltage shifts due to channel hot-electron injection between devices fabricated by electron-beam lithography and optical lithography.

orders of magnitude due to electron-beam radiation. This is due positively charged centers created by the radiation, which act as electron traps.

By using the substrate hot-electron injection method, the effective trap concentration created in the oxide by electron-beam radiation and its cross section are obtained. Figure 4.23 is a plot of effective trapping efficiency versus the number of injected electrons per unit area for both devices. From Figure 4.23, the effective trap concentration and its cross section in electron-beam lithography are determined to be about 5×10^{10} and 2×10^{-15} cm^2, respectively. These values show an increase of more than an order of magnitude, compared with those of devices fabricated by optical lithography.

Electron-beam radiation–induced electron traps further aggravate device degradation due to hot-electron injection by increasing the probability that an injected electron may become trapped. Thus electron-beam radia-

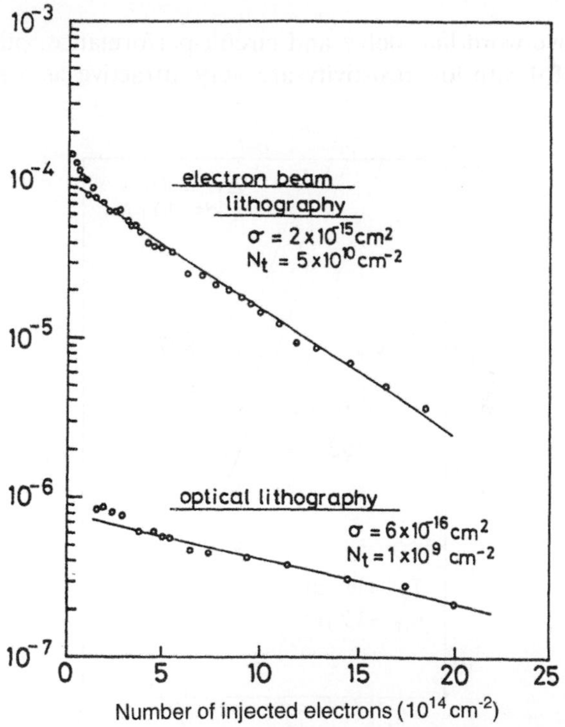

Figure 4.23 Plots of effective trapping efficiency vs. the number of injected electrons per unit area for both devices fabricated by electron-beam lithography and optical lithography.

tion can bring about a lowering of about 1 V in the highest applicable voltage, BV_{DC}. Therefore, it is important to take into account the influence of radiation damage on ULSI design [Ning, 1976].

4.3.2 Isolation Effects on Hot-Carrier Degradation

The dependence of lifetime on gate width, W_{eff}, is shown in Figure 4.24 for the degradation condition of $\Delta V_{th} = 10$ mV. The dependence of interface trap density, N_{it}, which was measured using the charge pumping technique, on W_{eff} is shown in Figure 4.25. The N_{it} vs. W_{eff} relationship is consistent with the lifetime behavior. In the narrower channel, interface trap density induced during the fabrication process is larger. The isolation process used was local oxidation of silicon (LOCOS).

4.4 Materials Effects on Hot-Carrier Degradation

To improve word line delay and circuit-performance, pure refractory metals (W, Mo) with low resistivity are very attractive as a substitute for

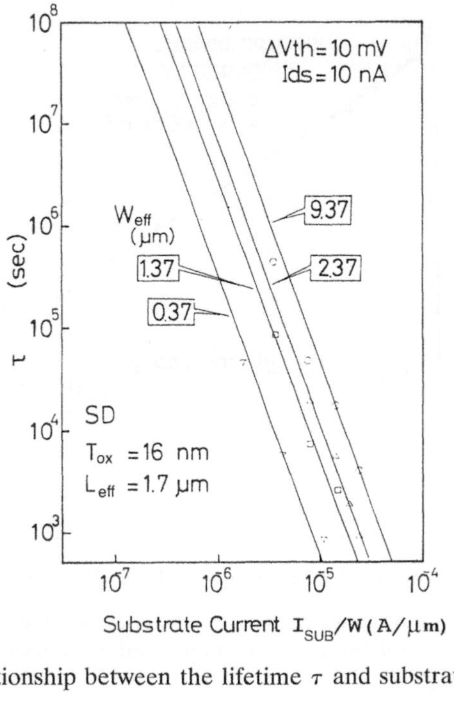

FIGURE 4.24 Relationship between the lifetime τ and substrate current I_{SUB}.

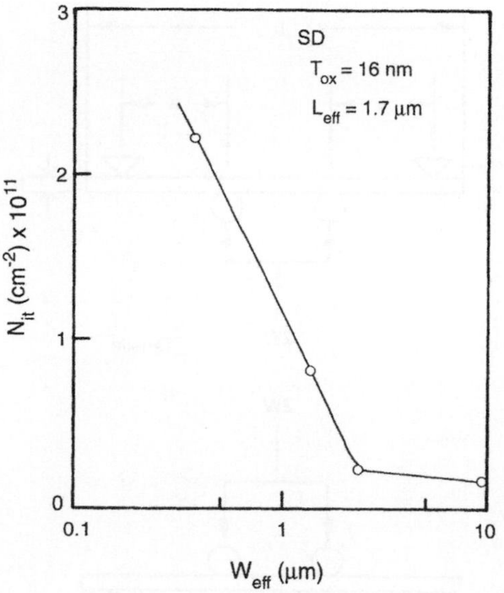

FIGURE 4.25 Relationship between the interface trap density N_{it} and channel width W.

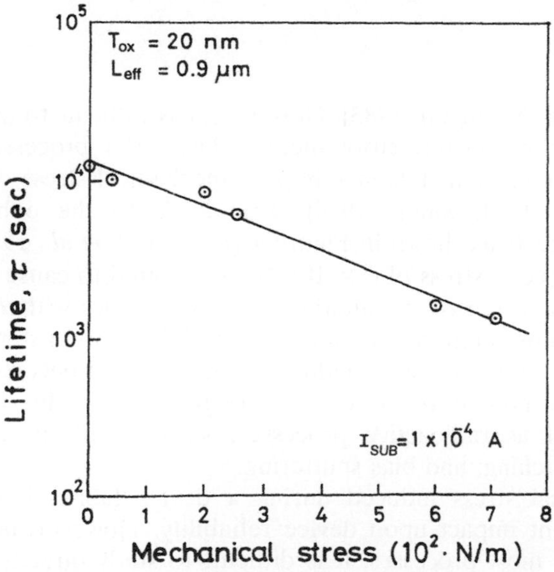

FIGURE 4.26 Influence of mechanical stress on the lifetime of MOS devices with W-gate.

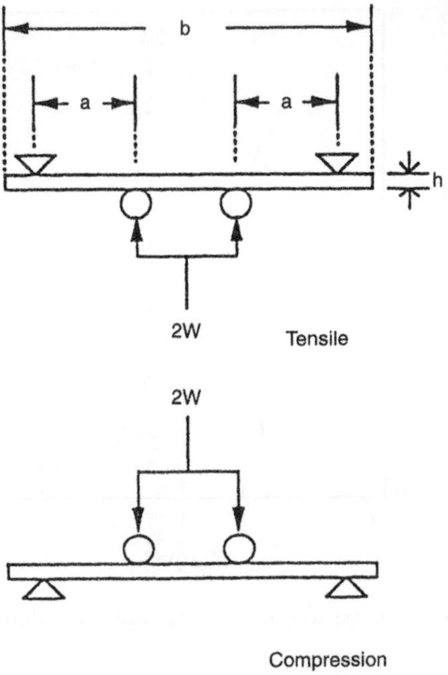

FIGURE 4.27 Schematic diagram of measurement setup: stress $\sigma_x = 6W/(a/bh^2)$.

poly-Si [Itoh and Sunami, 1983]. However, it is difficult to apply standard poly-Si processes to refractory metals. Thus, the processes which are specific to refractory metals may impose mechanical stress on gate oxides [Iwata *et al.*, 1984; Sinha, 1974]. This leads to the enhancement of hot-carrier effects as shown in Figure 4.26 [Takeda *et al.*, 1985a].

For example, a stress of 8×10^2 N/m is found to cause a decrease in lifetime by one order of magnitude for a MOS device with $L_{\text{eff}} = 0.8\ \mu\text{m}$ and $t_{\text{ox}} = 20$ nm. Similarly, when new materials, process steps, or thermal processes are adopted, oxide-quality degradation will become increasingly important [Arimoto *et al.*, 1985]. This degradation is due to mechanical stress as well as radioactive processes, such as advanced lithography, reactive ion etching, and bias sputtering.

Mechanical stress induced during a device fabrication process can have significant impact upon device reliability. However, because of the complexity of most processes, it is difficult to study directly the effect of process-induced mechanical stress. In order to investigate the mechanical stress effect on device electrical characteristics, an external force is directly applied on a silicon chip to simulate the process-induced stress.

4.4.1 MECHANICAL STRESS EFFECT

Both *n*- and *p*-channel single-drain MOSFETs fabricated on an (100)-oriented substrate were evaluated with t_{ox} = 10 nm, drain length L_D = 0.5–10 μm, and drain width W_D = 10 or 100 μm. An external stress ranging from −75 to 75 MPa was applied using the four-point-bending technique, as shown in Figure 4.27. The surface stress (σ_x) was calculated from beam analysis using weight (W) precisely controlled by the standard tensile testing machine. The arrangement used in this study was (a) longitudinal, $\sigma_x \parallel J \perp (011)$, and (b) transverse, $\sigma_x \perp J \perp (011)$, where J

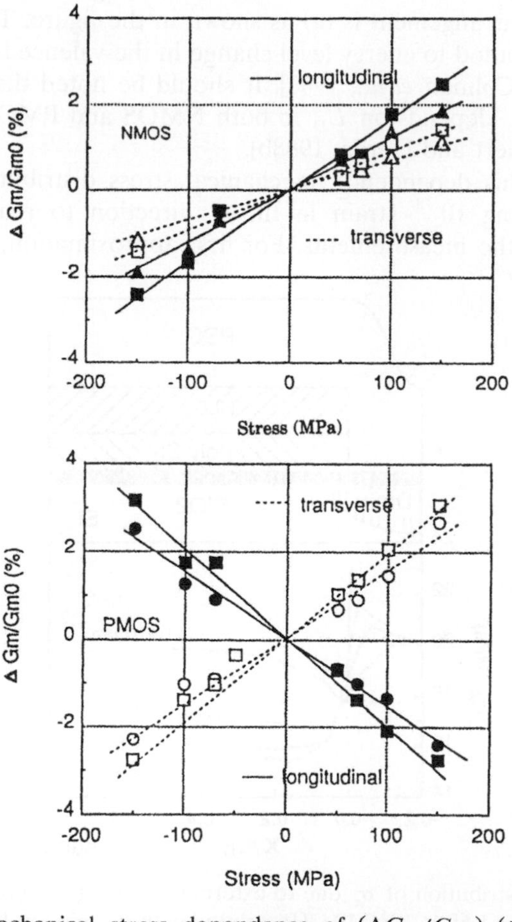

FIGURE 4.28 Mechanical stress dependence of ($\Delta G_m/G_{m0}$) (□, ○, △: L = 2, 1, 0.8 μm). (Reprinted with permission from A. Hamada, T. Furusawa, N. Saito, and E. Takeda, "A New Aspect of Mechanical Stress Effects in Scaled MOS Devices," *IEEE Trans. Electron Devices* **38**, 895–900 (© 1991 IEEE).)

indicates the direction of current flow [Hamada *et al.*, 1990]. A stress analysis program named SIMUS 2D/F, which can analyze the stress state of thin multilayer structures such as LSI devices through their manufacturing processes, was used.

Figure 4.28 shows external mechanical stress dependence of $\Delta G_m/G_{m0}$ in the linear region with L_D as a parameter. G_{m0} is defined as the value without external mechanical stress. For NMOS technology, as is well known, $\Delta G_m/G_{m0}$ decreases under compressive stress and increases under tensile stress due to changes in energy band structure [Dorda, 1971]. For PMOS, on the other hand, $\Delta G_m/G_{m0}$ decreases under compressive stress and increases under tensile stress when the arrangement is (b) and vice versa when the arrangement is (a) as shown in the figure. These phenomena can be attributed to energy level change in the valence band caused by external stress [Colman *et al.*, 1968]. It should be noted that the slope of $\Delta G_m/G_{m0}$ vs. σ_x depends on L_D in both NMOS and PMOS, as reported by others [Borchert and Dorda, 1988b].

To clarify this dependence, mechanical stress distribution was simulated by imposing 10^{-4} strain in the x direction to realize the same condition as in the measurements. For first approximation, it is sufficient

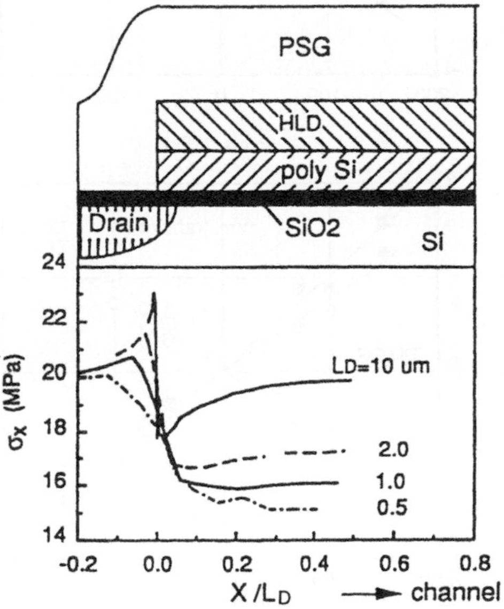

FIGURE 4.29 Distribution of σ_x due to external tensile strain (10^{-4} strain) along the channel. $E_{Si} = 1.68 \times 10^5$ MPa. (Reprinted with permission from A. Hamada, T. Furusawa, N. Saito, and E. Takeda, "A New Aspect of Mechanical Stress Effects in Scaled MOS Devices," *IEEE Trans. Electron Devices* **38**, 895–900 (© 1991 IEEE).)

to take the surface component only into account because the thickness of the inversion layer is in the order of 100 A. The distribution of σ_x along the channel shown in Figure 4.29 reveals the gate length dependence. The gate edge is defined as the origin and the x axis is described in units of L_D. For the long-channel device, σ_x concentrates at the gate edge and then tends to become stable toward the middle of channel. However, for the shorter L_D, no local concentrated component appears even at the gate edge, where σ_x is smaller than in the long channel. From these simulated results, the stress distribution due to external stress is strongly influenced by L_D, which will cause different electrical behavior.

In order to examine the correlation between $\Delta G_m/G_{m0}$ measured electrically and σ_x calculated analytically, the gate length dependence was plotted as shown in Figure 4.30. The maximum σ_x along the channel region was chosen. $\Delta G_m/G_{m0}$ was measured under a tensile stress (σ_x = 75 MPa) in the longitudinal arrangement. In the region where $L_D > 2$ μm, neither $\Delta G_m/G_{m0}$ nor σ_x seems to depend on L_D. However, in the region where $L_D < 2$ μm, both $\Delta G_m/G_{m0}$ and σ_x decrease as L_D decreases. Therefore, under external stress, the dependence of $\Delta G_m/G_{m0}$ on L_D results from the dependence of σ_x distribution on L_D. This suggests that in the deep submicrometer region, the external mechanical stress effect, such as that due to chip mold, tends to be less.

Figure 4.31 shows the time dependence of device degradation due to drain avalanche hot-carrier injection in both NMOS and PMOS under

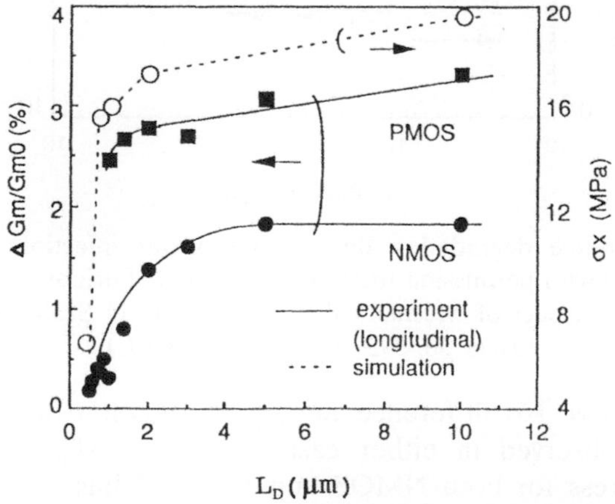

FIGURE 4.30 Correlation between the experimental results ($\Delta G_m/G_{m0}$) and the simulated results of σ_x maximum due to external stress. (Reprinted with permission from A. Hamada, T. Furusawa, N. Saito, and E. Takeda, "A New Aspect of Mechanical Stress Effects in Scaled MOS Devices," *IEEE Trans. Electron Devices* **38**, 895–900 (© 1991 IEEE).)

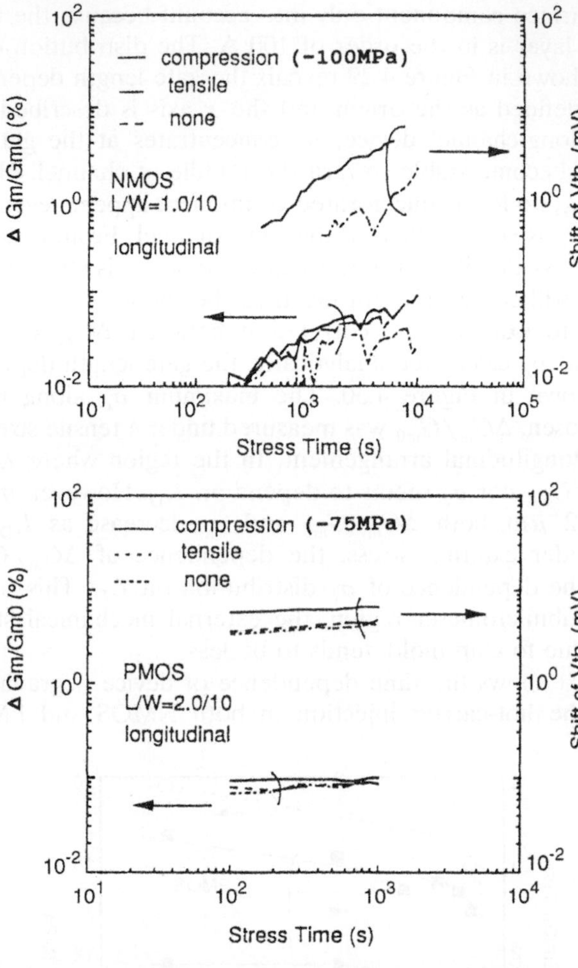

FIGURE 4.31 Device degradation due to hot-carrier injection under external stress. (Reprinted with permission from A. Hamada, T. Furusawa, N. Saito, and E. Takeda, "A New Aspect of Mechanical Stress Effects in Scaled MOS Devices," *IEEE Trans. Electron Devices* **38**, 895–900 (© 1991 IEEE).)

longitudinal stress. No difference in G_m degradation due to hot-carrier injection was observed in either case. However, ΔV_{th} is larger under compressive stress for both NMOS and PMOS. Thus, the interface trap generation induced by hot carriers is not increased by external stress. On the other hand, the capture rate of hot electrons in SiO_2 is found to be extremely sensitive to external compressive stress. This is mainly because the bond angle and distance between Si and SiO_2 are directly related to the electron trapping process.

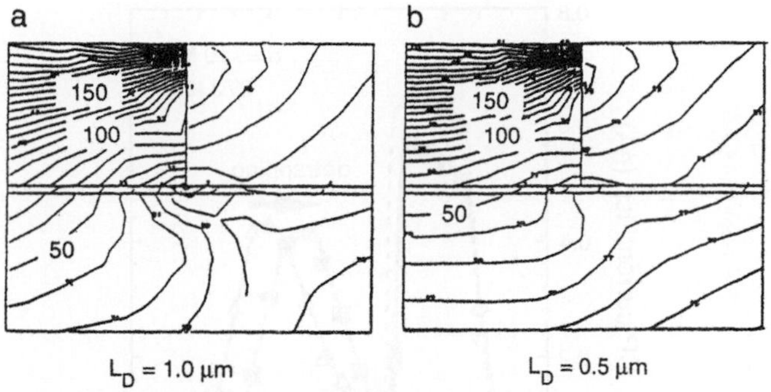

FIGURE 4.32 Distribution of σ_x (MPa) for (a) $L_D = 1.0$ μm and (b) $L_D = 0.5$ μm. (Reprinted with permission from A. Hamada, T. Furusawa, N. Saito, and E. Takeda, "A New Aspect of Mechanical Stress Effects in Scaled MOS Devices," *IEEE Trans. Electron Devices* **38**, 895–900 (© 1991 IEEE).)

Figure 4.32 shows process-induced stress profile. Although the concentrated component is seen at the gate edge for $L_D = 1$ μm, it vanishes when $L_D = 0.5$ μm. The magnitude of σ_x gets smaller as L_D gets shorter. The results of Figures 4.31 and 4.32 imply that in scaled MOS devices the effect of external mechanical stress on hot-carrier degradation tends to be less.

In order to clarify the effect of mechanical stress on electron trapping efficiency, hot-electron detrapping behavior in PMOSFETs after hot-electron injection was investigated under applied compressive mechanical stress. The used devices were $t_{ox} = 15$ nm, $L_{eff} = 0.5$, 0.9, 1.5 μm fabri-

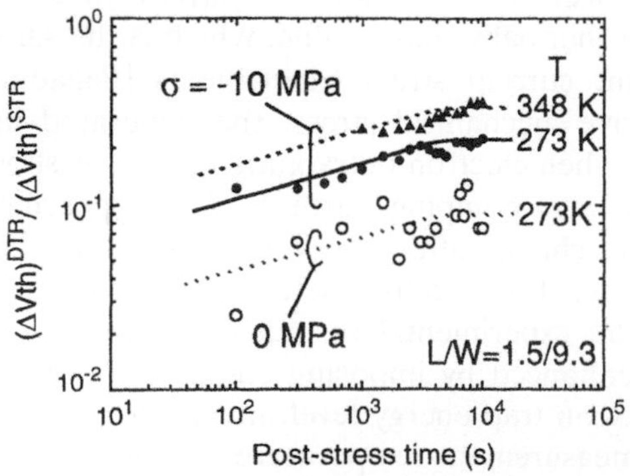

FIGURE 4.33 The dependence of $(\Delta V_{th})^{DTR}/(\Delta V_{th})^{STR}$ on monitor time for PMOSFETs. $(\Delta V_{th})^{DTR} = \Delta V_{th}$ after stress is removed with all the terminals connected to the ground. $(\Delta V_{th})^{STR} = \Delta V_{th}$ after hot-electron injection for 500 s at $V_D = -12$ V, $V_G = -1.5$ V, and $V_{SUB} = 0$ V.

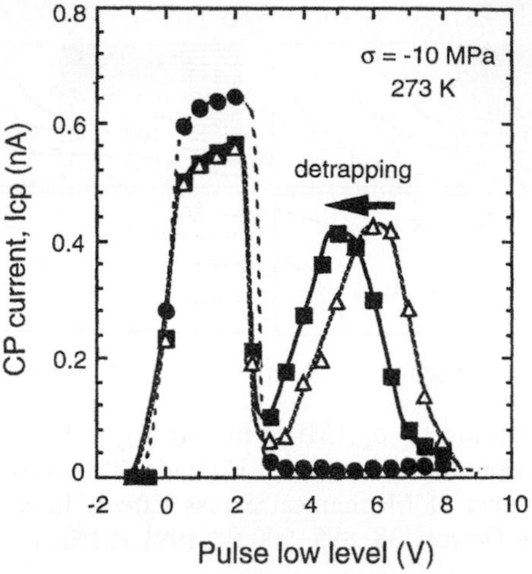

FIGURE 4.34 Charge-pumping current characteristics under compressive mechanical stress of $\sigma = -10$ MPa. Circles indicate initial I_{cp}, triangles for I_{cp} after hot-electron injection for 500 s and squares for I_{cp} after stress is removed.

cated on (100)-oriented silicon substrate. Figure 4.33 shows the threshold voltage change after hot-electron injection. $(\Delta V_{th})^{DTR}$ corresponds to the V_{th} shift after stress is removed, whereas $(\Delta V_{th})^{STR}$ corresponds to the shift due to hot-electron injection for 500 s. Although compressive mechanical stress enhances electron detrapping, a larger $(\Delta V_{th})^{DTR}/(\Delta V_{th})^{STR}$ was observed under compressive mechanical stress mechanical stress ($\sigma = -10$ MPa) than that under no stress. This result indicates that electron detrapping is also enhanced under compressive stress. In every case, $(\Delta V_{th})^{DTR}$ reaches a certain thermally stable value, which is the same as the behavior found in constant current stress experiments [Pagaduan *et al.*, 1989a]. Under compressive mechanical stress, the generated interface traps remain unchanged when electron detrapping occurs, as shown in Figure 4.34. The maximum charge-pumping current (I_{cp}) appearing at higher pulse voltages shows no change after stress was removed, whereas there was a voltage shift as a result of electron detrapping in that region. One possible explanation of the experimental results that both electron trapping and detrapping are enhanced by imposing compressive stress is that changes occur in the electron trap energy level, not in the interface trap density.

By varying measurement temperature, an electron trapping activation energy, E_a, can be obtained from the relation $(\Delta V_{th})^{DTR}/(\Delta V_{th})^{STR} \propto \exp(-E_a/kT)$ as in Figure 4.35. $E_a = 80$ meV was obtained after hot-electron injection with the conventional measurement method without me-

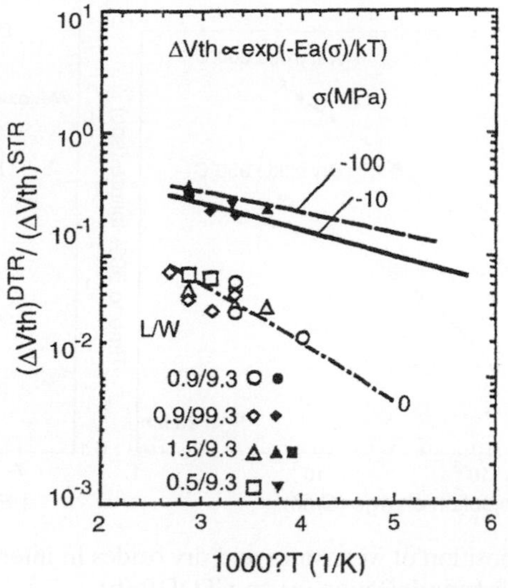

FIGURE 4.35 Plot of $(\Delta V_{th})^{DTR}/(\Delta V_{th})^{STR}$ vs. $1/T$ for different stresses.

chanical stress. One can see that E_a decreases with the applied compressive mechanical stress. Figure 4.36 shows the relationship between E_a and the external mechanical stress. The relationship is not linear, and E_a changes continuously with compressive stress. These experimental results can be understood by considering the dependence of electron trap level on bond lengths and angles [Sakurai and Sugano, 1981], which change contin-

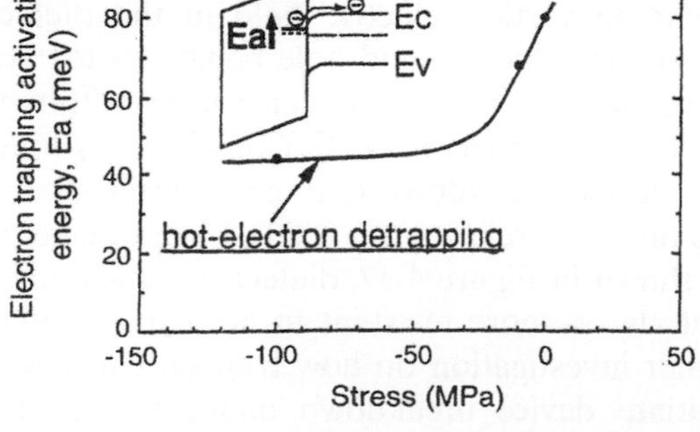

FIGURE 4.36 Relationship between electron trapping activation energy and the external mechanical stress.

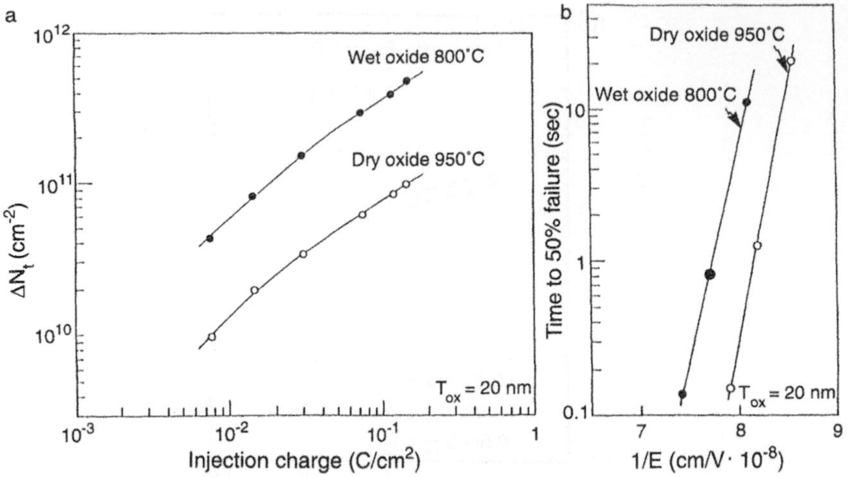

FIGURE 4.37 Composition of wet oxides and dry oxides in interface state variation due to avalanche substrate injection (a) and TDDB (b).

uously with external mechanical stress. Thus the hot-electron trapping activation energy decreases as the electron trap level changes continuously with compressive mechanical stress.

4.4.2 HIGH-QUALITY GATE DIELECTRICS

Hot-carrier effects are related to the channel lateral electric field, and time-dependent dielectric breakdown (TDDB) in a thin gate dielectric is strongly dependent on vertical electric field in the dielectric. However, there seem to be similar electron and hole behaviors for both phenomena in carrier trapping and interface trap generation within thin dielectrics [Hirayama *et al.*, 1981; Tsubouchi, 1983]. Consequently long-term hot-carrier effects cause device breakdown (e.g. gate–drain short), which means that these effects might be related to a TDDB-like breakdown mechanism. Furthermore, as shown in Figure 4.37, dielectrics which are more resistant to TDDB are not always more resistant to hot-carrier effects (e.g. wet vs. dry oxides). Further investigation on how trapped carriers due to hot-carrier injection initiate device breakdown under long-term stress will be important in elucidating the TDDB phenomenon [Raider, 1973; Chen *et al.*, 1985; Kusaka *et al.*, 1986].

4.5 Summary

Device degradation due to dynamic stress has been treated extensively in this chapter. Carrier trapping and trap generation models for enhanced degradation (compared with DC stress) were presented. Process and materials effects on hot-carrier-induced degradation were reviewed and explained with similar analyses. Other effects such as ambient temperature will be discussed in the next chapter.

CHAPTER

5

HOT-CARRIER EFFECTS AT LOW TEMPERATURE AND LOW VOLTAGE

5.1 Introduction

In Chapter 5, device degradation due to hot carriers having energies below the Si–SiO_2 energy barrier, is discussed. In the past, when the supplied voltage was under 3 V, hot-carrier device degradation was assumed to be absent because hot carriers cannot surmount the Si–SiO_2 barrier height (3.2 eV). However, impact ionization can occur even at $V_D = 1$ V, which is close to the Si band gap (1.12 eV), if the electric field is high enough [Eitan *et al.*, 1982]. Several reports on hot-carrier degradation in the region of low drain voltage are listed as follows.

1. Gate current observation at $V_D = 2.35$ V, $V_D = 3$ V [Tam *et al.*, 1983]
2. Gate current observation by channel hot electron (CHE) injection at $V_D = 1.6$ V [Ricco *et al.*, 1984]
3. G_m degradation at $V_D = 2.5$ V, $V_G = 1$ V [Takeda, 1984]

These experimental results cannot be explained by a simple "lucky electron" approximation. Also, substrate current was observed experimentally in the region of energies below the Si band gap ($V_D < 1$ V).

The obtained results will serve as a guide for the development of future very large-scale integrated (VLSI) submicrometer metal-oxide semiconductor (MOS) devices.

5.2 Hot-Carrier Effects at Low Temperature

At low temperature, hot-carrier effects are enhanced. As shown in Figure 5.1a and b, substrate current at 77 K is five times greater than that at room temperature (RT) and CHE gate current is about 1.5 orders of magnitude greater than that at RT. This is mainly due to an increase in impact ionization rate and electron mean free path at low temperature. Under the condition of constant V_D at 77 K, a degradation enhancement of two orders of magnitude occurs compared with RT. In addition, G_m degradation at 77 K is greater than at RT by about one order of magnitude even at I_{SUB} = constant as shown in Figure 5.2 This is because electron trapping efficiency increases with decreasing temperature and the effect of fixed charges becomes large at low temperature. The CHE mode also causes more severe degradation than the drain avalanche hot carrier (DAHC) mode at 77 K, while at RT the CHE mode is less severe. These results are shown in Figure 5.3. These experimental results imply that at low temperature ($\sim$ 77 K), it is important to account for the electron-trapping efficiency and the influence of many trapped electrons on G_m degradation [Estreich, 1978; Tsuchiya *et al.*, 1985].

5.3 Device Performance Degradation

5.3.1 G_m DEGRADATION

Figure 5.4 shows the stress-time variation in G_m degradation and V_{th} shift for a device with $L_{eff} = 0.3$ μm and $t_{ox} = 5$ nm. The G_m was measured at $V_D = 0.1$ V and $V_G = 0.4$ V. A gate voltage of 0.4 V provides the maximum transconductance in the triode region. The V_{th} shown in this figure was taken at a rather high channel current, $I_D = 1$ mA. This V_{th} shift at high channel current, in fact, implies G_m degradation because the tailing coefficient in the subthreshold region increases due to DC stress. The stress condition is that which provides the peak I_{SUB} in an I_{SUB} vs. V_G plot. Back bias (V_{SUB}) was zero.

It can be seen from this figure that device degradation indeed occurs as low as $V_D = 25$ V. The Si–SiO_2 barrier height at the drain will not be

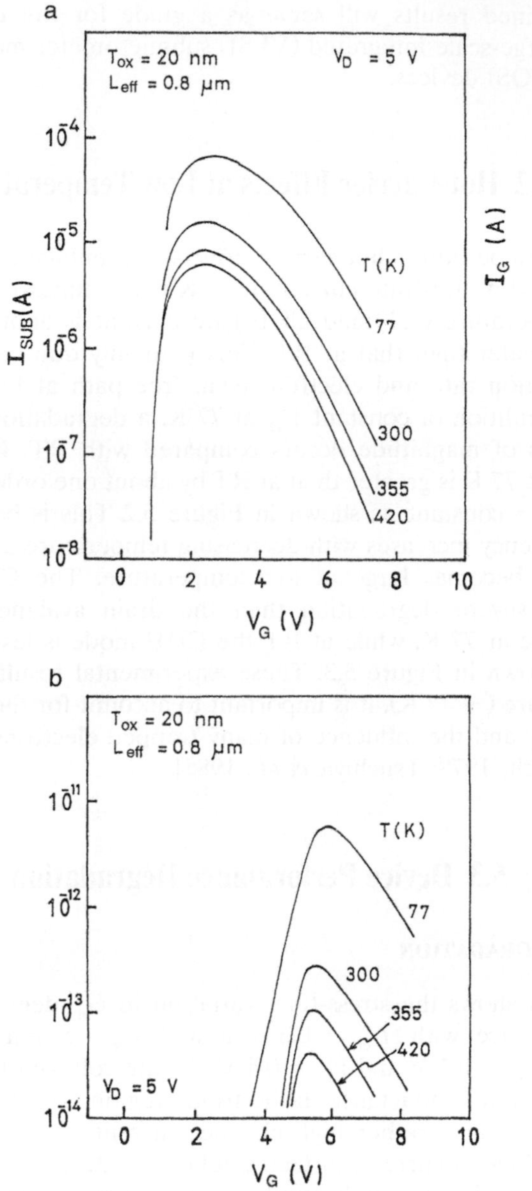

FIGURE 5.1 (a) Substrate current characteristics with temperature as a parameter. (b) Gate current characteristics with temperature as a parameter.

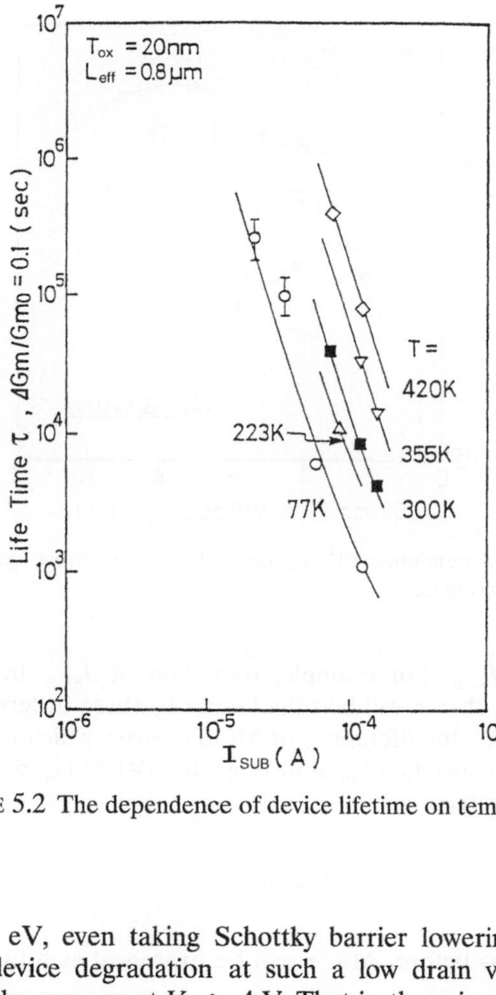

FIGURE 5.2 The dependence of device lifetime on temperature.

lower than 2.7 eV, even taking Schottky barrier lowering into account. Nevertheless, device degradation at such a low drain voltage seems to behave almost the same as at $V_D > 4$ V. That is, there is no sharp cutoff in the behavior of device degradation from $V_D = 4$ V to 2.5 V. It should also be noted that the time dependence of device degradation in such a small MOS device is the same as that for long-channel MOS devices.

Figure 5.5 shows the V_D dependence of lifetime, which is the time it takes for V_{th} at $I_D = 1$ mA to shift 10 mV in a small device with $L_{eff} = 0.3$ μm and $t_{ox} = 5$ nm. It is obvious that there is no deviation from the τ vs. V_D relationship given by Eq. (3.8) in the V_D range of 4–2.5 V. In addition, Figure 5.7 shows the dependence of lifetime on I_{SUB}. It can be seen that the τ vs. I_{SUB} relationship given by Eq. (5.1) is quite valid over a wide range of device parameters. It is possible to predict the lifetime by

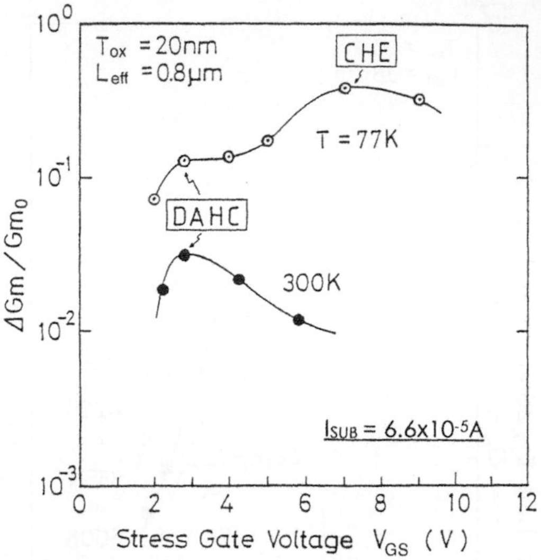

FIGURE 5.3 The dependence of G_m degradation on stress gate voltage with temperature as a parameter.

measuring only I_{SUB}. For example, reduction of I_{SUB} by one order of magnitude means that τ will become longer by three orders of magnitude.

Using Eq. (3.8), the lifetime τ of MOS devices, which is defined as the time it takes for a specified V_{th} shift (e.g., 10 mV) or G_m degradation (e.g., 10%) can be expressed as

$$\tau \propto \exp(b/V_D), \tag{5.1}$$

where $b = \alpha/n$ as before. Also τ can be expressed as a function of I^m_{SUB}:

$$\tau \propto (I^m_{SUB})^{-l}, \tag{5.2}$$

where $l = (\alpha/\beta)/n$ as before, and it has a value of 3.1–3.4 over a wide range of device parameters. The experimental evidence verifying Eqs. (3.8) and (3.10) is shown in Figures 5.6 and 5.7. It can be seen that these relations are valid over a wide range of device parameters. Since τ is related simply to substrate current, it is possible to predict the lifetime by measuring only I_{SUB}. Furthermore, it is significant that there is no deviation from this tendency in hot-carrier phenomena for bias voltages less

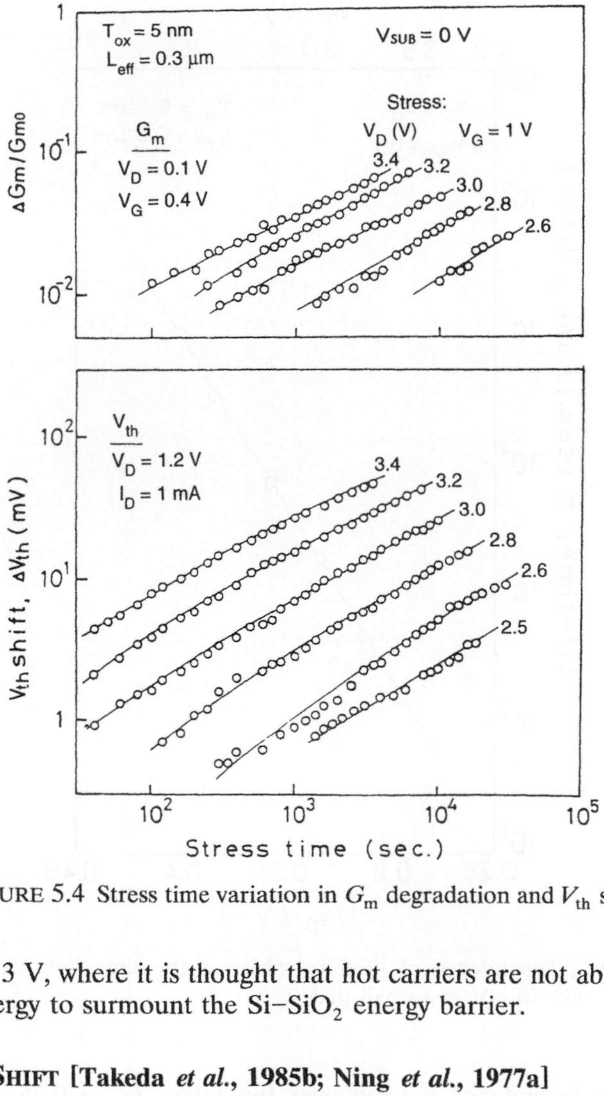

FIGURE 5.4 Stress time variation in G_m degradation and V_{th} shift.

than $V_D = 3$ V, where it is thought that hot carriers are not able to obtain enough energy to surmount the $Si-SiO_2$ energy barrier.

5.3.2 V_{th} SHIFT [Takeda *et al.*, 1985b; Ning *et al.*, 1977a]

To obtain quantitative guidelines for submicrometer device design and to investigate hot-carrier phenomena under a bias of less than 3 V, an empirical model for device degradation was constructed. It was found from Figure 5.8 that G_m degradation and/or V_{th} shift at high channel current can be expressed as

$$\Delta G_m/G_{m0}(\Delta V_{th}) = A \cdot t^n. \tag{5.3}$$

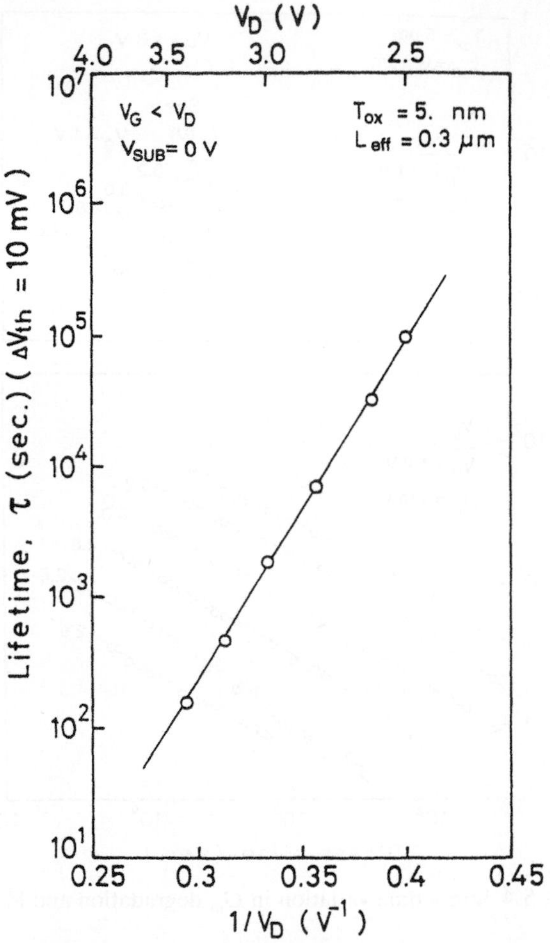

FIGURE 5.5 V_D dependence of lifetime, which is the time it takes for V_{th} at $I_D = 1$ mA to shift 10 mV in a small device.

Such a time dependence is valid over the range $L_{eff} = 0.25\text{–}5$ μm and $t_{ox} = 5\text{–}35$ nm. The slope, n, depends mainly on V_G and has a value of 0.5–0.6 in the case of the DAHC injection mode. The magnitude of degradation, A, has been known to have the V_D dependence

$$A \propto \exp(-\alpha/V_D). \tag{5.4}$$

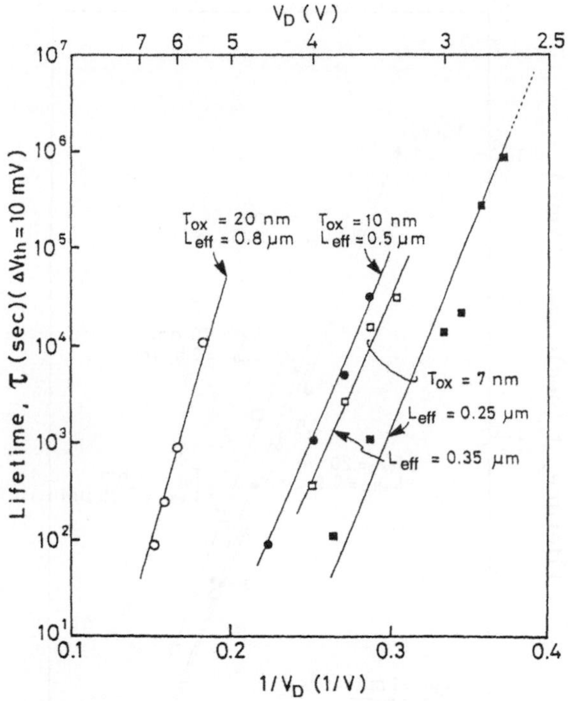

FIGURE 5.6 Relationship between lifetime of MOS devices and drain voltage.

That is, A does not depend on V_G. The peak substrate current (I_{SUB}^{p}) has also the same exponential dependence

$$I_{SUB}^{p} \propto \exp(-\beta/V_D). \tag{5.5}$$

This is because the impact ionization rate (α) has the same dependence on electric field ($\mathscr{E}$). A relationship between $\mathscr{E}$ and V_D will be explained later. It is important to point out that A is related to the number of excess carriers generated by impact ionization, by considering Eqs. (5.4) and (5.5).

Using Eqs. (5.2), (5.3), and (5.4), the lifetime (τ) of MOS devices can be expressed as

$$\tau \propto \exp(b/V_D), \qquad b = \alpha/n. \tag{5.6}$$

Also, τ can be expressed as a function of I_{SUB}^{p}, yielding

$$\tau \propto (I_{SUB}^{p})^{-l}, \qquad l = (\alpha/\beta)/n, \tag{5.7}$$

where l has a value of 3.2–3.3 over a wide range of parameters.

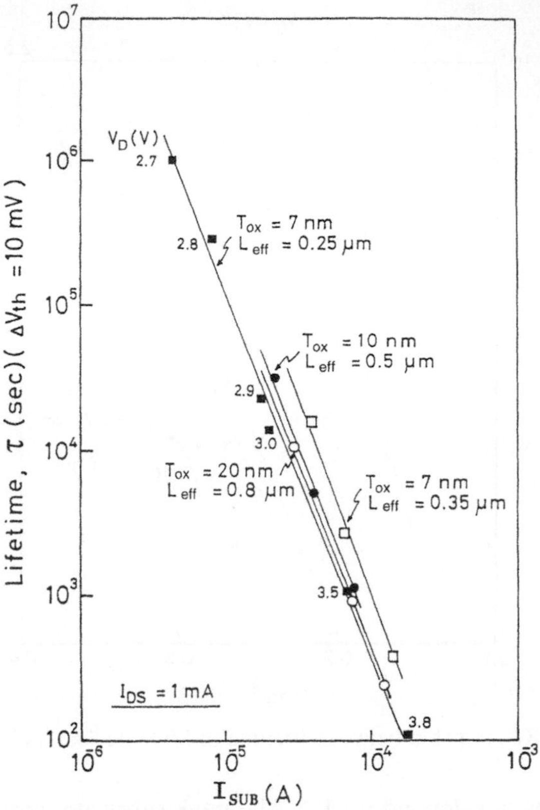

FIGURE 5.7 I_{SUB}^{P} dependence of lifetime for various kinds of MOS devices.

5.4 Device Degradation Mechanisms [Takeda *et al.*, 1985a; Smith, 1978; Moslehi and Saraswat, 1984]

Figure 5.9 shows a significant relation between lifetime and the vertical component of the peak electric field ($\mathscr{E}_p$) at the drain for MOS devices with various device parameters. The electric field was calculated using the process/device simulators SUPREM [Antoniadis *et al.*, 1978] and CADDET [Toyabe and Asai, 1979]. It is clear that τ can be expressed as

$$\tau \propto \exp(c/\mathscr{E}_p), \tag{5.8}$$

which can be explained as follows.

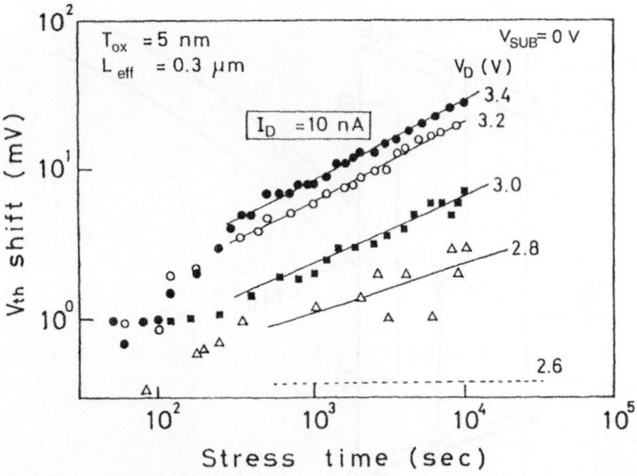

FIGURE 5.8 Stress time variation of V_{th} shift at $I_D = 10$ nA for a device with $L_{eff} = 0.3$ μm and $t_{ox} = 5$ nm.

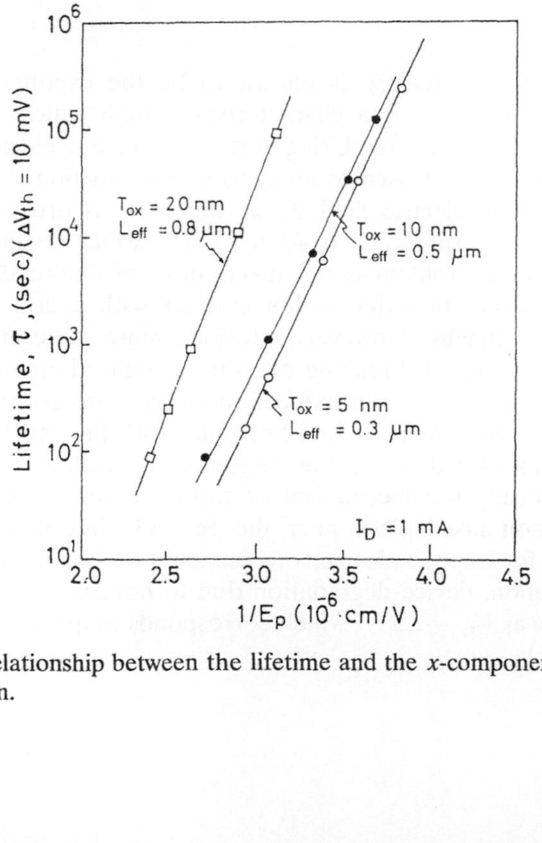

FIGURE 5.9 Relationship between the lifetime and the x-component peak electric field at the drain.

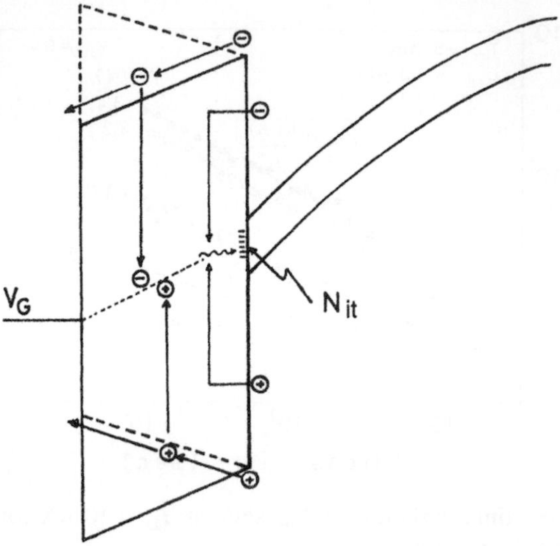

FIGURE 5.10 Mechanism of interface state generation.

The surface potential ψ_s is known to be the exponential function, $\psi_s \propto \exp(x/l_0)$, where l_0 is a characteristic length which can be determined from device structure. Using this relation, the electric field $\mathscr{E} = \partial\psi_s/\partial x$ can also be expressed as an exponential function, $\mathscr{E} \propto \exp(x/l_0)$. Therefore the peak electric field $\mathscr{E}_p$ at the drain is proportional to the drain voltage, $\mathscr{E}_p \propto \exp(x_D/l_0) \propto \psi_s(x_D) = V_D$, which results in Eq. (5.8).

One remaining problem is the mechanism of device degradation or interface state generation due to hot carriers with energy less than the Si–SiO_2 barrier height. However, interface state generation near the Si–SiO_2 interface due to breaking of some interfacial chemical bonds by holes or, more likely, hydrogen-induced interface state generation through recombination of hot electrons and hot holes with the assistance of direct tunneling is considered to be the degradation mechanism. Figure 5.10 shows schematically the mechanism of interface state generation. Since this phenomenon takes place near the Si–SiO_2 interface, it would be relatively easy for trapped electrons to escape into the channel. According to this explanation, device degradation due to hot-carrier injection would occur at as low as $V_D = 2.2$ V, which corresponds to half of the H-H bond energy (4.5 eV).

5.5 Summary

Device degradation due to hot carriers having energies below the Si–SiO_2 barrier height is reviewed using scaled MOS devices. G_m degradation occurred under the condition of V_D smaller and larger than 3 V. On the contrary, no V_{th} shift is observed at $V_D \simeq 2.5$ V. The hot-carrier effect will remain a significant reliability problem for submicrometer MOS devices even when the supplied voltage is reduced to 3.3 V or below.

CHAPTER

6

DEPENDENCE OF HOT-CARRIER PHENOMENA ON DEVICE STRUCTURE

6.1 Introduction

The dependence of hot-electron injection upon metal-oxide-semiconductor field-effect transistor (MOSFET) structures has been investigated, both experimentally and theoretically, in the search for structures suitable for scaling down to the submicrometer level. Hot-electron injection was evaluated directly by measuring the gate current, and modeling was undertaken in order to account quantitatively for its behavior. The scaling down of each MOSFET structural parameter proved to intensify hot-electron injection.

Several MOSFET structures for minimizing hot-electron injection are proposed in this chapter, on the basis of the injection model, and experimentally evaluated. The results demonstrate the feasibility of the proposed "hot-electron-resistant" structures in VLSI components.

6.2 Variations of Device Structure

Various kinds of n-channel MOSFET structures were tested. With the conventional arsenic (As)-drain structure, structural parameters were varied over a wide range in order to evaluate the effect of scaling down upon hot-carrier injection. Channel length was varied over the range 0.5–5 μm, gate oxide thickness 20–50 nm, junction depth for source and drain diffusion layers 0.2–0.5 μm, channel implant dose 0–2 $\times 10^{12}$ ions/cm^2, and substrate resistivity 1–100 Ω-cm.

In improving the resistance of devices against hot-electron injection, drain junction structure and channel doping are significant. A phosphorus (P)-drain structure, an As–P(n^+–n^-) double diffused drain structure, an offset gate structure, and a buried channel structure were evaluated, as compared with the conventional As-drain structure for 1-μm devices.

6.3 Device Parameter Dependence

6.3.1 Effective Channel Length

As described in Section 3.3, the hot-carrier breakdown voltage is related to the effective gate length L_{eff} according to

$$BV_{\text{DC}} = BV_{\text{HC}} = C_1 L_{\text{eff}}/(L_{\text{eff}} + C_2). \tag{6.1}$$

This equation is valid for any device parameter and structure.

6.3.2 Channel Dose

In order to avoid undesirable short-channel effects, as well as to improve device performance, the scaling principle proposed by Dennard *et al.* (1974), has been used for MOSFET structures. According to this principle, such vertical dimensions in MOSFET as gate oxide thickness and junction depth for source and drain diffusion layers should be scaled down, along with the horizontal dimensions. Inverse substrate doping density should also be decreased proportionately. Here, it would be useful to discuss the effects of the scaling principle upon hot-electron injection. Shrinking of effective channel length drastically lowers the highest applicable voltage (breakdown voltage), BV_{DC}. The scaling down of other structural parameters, such as gate oxide thickness, junction depth, and sub-

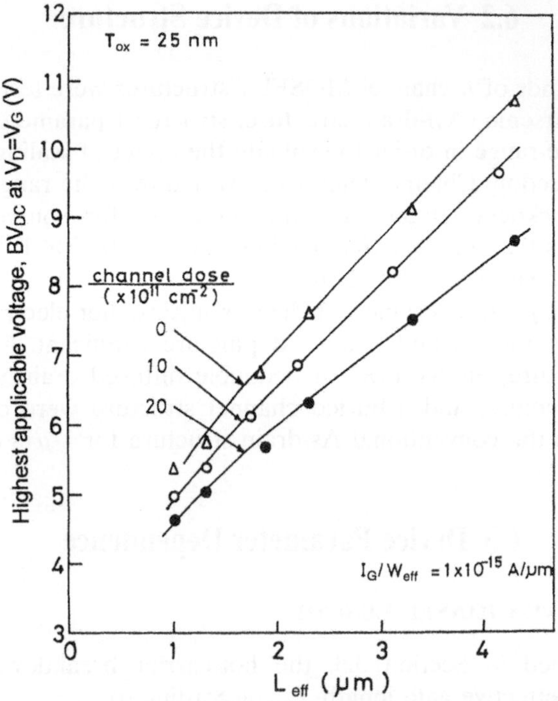

FIGURE 6.1 Dependence on channel implant dose of highest applicable voltage BV_{DC}, which is limited by hot-electron injection. Source/drain junction depth is 0.3 μm.

strate doping density, also intensified the injection of hot electrons. Examples of these dependences are shown in Figure 6.1, which shows that an increase in channel implant dose of boron appreciably lowers BV_{DC}.

Figure 6.2 shows BV_{DC} and BV_{DS} as a function of effective channel length when both gate-oxide thickness and inverse channel doping are scaled down simultaneously so as to make the threshold voltage constant. It can be seen clearly that hot-carrier generation imposes the most severe limitations on device design. It should be noted that, for devices with smaller effective channel lengths than 1 μm, the highest applicable voltage determined by hot-electron injection, BV_{DC}, drops to as low as 3.5 V.

Therefore, the influence of scaling down device dimensions on hot-carrier generation can be fatal for submicrometer devices. This demonstrates the urgent need for strengthening the resistance of scaled-down devices against hot-carrier generation.

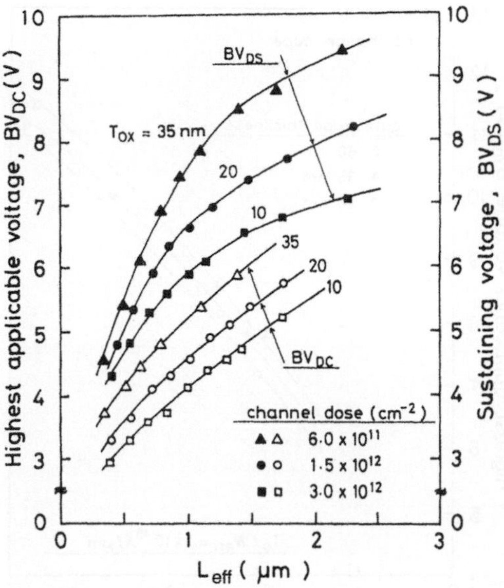

FIGURE 6.2 Examples of lowering in BV_{DC} and BV_{DS} as a function of L_{eff}, in the case of simultaneous scaling down of both gate oxide thickness and the inverse channel doping.

6.3.3 GATE OXIDE THICKNESS

Figure 6.3 shows the t_{ox} dependence of BV_{DC} in devices with no channel doping. The dependence on t_{ox} is less than that for devices with channel doping. This implies that a high electric field is not necessary to pull hot carriers toward the gate. The t_{ox} dependence of gate current, however, causes less device degradation than expected, especially as seen in the threshold voltage shift, ΔV_{th}, where the following relation was reported [Yoshida *et al.*, 1985]:

$$\Delta V_{th} \propto t_{ox}^2. \tag{6.2}$$

Equation (6.2) indicates that degradation gets worse as t_{ox} becomes thicker, which is opposite to the result which is deduced from considering gate current. Therefore Eq. (6.2) can be interpreted as follows:

$$\Delta V_{th} = \Delta Q / C_{ox} \propto t_{ox} \, \Delta Q. \tag{6.3}$$

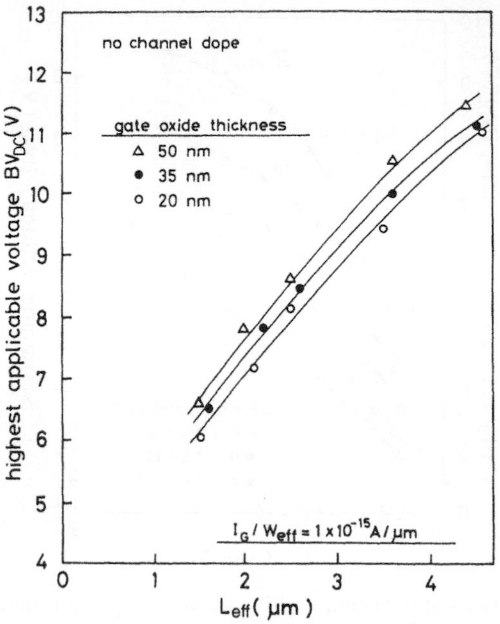

FIGURE 6.3 Dependence on gate oxide thickness of highest applicable voltage, BV_{DC} which is limited by hot-electron injection. Source/drain junction depth is 0.3 μm.

If carriers are trapped uniformly in the gate oxide, ΔQ can be described by

$$\Delta Q \propto t_{ox}. \tag{6.4}$$

Equations (6.3) and (6.4) lead to Eq. (6.2). Further study is expected on the validity of these equations for thin gate oxides and for the regions where tunneling current is dominant.

From Figures 6.1–3, it is found that hot-carrier effects most strongly depend on L_{eff}; however, it is impossible to control these effects using L_{eff} as long as MOS devices are scaled. The key factor in reducing the electric field near the drain is improvement of the drain–substrate p–n junction. With this idea in mind, the new device structure was proposed.

6.3.4 GATE TO DRAIN–SOURCE OVERLAPPED LENGTH

The drain structure in lightly doped drains (LDDs) can be described completely by three parameters, length of n^- region L_n, gate–drain/

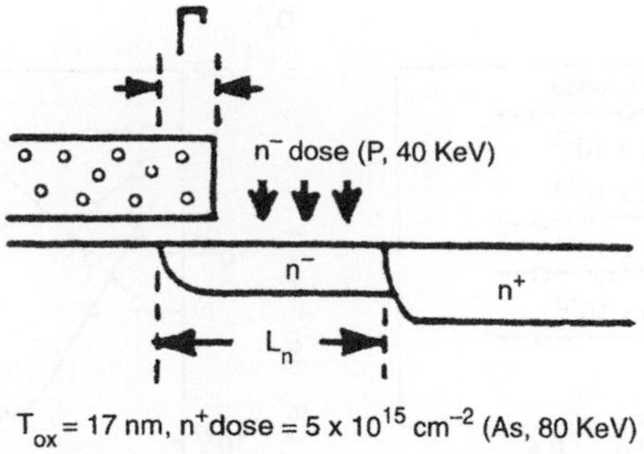

FIGURE 6.4 Schematic cross section of an LDD device; n^- length L_n, gate–drain/ source overlap length Γ, and n^- dose N_D are important device parameters.

source overlap length Γ, and n^- dose N_d, as shown in Figure 6.4. The influence of these parameters on LDD device characteristics is studied using a three-dimensional (3-D) device simulator (CADDETH). However, the simulation results shown here seem to have little actual 3-D effect. The devices have an n^- length L_n from 0.1 to 0.6 μm, overlap length Γ from 0.05 to 0.4 μm, n^- dose N_d from 1×10^{12} to 4×10^{13} cm^{-2} (P, 40 keV), effective gate length L_{eff} from 0.3 to 0.8 μm, and gate oxide thickness $t_{ox} = 17$ nm with an oxide sidewall spacer. The substrate peak acceptor concentration is 1×10^{17} cm^{-3}.

Figure 6.5a illustrates $\mathscr{E}_m$ versus Γ and Figure 6.5b shows GR_m versus Γ, where $L_{eff} = 0.8$ μm and $L_n = 0.2$ μm. It is found from this behavior that $\mathscr{E}_m$ and GR_m decrease with increasing Γ until $\Gamma = W_d$ for the range of $n^- = 5 \times 10^{12}$ to 4×10^{13} cm^{-2} just as for L_n. This overlap effect can be explained by the following differential equation, which relates the lateral electric field $\mathscr{E}_y$ to the n^- doping $N_d(y)$:

$$\frac{X_J}{\eta}\frac{d\mathscr{E}_y(0,y)}{dy} = (qN_{SUB}X_J - Q_n - \epsilon_{ox}\mathscr{E}_x(0,y) - qN_d(y)X_J)/\epsilon_s, \quad (6.5)$$

where X_J is n^- depth, η is a fitting parameter, N_{SUB} is the (average) channel doping concentration, and Q_n is the mobile carrier density per unit area. In this equation, the term, $\epsilon_{ox}\mathscr{E}_x(0,y)$, corresponds to the gate-induced charge due to the overlap effect. Because this term reduces the total charge, the drain electric field decreases as the overlap length becomes larger. That is, the depletion-layer width W_d widens in order to drop the drain bias across a region with reduced field. Although Equation (6.5) shows the overlap effect, it does not include such low-dimensional

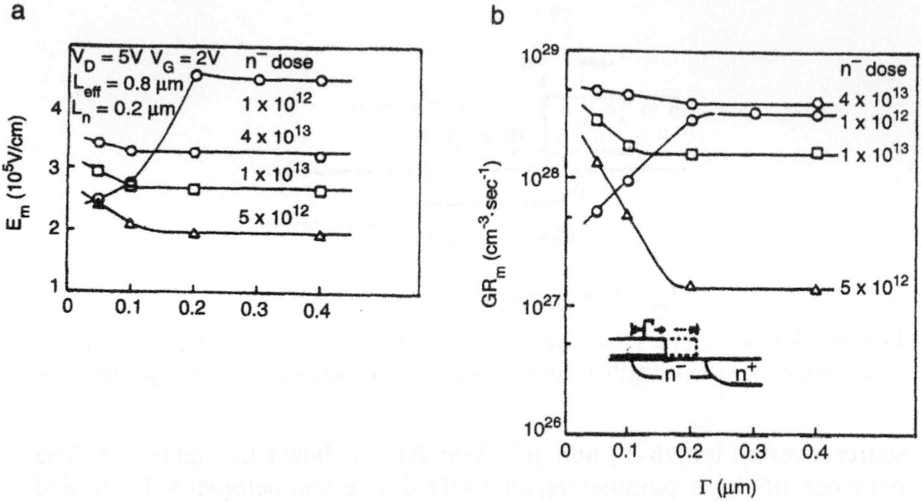

FIGURE 6.5 (a) Maximum channel electric field E_m and (b) maximum hot-carrier generation rate GR_m versus gate–drain/source overlap length Γ characteristics calculated from CADDETH, where $L_{eff} = 0.8$ μm and $L_n = 0.2$ μm.

(2-D) effects as impurity profile and gate-edge effect. On the other hand, for $N_d = 1 \times 10^{12}$ cm^{-2}, device characteristics are similar to those of gate-offset MOS devices. Then n^- region is completely depleted, and there are induced carriers underneath the gate. As the gate edge approaches the n^+ region with increasing Γ, the effective depletion width becomes smaller. This reduction results in increased $\mathscr{E}_m$.

6.4 Device Structure Dependence

6.4.1 DRAIN STRUCTURES

We investigated four kinds of MOSFET structures as "hot-carrier-resistant" structures. These include As–P (n^+–n^-) double diffused drain, P-drain, offset gate, and buried-channel devices, as shown in Figure 6.6a–d. The main fabrication processes in these devices are as follows.

1. As–P (n^+–n^-) double diffused drain: where narrow, self-aligned n^- regions are introduced in the periphery of n^+ source–drain diffusions by using double implantation of As and P. The optimized impurity profiles are shown in Figure 6.7. This was calculated using SUPREM. The total junction depth is 0.3 μm and is reasonable for realizing VLSI.

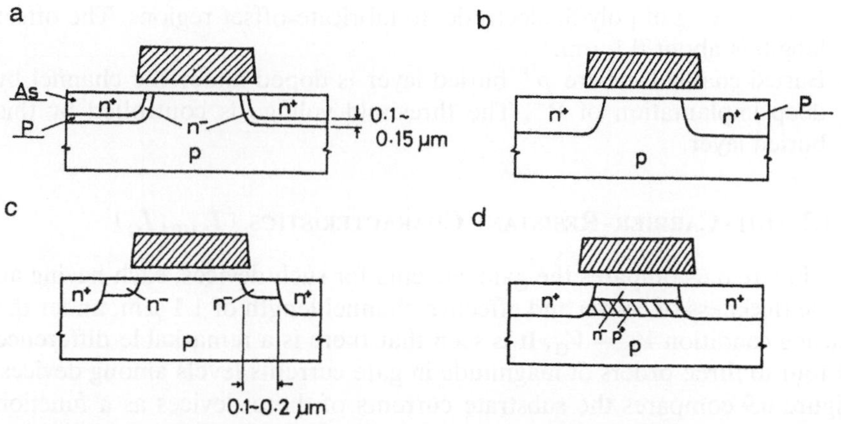

FIGURE 6.6 Cross section of a few of MOSFET structures investigated. (a) As–P (n^+–n^-) double diffusion; (b) P-drain; (c) LDD; (d) Buried channel.

2. P-drain: where phosphorus diffusant is used as source–drain diffusion in order to fabricate a graded drain junction. The junction depth is about 0.3–0.35 μm.
3. LDD: where self-aligned n^- regions are introduced between the channel and n^+ junctions. The n^+ regions are implanted after selective

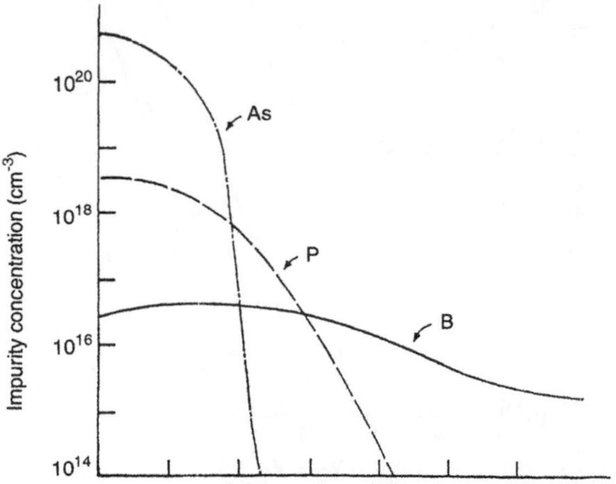

FIGURE 6.7 Impurity profiles in As–P (n^+–n^-) double diffused drain structure.

oxide coating of poly-Si electrode, to fabricate offset regions. The offset length is about 0.1 μm.

4. Buried channel: where p^+ buried layer is doped under the channel by deep implantation of B^+. The threshold voltage is controlled by this buried layer.

6.4.2 HOT-CARRIER-RESISTANT CHARACTERISTICS (I_{SUB}, I_G)

Figure 6.8 compares the gate currents for such devices, each having an oxide thickness of 35 nm and effective channel length of 1.1 μm, under the biasing condition $V_D = V_G$. It is seen that there is a remarkable difference of four to three orders of magnitude in gate currents levels among devices. Figure 6.9 compares the substrate currents of these devices as a function of V_G, with V_D as a parameter such that the peak substrate current occurs at the same V_G for every device. There are also appreciable difference among these devices.

Graded drain junctions such as the P-drain and As–P double diffused drain are obviously effective in reducing hot-carrier generation. The device

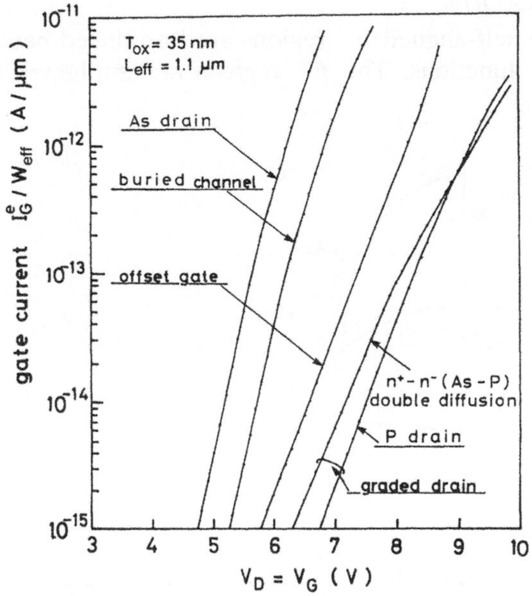

FIGURE 6.8 Comparison of gate currents for various kinds of MOSFET structures, each having an oxide thickness of 35 nm and an effective channel length of 1.1 μm.

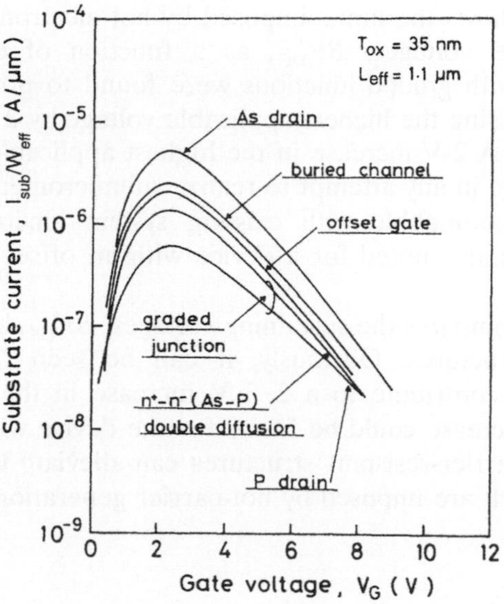

FIGURE 6.9 Comparison of a substrate currents between various kinds of device structures as a function of V_G.

with a P-drain is found to have the lowest gate and substrate currents. This is rather striking, considering the present prevalence of As-drain devices. It may be necessary to reconsider the advantages of the P-drain junction, in spite of the suitability of the shallow As-drain junction for VLSI miniaturization. The problems of relatively large junction depth and high sheet resistance for the P-diffused layer were solved by introducing As as a codiffusant in a manner in which a shallow but abrupt As implant layer, n^+, is surrounded by a gradual P implant layer, n^-, as shown in Figure 6.6a.

The device with an offset gate also has a rather low gate current. This indicates that the separation of the gate electrode from the maximum electric field is effective in reducing hot-electron injection, as expected from the calculation results. This structure is also effective for reducing substrate current because n^- regions are introduced near the drain.

On the basis of the experiments and hot-carrier generation models, two of the guiding principles for minimizing hot-carrier generation that we realize were that it is best to (1) use a graded drain junction for reducing the electric field and (2) use an offset gate for separating the gate electrode from the localized peak of the electric field.

Figure 6.10 shows the limits imposed by hot-electron injection on the highest applicable voltages, BV_{DC}, as a function of effective channel length. Devices with graded junctions were found to provide remarkable improvements, raising the highest applicable voltage by 2 V more than the As-drain devices. A 2-V increase in the highest applicable voltage is quite a considerable one in any attempt to realize submicrometer circuits having supply voltages compatible with existing system environments. A 1-V improvement was also noted for a device with an offset gate or a buried channel.

Figure 6.11 compares the sustaining voltages, BV_{DS}, between "hot-carrier-resistant" structures. Obviously, it can be seen that devices with graded junctions contribute to a 2–3 V increase in the drain sustaining voltage. A 1-V increase could be found for the device with an offset gate. Thus these hot-carrier-resistant structures can alleviate the limitations on VLSI design which are imposed by hot-carrier generation.

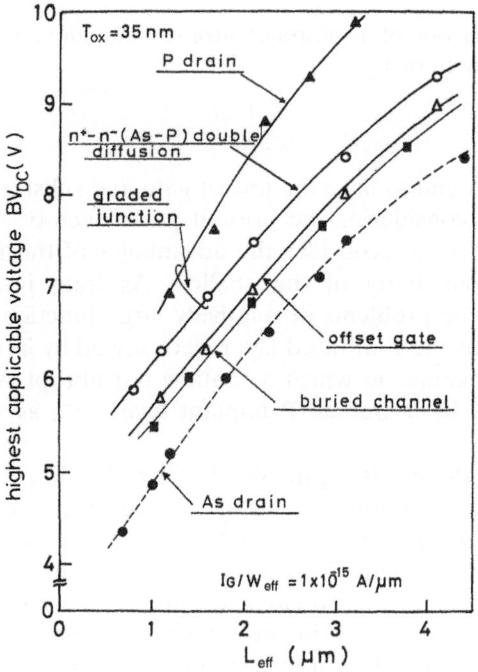

FIGURE 6.10 Comparison of a highest applicable voltage, BV_{DC}, for various kinds of MOSFET structures.

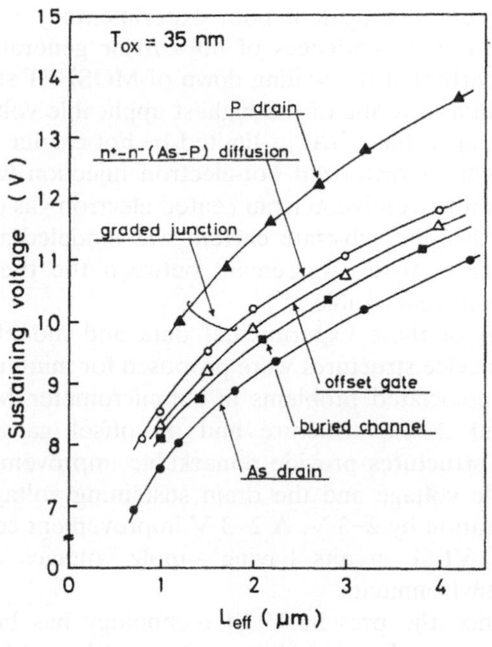

FIGURE 6.11 Comparison of the drain sustaining voltage, BV_{DS}, for various kinds of MOSFET structures.

In terms of short-channel effects such as threshold voltage lowering, these structures have almost the same characteristics as conventional MOSFETs. During fabrication, no additional masking steps are required. Wafer processing is quite compatible with conventional NMOS process. There is no transconductance degradation during wafer processing. Packing densities are the same as with conventional technologies.

Fabrication controllability could be a problem in these hot-carrier-resistant MOSFETs. Care should be taken in fabricating the n^- region of the As–P double diffused and offset gate devices in order to control smaller dimensions of 0.1-μm size. However, extensive use of self-aligning technologies would make these structures quite feasible, even at VLSI levels.

6.5 Summary

In the search for MOSFET structures suitable for scaling down to the submicrometer level, hot-carrier generation for various kinds of device structures has been investigated, both experimentally and theoretically. The device structure dependences of hot-carrier generation shown here demonstrated clearly that the scaling down of MOSFET structural parameters results in fatal lowering of the highest applicable voltage, BV_{DC}, and the drain sustaining voltage, BV_{DS}, limited by hot-carrier generation. The gate current which characterized hot-electron injection was modeled numerically as thermionic emission from heated electron gas over the Si–SiO_2 energy barrier. Also, the substrate current was modeled analytically using a two-dimensional analysis. Agreement between the theory and experimental results were remarkable.

On the basis of these experimental data and models of hot-carrier generation, two device structures were proposed for minimizing hot-carrier generation and associated problems in submicrometer MOSFET. These include a graded device structure and an offset gate structure. The proposed device structures provide remarkable improvements, raising the highest applicable voltage and the drain sustaining voltage as limited by hot-carrier generation by 2–3 V. A 2–3 V improvement could bring about a realization of VLSI circuits having supply voltages compatible with existing system environments.

However, since the present VLSI technology has been using many radiative process steps besides electron-beam lithography, the resultant devices are expected to be more susceptible to hot-carrier injection than devices fabricated without any radiative process steps. Therefore, it would be necessary to make reliable insulators with a smaller trapping density and capture cross section, as well as hot-carrier-resistant MOSFET structures, in order to realize devices and circuits with higher reliability.

CHAPTER

7

As–P Double Diffused Drain (DDD) Versus Lightly Doped Drain (LDD) Devices

7.1 Introduction

Device miniaturization achieved by means of advanced silicon fabrication technologies is approaching the physical limits predicted by operating principles. However, because reduction in device dimensions is not accompanied by a corresponding reduction in supply voltage, higher electric fields can be generated within the device. The so-called "distortion" of the simple scaling law proposed by Dennard *et al.* [1974] that results from this nonscalability of the supply voltage is strongly manifested in devices with an effective channel length of 1 μm or less. This distortion imposes various kinds of constraints on device design. Among these constraints, short-channel effects and hot-carrier effects, which seem to have a design trade-off relation with each other, are the most important problems to overcome in realizing submicrometer very large-scale integration (VLSI).

Figure 7.1 shows the minimum feature sizes, effective channel length L_{eff}, and oxide thickness t_{ox} of metal-oxide semiconductor (MOS) devices,

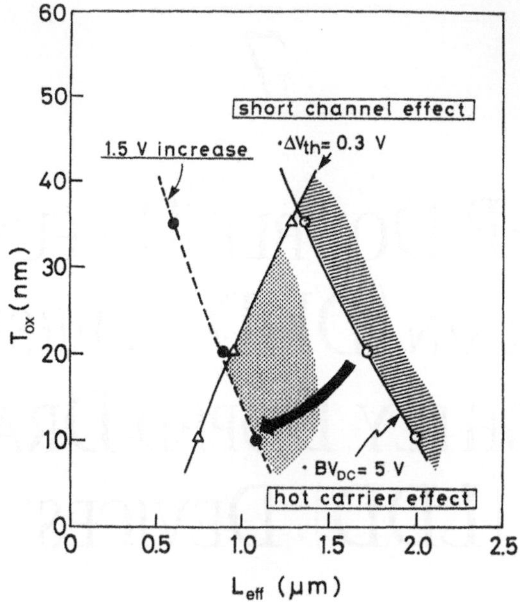

Figure 7.1 Minimum possible feature size, effective channel length, and oxide thickness of MOS devices as allowed by a conventional As drain structure. (Reprinted with permission from E. Takeda, H. Kume, Y. Nakagome, T. Makino, A. Shimizu, and S. Asai, "An As–P(n^+–n) Double Diffused Drain MOSFET for VLSI's," *IEEE Trans. Electron Devices* **30**, 652–657 (© 1983 IEEE).)

as allowed by a conventional As drain structure, where some reasonable criteria for device characteristics, such as V_{th} lowering and hot-carrier effects, are assumed. The allowable regions with the use of a conventional As drain are the shaded portion shown in the figure and are determined by hot-carrier and short-channel limits. It is obvious that the minimum feature sizes in this shaded portion correspond to a 2- or 3-μm design rule (L_{eff} = 1.5 μm and t_{ox} = 35 nm).

Therefore, to make the submicrometer design rule feasible, it is vital to raise the highest applicable voltage BV_{DC} as limited by hot-carrier effects. This requires at least 1.5 V without deterioration of short-channel effects [Ogura *et al.*, 1980; Hsu and Grinolds, 1984]. In particular, hot-carrier effects would be fatal in submicrometer MOSFETs even at a reduced operating voltage (3 V), unless a hot-carrier-resistant device structure or a trap-free gate insulator were developed [Hu *et al.*, 1983; Katto *et al.*, 1984; Takeda *et al.*, 1986c; El-Mansy, 1982].

In this chapter, an As–P (n^+–n^-) double diffused drain is characterized as the most feasible device structure from the viewpoint of device design. Narrow, self-aligned n^- regions introduced at the periphery of n^+ source–drain diffusions are a unique feature of this device structure. This device also makes good use of both As and P in realizing a graded junction.

Comparison between this double diffused drain and a conventional As drain is made for the wide range of $L_{eff} = 0.5–5$ μm to demonstrate short-channel effects. These phenomena are explained by discriminating between channel hot-electron injection due to "lucky electrons" and avalanche hot-carrier injection due to impact ionization at the drain.

By employing experiments and simulations using the two-dimensional process/device simulators SUPREM and CADDET [Antoniadis *et al.*, 1978; Toyabe *et al.*, 1978], it is proven that this device structure provides remarkable improvements, not only in terms of channel hot-electron effects but also in terms of avalanche hot-carrier effects, which are more responsible for hot-carrier-related device degradation. We have also succeeded in directly measuring hot-hole gate current related to the creation of surface states or trapping centers. This current seems to play a significant role as well.

7.2 DDD Structure and Its Fabrication Process

Figure 7.2 shows a cross section of an As–P double diffused drain and its drain impurity profile. The self-aligned n^- regions introduced at the periphery of n^+ source–drain diffusions are obtained by double implantation of As and P. This helps to reduce the electric field at the drain.

The n^- peak concentration is the most important parameter for determining both lateral and vertical impurity profiles which influence device characteristics, and it was varied over the range from 1×10^{18} to 8×10^{19} cm^{-3}. Figure 7.3. shows the dependence of both the x component of the electric field at the drain and the junction depth on n^- peak concentration. The devices used in this simulation have $L_{eff} = 1.1$ μm and $t_{ox} = 20$ nm. It was found that the electric field under the bias condition of $V_D = V_G = 5$ V and $V_{SUB} = -2$ V decreases abruptly with the increase of n^- peak concentrations from 1×10^{17} to 5×10^{18} cm^{-3} and then saturates at about 5×10^{18} cm^{-3}. The junction depth also increases monotonously as the n^- concentrations increases. Therefore, the n^- peak concentration can be adjusted to optimize both the electric field and junction depth, which may have an important influence on packing density and short-channel effects.

The effective channel length was varied over the range $L_{eff} = 0.5–5$ μm to facilitate investigation of the effects of MOS device downscaling. Oxide thickness was also taken as a parameter. The channel implant was

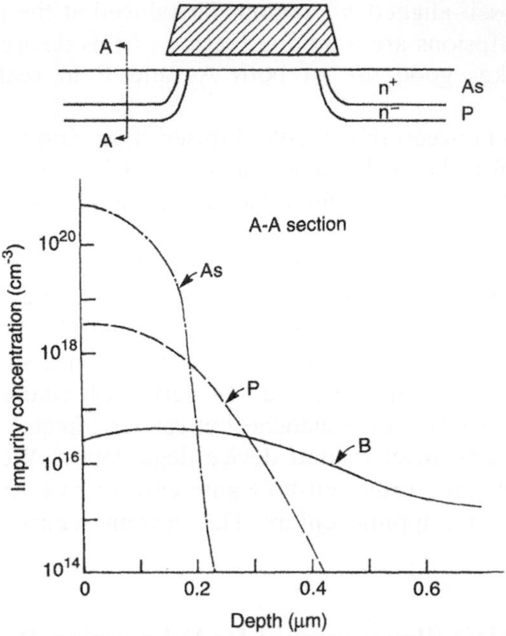

FIGURE 7.2 A cross section of an As–P double diffused drain and a typical drain impurity profile. (Reprinted with permission from E. Takeda, H. Kume, Y. Nakagome, T. Makino, A. Shimizu, and S. Asai, "An As–P(n^+–n) Double Diffused Drain MOSFET for VLSI's," *IEEE Trans. Electron Devices* **30**, 652–657 (© 1983 IEEE).)

controlled to tailor the threshold voltage to about 0.5 V. A typical boron channel impurity profile is also shown in Figure 7.2.

Figure 7.4 illustrates the fabrication process in simplified form [Takeda *et al.*, 1982c]. First, after patterning the gate electrode with poly-Si, the n^- source–drain is implanted with P. Next, annealing of the n^- region is carried out in an N_2 atmosphere at 1000° C. Then, the n^+ source–drain is implanted with As. After that, the conventional processing steps take over. It should be noted that no additional masking or process steps, such as oxide sidewall fabrication, are required. Therefore, it can be confidently said that packing density and reproducibility for the process are the same as with conventional technologies [Tsang *et al.*, 1982; Hanafi, 1985].

7.3 DDD Device Characteristics

7.3.1 I_D–V_D Characteristics

I–V characteristics for both As drain and As–P double diffusion structures having an effective channel length of 1 μm are shown in Figure

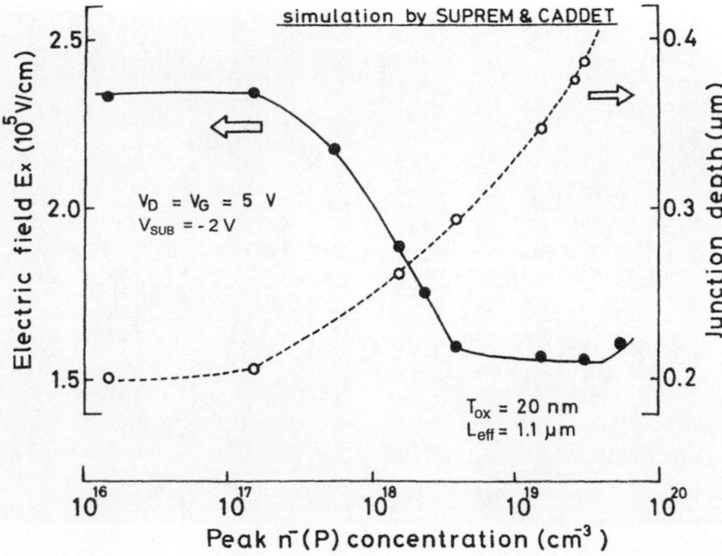

FIGURE 7.3 Dependence of both the x-component electric field at the drain and the junction depth on n^- peak concentration. (Reprinted with permission from E. Takeda, H. Kume, Y. Nakagome, T. Makino, A. Shimizu, and S. Asai, "An As–P(n^+–n) Double Diffused Drain MOSFET for VLSI's," *IEEE Trans. Electron Devices* **30**, 652–657 (© 1983 IEEE).)

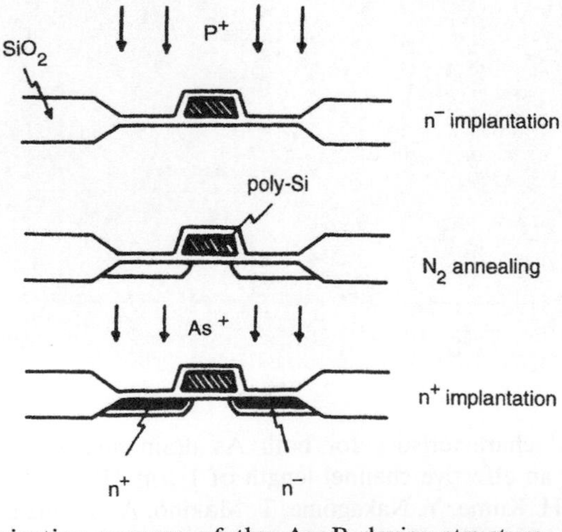

FIGURE 7.4 Fabrication process of the As–P device structure. (Reprinted with permission from E. Takeda, H. Kume, Y. Nakagome, T. Makino, A. Shimizu, and S. Asai, "An As–P(n^+–n) Double Diffused Drain MOSFET for VLSI's," *IEEE Trans. Electron Devices* **30**, 652–657 (© 1983 IEEE).)

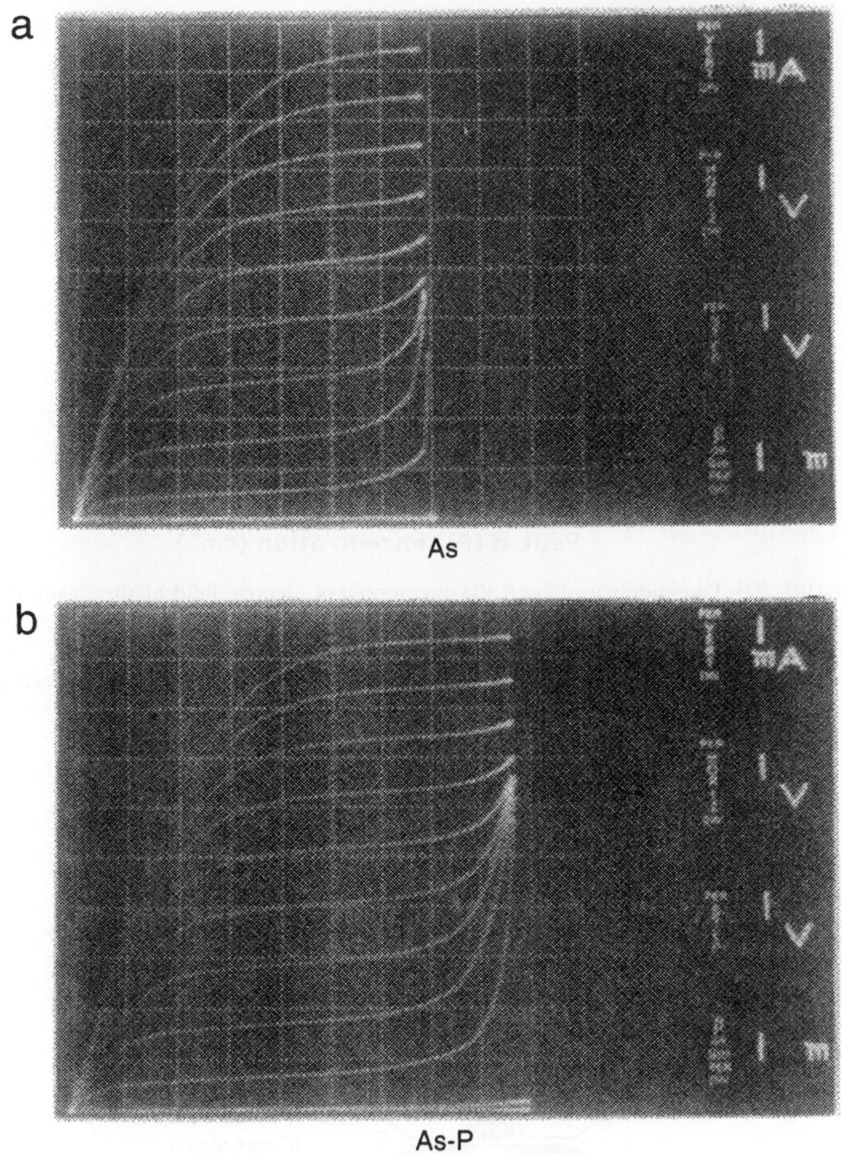

FIGURE 7.5 *I–V* characteristics for both As drain and As–P double diffused structures having an effective channel length of 1 μm. (Reprinted with permission from E. Takeda, H. Kume, Y. Nakagome, T. Makino, A. Shimizu, and S. Asai, "An As–P(n^+–n) Double Diffused Drain MOSFET for VLSI's," *IEEE Trans. Electron Devices* **30**, 652–657 (© 1983 IEEE).)

7.5. From the figure, it can be deduced that no degradation of transconductance in the As–P device structure is found as compared to the As device. The As–P structure contributes to about a 1.5-V increase in drain sustaining voltage.

7.3.2 DRAIN SUSTAINING VOLTAGE (BV_{DS})

Figure 7.6 shows drain sustaining voltage as a function of effective channel length, with the n^- peak concentration as a parameter. It can be seen that the drain sustaining voltages are strongly dependent on the n^- peak concentration. Even in the submicrometer region, the drain sustaining voltage increases by as much as 2 V. In particular, for a 0.6-μm device, BV_{DS} is 7.8 V for an As–P drain at an n^- peak concentration of 7.5×10^{18} cm^{-3} compared to only 5.5 V for a conventional drain structure.

7.3.3 SHORT-CHANNEL EFFECTS—V_{TH} LOWERING

The As–P double diffused structure does not enhance the falloff in threshold voltage that accompanies a reduction in channel lengths. Figure

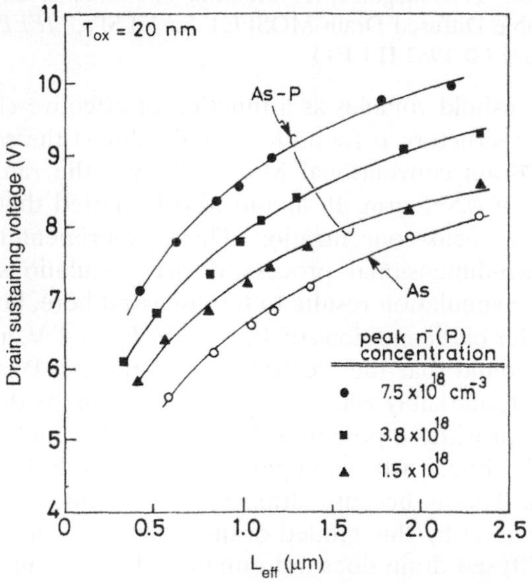

FIGURE 7.6 Drain sustaining voltages as a function of effective channel length with n^- peak concentration as a parameter. (Reprinted with permission from E. Takeda, H. Kume, Y. Nakagome, T. Makino, A. Shimizu, and S. Asai, "An As–P(n^+–n) Double Diffused Drain MOSFET for VLSI's," *IEEE Trans. Electron Devices* **30**, 652–657 (© 1983 IEEE).)

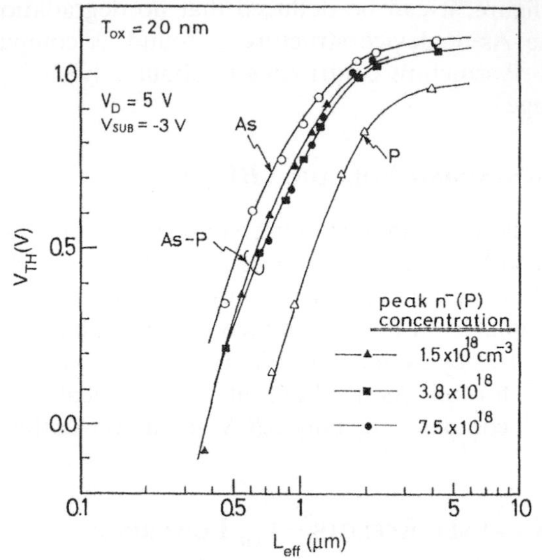

FIGURE 7.7 V_{th} lowering characteristics for As–P and As drain structures with the P peak concentration as a parameter. (Reprinted with permission from E. Takeda, H. Kume, Y. Nakagome, T. Makino, A. Shimizu, and S. Asai, "An As–P(n^+–n) Double Diffused Drain MOSFET for VLSI's," *IEEE Trans. Electron Devices* **30**, 652–657 (© 1983 IEEE).)

7.7 shows the threshold voltages as a function of effective channel length. The As–P device structure is found to provide almost the same characteristics as an As drain conventional MOSFET over the range of effective channel lengths of 0.5–5 μm. It should also be noted that there is little dependence on n^- peak concentration. These experimental results can be explained by two-dimensional process/device simulations as shown in Figure 7.8. In the simulation results to be discussed here, both devices had $L_{eff} = 1$ μm under bias conditions of $V_D = 6$ V, $V_G = 1$ V, and $V_{SUB} = -3$ V. First, it was found that the electric field of the As–P structure at the drain decreased remarkably when compared with the As drain. The As–P device structure provides a potential distribution that makes V_{th} lowering or punchthrough difficult when compared with an As drain having a rather abrupt junction. This is because drain-induced surface potential barrier lowering is alleviated by this graded drain structure. This proves that the As–P double diffused drain does not enhance short-channel effects.

7.3.4 Tail Coefficient and V_{TH} Scattering

The device characteristics in the subthreshold region are also important factors affecting switching properties or retention of dynamic memo-

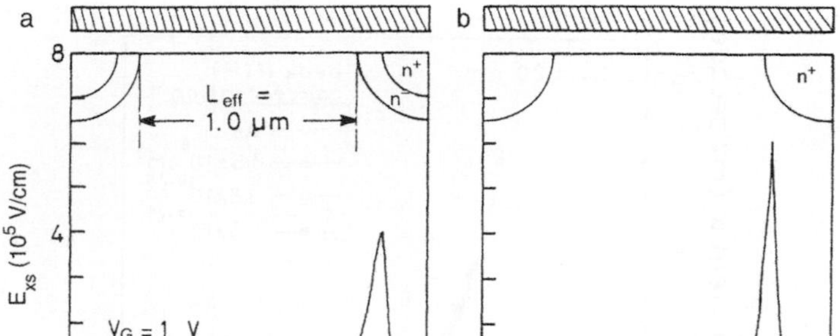

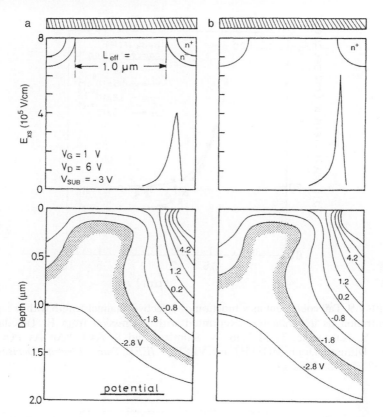

FIGURE 7.8 Simulation results by two-dimensional process/device analysis programs, SUPREM and CADDET: electric fields and potential distributions for As and As–P drains. (a) As–P drain; (b) As drain. (Reprinted with permission from E. Takeda, H. Kume, Y. Nakagome, T. Makino, A. Shimizu, and S. Asai, "An As–P(n^+–n) Double Diffused Drain MOSFET for VLSI's," *IEEE Trans. Electron Devices* **30**, 652–657 (© 1983 IEEE).)

ries. Figure 7.9 presents tailing coefficient as a function of effective channel length, with the n^- peak concentration as a parameter. It can be seen that there is little dependence of tailing coefficient on n^- peak concentration over the range of $L_{eff} = 0.5–5\ \mu m$.

Figure 7.10 compares the standard deviation, σ, of V_{th} scattering for As, As–P and P drains and displays the results as a function of effective channel length. It should be pointed out that the As–P devices provide almost the same standard deviation, $\sigma < 100$ mV, as conventional drains do over the range $L_{eff} = 0.5–5\ \mu m$. Although P drains do not have very good V_{th} scattering characteristics, this is mainly because of a sharp V_{th} lowering for the P drains, as can be seen in Figure 7.7.

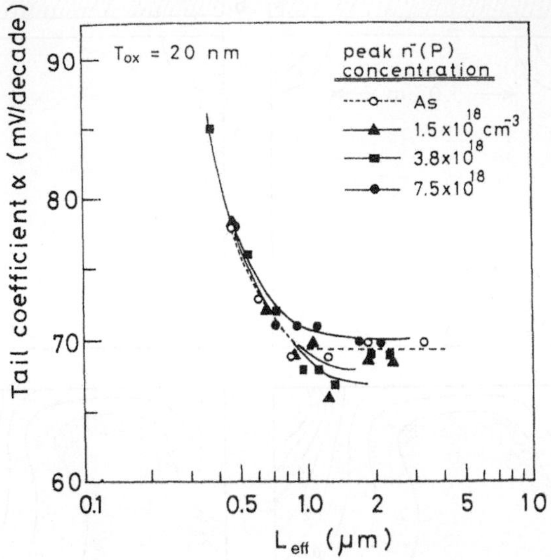

Figure 7.9 Tail coefficient as a function of effective channel length with n^- peak concentration as a parameter. (Reprinted with permission from E. Takeda, H. Kume, Y. Nakagome, T. Makino, A. Shimizu, and S. Asai, "An As–P(n^+–n) Double Diffused Drain MOSFET for VLSI's," *IEEE Trans. Electron Devices* **30**, 652–657 (© 1983 IEEE).)

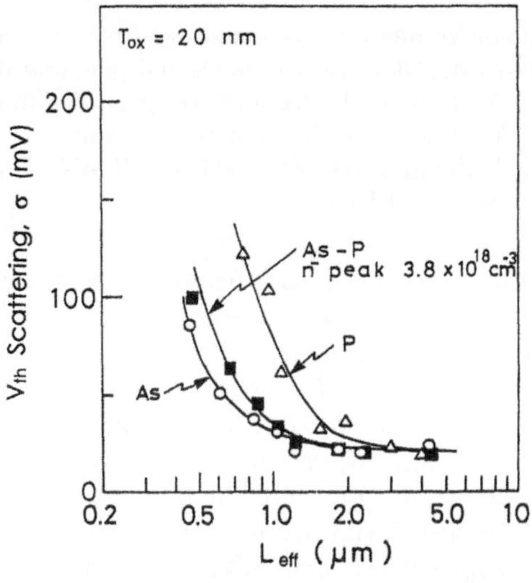

Figure 7.10 V_{th} scattering for As–P drain and As drain as a function of effective channel length.

7.3.5 HOT-CARRIER BREAKDOWN VOLTAGE OR HIGHEST APPLICABLE VOLTAGE (BV_{DC})

Figure 7.11 shows limits placed by channel hot-electron injection on the highest applicable voltage BV_{DC} for the devices used in the present work. These can be seen as a function of effective channel length. This highest applicable voltage was determined by taking an extrapolated threshold shift of 10 mV over 10 years as being allowable.

From the figure, it can be seen that As–P devices with n^- peak concentration from 1.5×10^{18} to 7.5×10^{18} cm^{-3} provide remarkable improvements, achieving a 2-V increase in the highest applicable voltage, when compared with an As drain device. It should also be noted that the highest applicable voltage is rather dependent on n^- peak concentration. A 2-V increase in the highest applicable voltage will make it possible to operate MOS devices with 1 μm or smaller dimensions at a 5-V supply voltage.

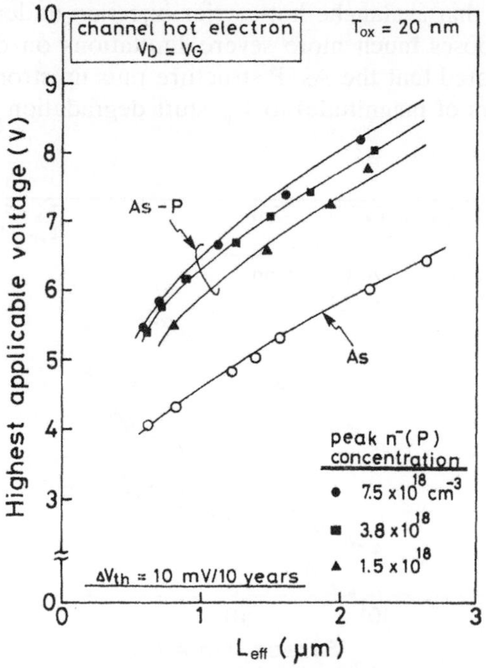

FIGURE 7.11 Highest applicable voltage determined by channel hot-electron injection with n^- peak concentration as a parameter. (Reprinted with permission from E. Takeda, H. Kume, Y. Nakagome, T. Makino, A. Shimizu, and S. Asai, "An As–P(n^+–n) Double Diffused Drain MOSFET for VLSI's," *IEEE Trans. Electron Devices* **30**, 652–657 (© 1983 IEEE).)

With avalanche hot-carrier injection under a bias region of $V_G < V_D$, on the other hand, degradation of transconductance rather than threshold-voltage shift is more pronounced. The time dependence of transconductance degradation for an As–P device with an effective channel length of 1 μm is shown in Figure 7.12, where it is compared with that for a conventional MOSFET. The bias condition of $V_G = 3$ V was chosen so that degradation might be the most severe. This bias condition also corresponds to the gate voltage that gives a peak value for the bell-shaped curve of substrate current versus gate bias.

The device with the As–P structure is found to provide excellent resistance against avalanche hot-carrier degradation. There is a remarkable difference between these devices, of more than two orders, in the stress time leading to specific degradation.

Figure 7.13 also demonstrates a significant difference in V_{th} shifts between those due to channel hot-electron injection and those due to avalanche hot-carrier injection, for both As–P and As drains having an effective channel length of 1.2 μm. Stress time was 500 s.

It is obvious that avalanche hot-carrier injection under bias conditions of $V_G < V_D$ imposes much more severe limitations on device design. It should also be noted that the As–P structure puts up strong resistance (by one or two orders of magnitude) to V_{th} shift degradation when compared

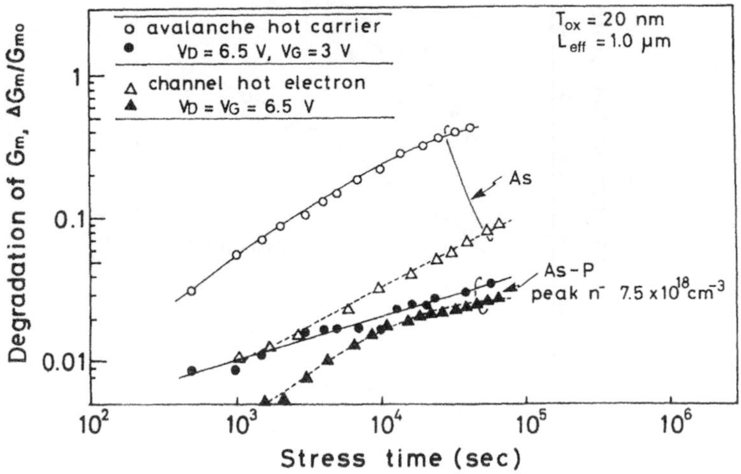

FIGURE 7.12 Time dependence of transconductance degradation for an As–P device with $L_{eff} = 1.0$ μm. (Reprinted with permission from E. Takeda, H. Kume, Y. Nakagome, T. Makino, A. Shimizu, and S. Asai, "An As–P(n^+–n) Double Diffused Drain MOSFET for VLSI's," *IEEE Trans. Electron Devices* **30**, 652–657 (© 1983 IEEE).)

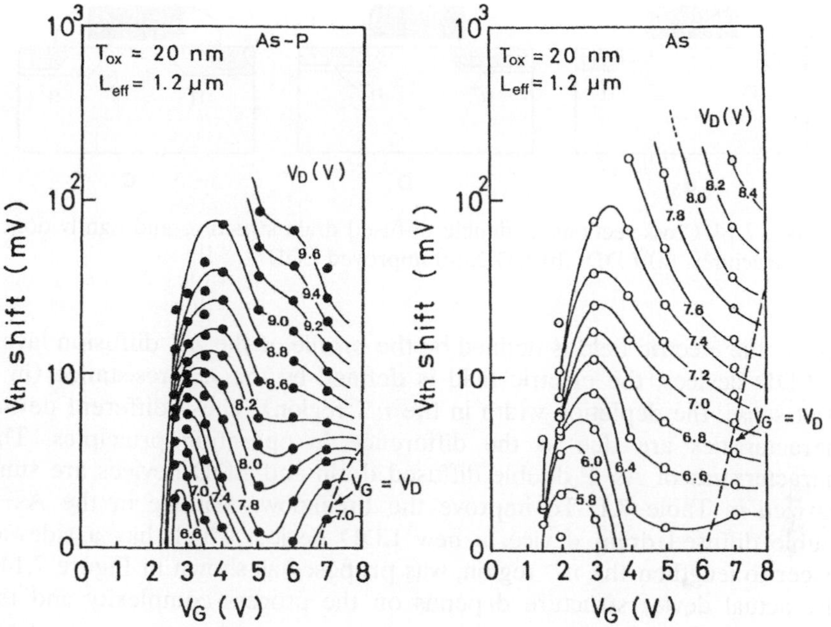

FIGURE 7.13 V_{th} shifts due to both channel hot-electron and avalanche hot-carrier injections for As and As–P drains having $L_{eff} = 1.2$ μm. (Reprinted with permission from E. Takeda, H. Kume, Y. Nakagome, T. Makino, A. Shimizu, and S. Asai, "An As–P(n^+–n) Double Diffused Drain MOSFET for VLSI's," *IEEE Trans. Electron Devices* **30**, 652–657 (© 1983 IEEE).)

with an As drain. The fact that avalanche hot-carrier injection imposed more stringent constraints is probably because hot holes having a larger effective mass than hot electrons are injected into the gate oxide at the same time as hot electrons, thereby creating interface traps.

7.4 DDD and LDD Device Operation Principles

From the viewpoint of mass production, the As–P double diffused drain structure and LDD structure are superior among the proposed high-performance device structures as shown in Figure 7.14. The different performance characteristics will be explained by simulation results.

Figure 7.15 shows the n^- concentration dependence of electric field at the drain edge ($\mathscr{E}_x^{max}$) for As–P drain and LDD devices. For the As–P drain device, $\mathscr{E}_x^{max}$ decreases with increasing n^- concentration, while the result is reversed for LDD devices. In the As–P double diffused drain

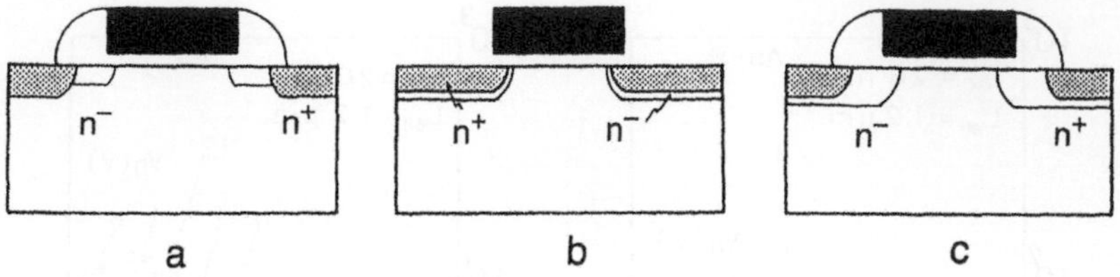

FIGURE 7.14 Cross sections of double diffused drain structure and lightly doped drain structures. (a) LDD; (b) DDD; (c) improved LDD.

device, the electric field is defined by the profile of the n^- diffusion layer. In LDD devices, the electric field is defined by the n^- resistance (in a strict sense, the depletion width in the n^- region). These different device characteristics are due to the difference in operating principles. The characteristics of As–P double diffused drain and LDD devices are summarized in Table 7.1. To improve the breakdown voltage in the As–P double diffused drain device, a new LDD device, which has a sidewall spacer to lengthen the n^- region, was proposed as shown in Figure 7.14c. The actual device structure depends on the process complexity and the supply voltage.

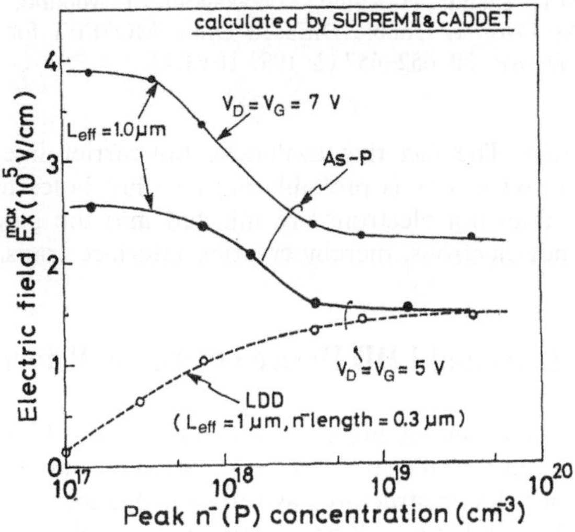

FIGURE 7.15 n^- peak concentration dependence of electric field in As–P double diffused drain and LDD device.

TABLE 7.1
Characteristics of As–P Double Diffused Drain Structure and LDD Structure

		As–P	LDD
Advantage	Process	Process simplicity Miniaturization	
	Device	Low G_m degradation High reproducibility	Sustain voltage control by n^- resistance Short-channel effect
Disadvantage	Process		Process complexity Miniaturization
	Device	Limit on improving sustaining voltage Short-channel effect	G_m degradation Initial stage G_m degradation

7.5 LDD Device Characteristics

7.5.1 DEVICE CHARACTERISTICS SPECIFIC TO LDD DEVICES

In LDD devices, the electric field is spread out in the n^- region, its peak value is reduced, and it is moved out from under the gate electrode. Hot carriers are generated by impact ionization in the n^- resistant region under the sidewall spacer, and in the case of NMOS, hot electrons are injected and trapped in the sidewall spacer causing pinch-off in the n^- region, as shown in Figure 7.16. This phenomenon was discovered by Hsu *et al.* [Hsu and Grinolds, 1984; Hsu and Chiu, 1984d]. This electron trapping within the sidewall spacer causes device degradation.

Specifically, LDD devices degrade as follows.

1. The initial stage G_m degradation. This degradation mode becomes worse with lower n^- concentration. Therefore, there must be an optimum n^- concentration.
2. The electron injection/trapping process in the sidewall, which occurs at the drain, also occurs at the source side when the n^- concentration is low and results in the new initial stage G_m degradation. This was found by Katto *et al.* [1984]. Figure 7.17 shows the simulation results for an LDD structure obtained using a 3-D hot-carrier simulator (H^2-CAST) [Hamada *et al.*, 1987]. When the n^- dose is 1×10^{13} cm^{-2}, the injected electron region is located under the gate electrode in the n^- layer. However for n^- dose $= 1 \times 10^{12}$ cm^{-2}, the injected region is located under the sidewall spacer on both the source and drain sides. Such a degradation phenomenon is specific to LDD and does not occur in

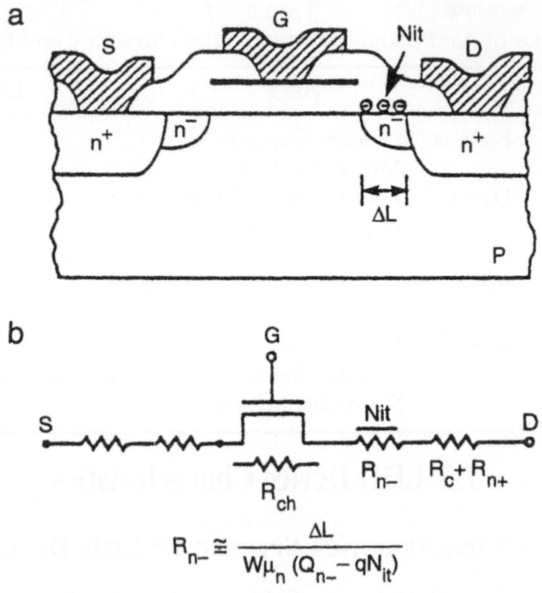

FIGURE 7.16 A schematic diagram of hole–electron trapping in an SiO_2 spacer specific to an LDD device. (Reprinted with permission from F.-C. Hsu and H. R. Grinolds, "Structure-Enhanced MOSFET Degradation Due to Hot-Electron Injection," *IEEE Electron Dev. Lett.* **5**, 71–74 (© 1984 IEEE).)

As–P double diffused drain devices. This is because, in the As–P structure, most of the n^- region is under the gate, so pinch-off does not occur.

7.5.2 Reduced Switchback Action Due to Source n^- Resistance

Characteristics specific to the LDD structure are shown in Figure 7.18. Here, reduced bipolar action due to n^- source resistance, called the switchback phenomenon, is described.

According to Hsu *et al.* [1982], the total source current I_s can be expressed as

$$I_s = \frac{(M-1)I_{ch}(R_{SUB}+R_{ext}) - V_{SUB} - 0.65}{[1-\gamma\alpha_T-(M-1)k\gamma\alpha_T](R_{SUB}+R_{ext})}, \tag{7.1}$$

where M is the avalanche multiplication factor, k is the fraction of the electrons collected by the drain that go through the high-field region, α_T

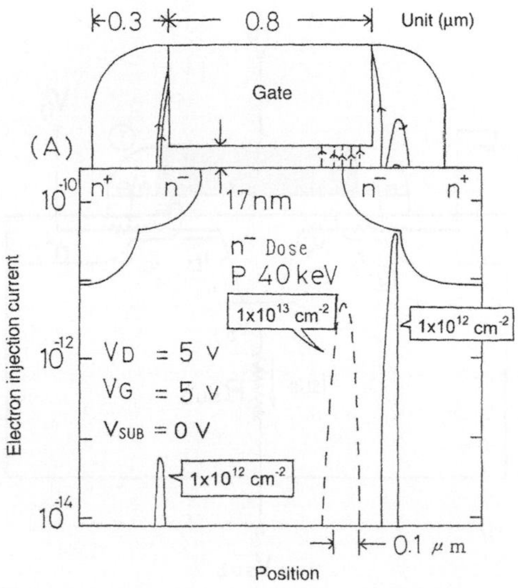

FIGURE 7.17 Simulated result of hot-carrier injection in an LDD device (H^2-CAST).

is the base transport factor, γ is the injection efficiency of the source junction, and I_{ch} is the channel current.

In general, switchback may occur under two conditions: (1) when the source–substrate junction is turned on ($V_b - V_e = 0.65$ V) or (2) when the bipolar loop (as shown in Figure 7.18) gain is bigger than one. If the switchback occurs under condition 1, from Eq. (7.1), substrate resistance (R_{SUB}) and external resistance (R_{ext}) are small and/or the gate voltage is low (small-channel current). On the other hand, the condition shown below is considered to correspond to 1. The peak value of substrate current (I_{SUB}) in a log I_{SUB} versus V_G plot is shown as a function of drain voltage (V_D) in Figure 7.19. The devices used have an effective channel length (L_{eff}) of 1.0 and 1.5 μm and a gate oxide thickness of 20 nm. Gate voltage that gives the I_{SUB} peak is about 3 V. This value is rather small and brings about a 0.1–0.2 mA/μm channel current. This condition apparently corresponds to 1 and causes drain soft breakdown (not complete switchback). In this case, $V_b - V_e = 0.65$ V and the bipolar loop gain would be smaller than one.

Figure 7.19 compares LDD with conventional MOS devices. The LDD is found to have a switchback voltage greater than conventional devices by about 3 V. Furthermore, there is a significant difference in I_{SUB} causing a drain soft breakdown between LDD and conventional devices. This difference (ΔI_b) is due to the n^- source resistance (r_s) as explained below.

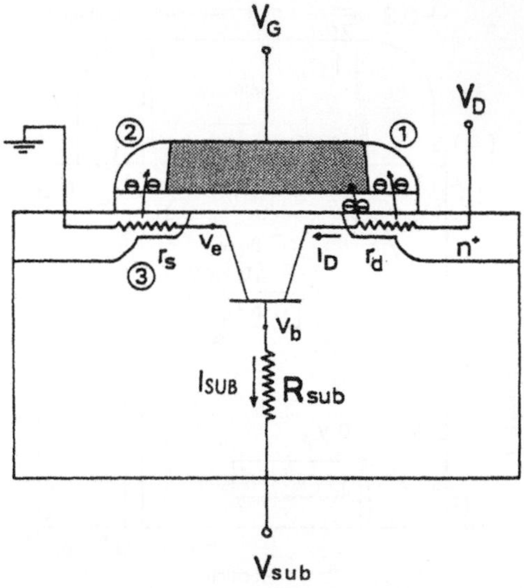

FIGURE 7.18 Device degradation mechanism in LDD MOSFETs. (1) Fast G_m degradation due to external channel pinch-off. (2) Different G_m degradation due to increased n^- source resistance. (3) Bipolar action reduction due to n^- source reduction.

From Figure 7.18, the potential V_b within the substrate is

$$V_b = R_{SUB} I_{SUB} + V_{SUB}. \tag{7.2}$$

The potential V_e at the source edge is

$$V_e = r_s I_{ch}. \tag{7.3}$$

Here, R_{SUB} is the effective substrate resistance, r_s is the n^- source resistance, and V_{SUB} is the substrate bias. Also, we did not consider the external series resistance R_{ext} and diffusion resistance plus contact resistance to simplify discussion. Using (7.2) and (7.3),

$$V_b - V_e = R_{SUB} I_{SUB} - r_s I_{ch} + V_{SUB}. \tag{7.4}$$

Drain switchback occurs when $V_b - V_e$ is about 0.65 V. This results in

$$\Delta I_b = (0.65 - V_{SUB} + r_s I_{ch})/R_{SUB} - I_{SUB0}, \tag{7.5}$$

$$I_{SUB0} = (0.65 - V_{SUB})/R_{SUB}, \tag{7.6}$$

where I_{SUB0} is the substrate current of a conventional MOSFET. Thus, the increased drain sustaining voltage in LDD structures is created by two factors. These are the reduced drain electric field due to n^- drain

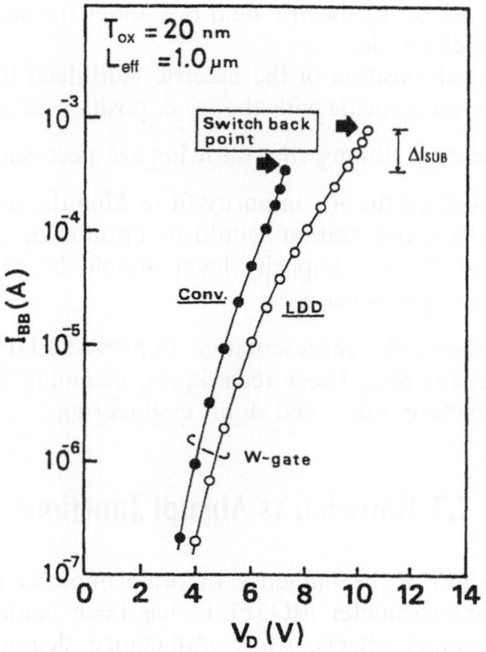

FIGURE 7.19 The peak value of substrate current as a function of drain voltage in both LDD and conventional devices.

resistance and the potential increase at the source edge due to n^- source resistance. In addition, the n^- source resistance r_s can be measured quantitatively by using Eqs. (7.5) and (7.6).

$$r_s = \Delta I_b(0.65 - V_{SUB})/(I_{ch} I_{SUB0}). \tag{7.7}$$

Therefore, the increase in n^- drain resistance due to pinch-off, which is caused by hot-electron injection, can be directly measured without many samples. Also, the channel current at the I_{SUB} peak and just before switchback does not vary with L_{eff}. The n^- source resistance in LDD devices used here with a channel width of 10 μm was about 60 Ω.

7.6 Improved LDD Devices

In order to reduce the device degradation specific to LDD structures and to improve their breakdown voltage, many improved device structures have been proposed. The objective of these is as follows.

1. To place the peak of the electric field not under the sidewall spacer but under the gate electrode.
2. To place the peak position of the electric field deep into the substrate so that it does not coincide with the peak position of current.

To satisfy these conditions, the following are necessary.

1. P should be used for the n^- impurity in making the graded drain.
2. The n^- impurity concentration should be optimized.
3. The position of the n^- impurity layer should be placed not at the surface but inside the substrate.

Figure 7.20 shows the representative improved LDD structures proposed by other companies. These techniques, including the As–P double diffused drain structure, are called drain engineering.

7.7 Gaussian vs Abrupt Junctions

A reexamination of the influence of drain impurity profile and junction depth on submicrometer MOSFETs has been made from the viewpoint of short-channel effects, transconductance degradation, and hot-carrier-related device degradation using 2-D simulation (SUPREM and CADDET) and experiments. It is shown that little scaling or even upscaling of junction depth, rather than the simple downscaling proposed by Dennard [1984], improves both short-channel effects and hot-carrier effects in the submicrometer regions. Also, significant guiding principles for device scaling in the realization of sophisticated submicrometer MOSFETs are proposed.

7.7.1 V_{th} Lowering

Figure 7.21 shows a comparison of V_{th} lowering characteristics between abrupt and gaussian junctions for MOS devices with $t_{ox} = 20$ nm, with x_j as a parameter. This figure shows that V_{th} lowering becomes more remarkable with increased X_j in the case of the abrupt junction, while, for the Gaussian junction, V_{th} lowering does not degrade at all. In fact, for $x_j = 0.6\ \mu$m with $L_{eff} = 0.5\ \mu$m, the result is improved. These simulation results are expressed as variations of the ratio of V_{th} lowering to effective channel length reduction, as shown in Figure 7.22. In the case of an abrupt junction, the ratio increases monotonically, as expected, with increase in x_j. On the other hand, for a Gaussian junction, the ratio decreases at x_j larger than 0.2–0.3 μm in the region $L_{eff} \leq 1\ \mu$m. Such a peculiar tendency becomes more remarkable in submicrometer regions.

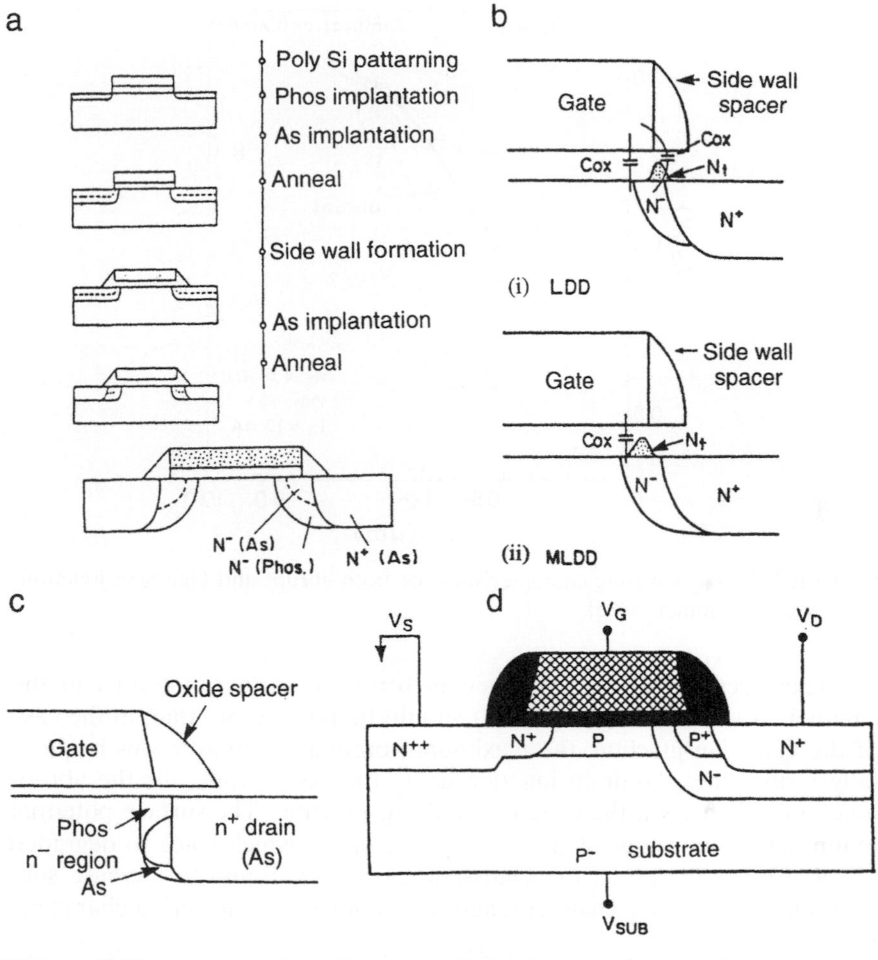

FIGURE 7.20 A cross section of improved LDD devices. (a) PLDD (reprinted with permission from Y. Toyoshima, N. Nishira, and K. Kanzaki, "Profiled Lightly Doped Drain (PLDD) Structure for High Reliable n-MOSFET's," *Tech. Digest VLSI Tech. Symp.*, 118–119 (© 1985 IEEE)); (b) MLDD (reprinted with permission from M. Kinugawa, M. Kakumu, S. Tokogawa, and K. Hashimoto, "Submicron MLDD n-MOSFET's for 5 *V* Operation," *Tech. Digest VLSI Tech Symp.*, 116–117 (© 1985 IEEE)); (c) BLDD (reprinted with permission from H. R. Grinolds, M. Kinugawa, and M. Kakumu, "Reliability and Performance of Submicron LDD MOSFET's with Buried-As n^- Impurity Profiles," *Tech. Digest IEDM*, 246–249 (© 1985 IEEE)); (d) JMOSFET (reprinted from S. Bampi and J. D. Plummer, "Modified LDD Device Structures for VLSI," *Tech. Digest IEDM*, 234–237 (© 1985 IEEE)).

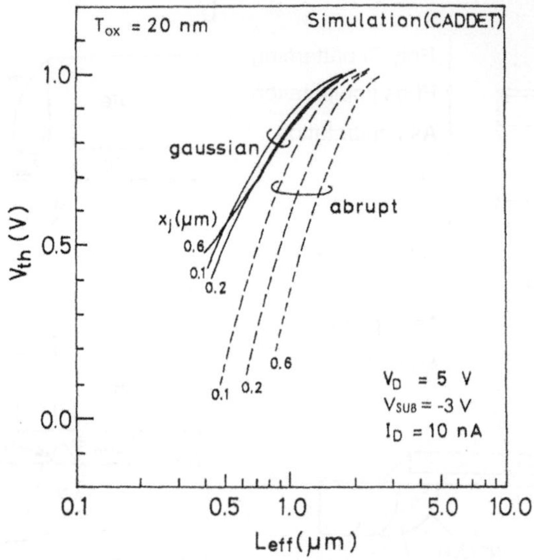

FIGURE 7.21 V_{th} lowering characteristics for both abrupt and Gaussian junctions [Bampi and Plummer, 1985].

These results can be explained in terms of surface potential in the channel, as shown in Figure 7.23. It should be pointed out that, in the case of the Gaussian junction, the maximum potential position moves horizontally farther into the drain junction as x_j increases, while, for the abrupt junction, it is fixed at the edge of the drain junction. The surface potential minimum in the center of the channel increases, which leads to degraded V_{th} lowering. Therefore, the Gaussian junction is found to increase substantially the effective channel length and improve V_{th} lowering characteristics.

These simulation results are also proved by experiments using phosphorus (P), which provides a nearly Gaussian profile, as a drain diffusant. Figure 7.24 shows the experimental results of V_{th} lowering for MOS devices with an As–P drain. They show little dependence of V_{th} lowering on junction depth, and at $x_j = 0.45\ \mu$m, V_{th} lowering even seems to become less in the region $L_{eff} \leq 0.5\ \mu$m.

7.7.2 G_m vs x_j Relation

Due to the substantial increase in L_{eff}, slight G_m degradation occurs in the case of the Gaussian junction with increased x_j, compared to that for the abrupt junction. Figure 7.25 shows the x_j dependence of transconductance, with L_{eff} as a parameter. It should be noted that the transcon-

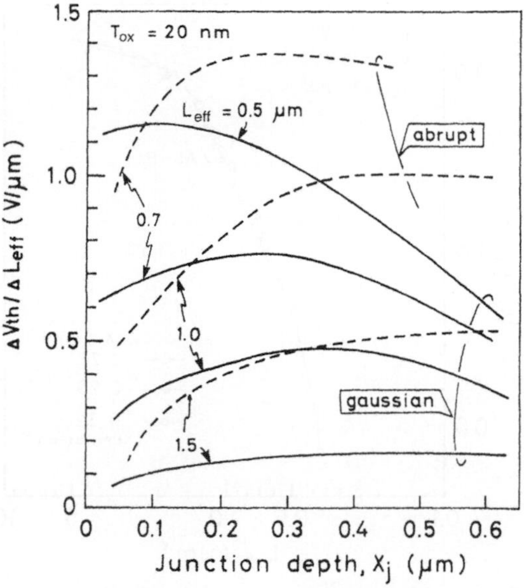

FIGURE 7.22 V_{th} lowering ratio to L_{eff} variation for both abrupt and Gaussian junctions as a function of junction depth [Grinolds *et al.*, 1985].

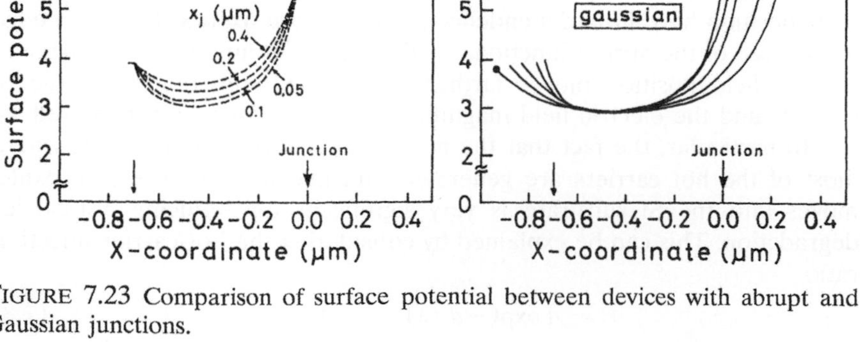

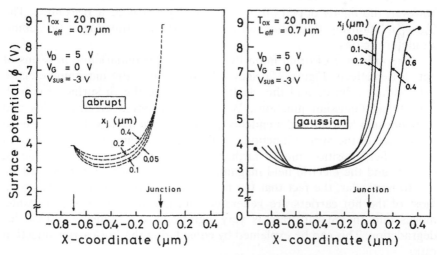

FIGURE 7.23 Comparison of surface potential between devices with abrupt and Gaussian junctions.

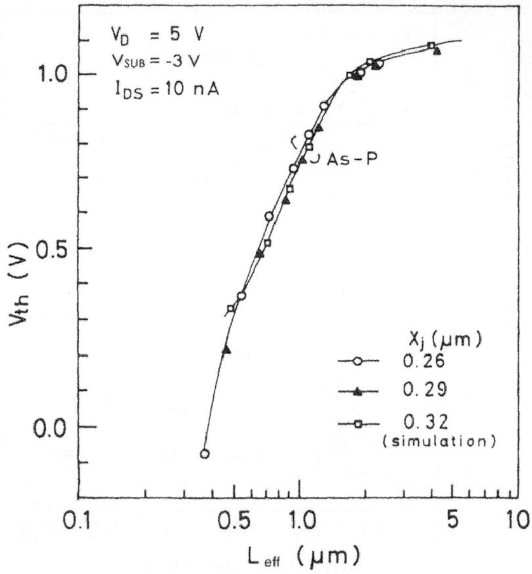

FIGURE 7.24 Experimental V_{th} lowering characteristics for As–P devices, with x_j as a parameter.

ductance for the Gaussian junction device gradually decreases by at most 10% as junction depth increases from less than 0.1 μm to 0.3 μm. This is because the gate electrode covers the drain diffusion layer, which leads to lessening of the increase in its diffusion resistance. In the LDD structure, however, the lightly doped offset diffusion layer acts as a resistance. The G_m–x_j relationship thus obtained is useful when optimizing x_j in submicrometer regions.

The influence of drain impurity profile is also remarkable in terms of hot-carrier effects. Figure 7.26 shows the position of the maximum electric field within a device and the electric field along the Si surface for both abrupt and Gaussian junctions. While the maximum electric fields and their position have little dependence on x_j and are fixed at the drain edge in the case of the abrupt junction, for the Gaussian junction the maximum electric field position moves farther into the Si substrate from the Si surface, and the electric field magnitude also decreases with increased x_j.

In particular, the fact that the maximum electric field position, where most of the hot carriers are generated and injected into the gate oxide, moves into the Si substrate is very significant for reducing hot-carrier degradation. This can be explained by considering the hot-carrier injection ratio

$$P = A \exp(-d/\lambda), \qquad A = 2.9, \tag{7.8}$$

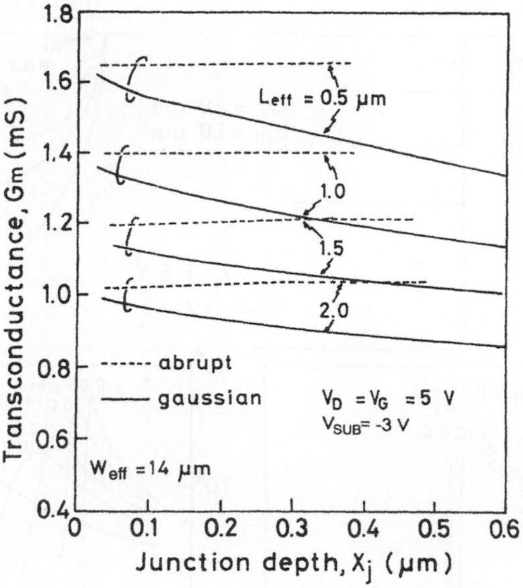

FIGURE 7.25 Transconductance characteristics as a function of x_j for abrupt and Gaussian junctions.

where λ is the carrier mean free path and d is the distance from the Si surface. For example, d at $x_j = 0.3$ μm in the Gaussian junction is about 0.1 μm, while λ is 5–10 nm. The impact of d on hot-carrier degradation, so-called "vertical offset", is considerable.

Figure 7.27 shows a stress-time variation in G_m degradation under the condition of substrate current $I_{SUB} = 1.6 \times 10^{-4}$ A for both As drain and As–P drain MOS devices with $L_{eff} = 0.8$ μm and $x_j = 0.3$ μm. It was found that an As–P nearly Gaussian junction provides stronger resistance to hot-carrier-related device degradation, compared with that for an As nearly abrupt junction. It is significant that the stress condition ($V_D = 7.5$ V) for the As–P drain is more severe than that ($V_D = 6.6$ V) for the As drain.

Figure 7.28 compares lifetime, τ, to hot-carrier effects for both As drains and As–P drains as a function of substrate current. The lifetime was defined as the time it takes for V_{th} to shift 10 mV. It is clear from this figure that As–P drains have longer lifetimes by more than one order of magnitude than As drains, in spite of the fact that the substrate current, which is considered the most important criterion when diagnosing hot-carrier injection, is constant. Constant substrate current also means that the electric field or ionization integral is approximately constant. The experi-

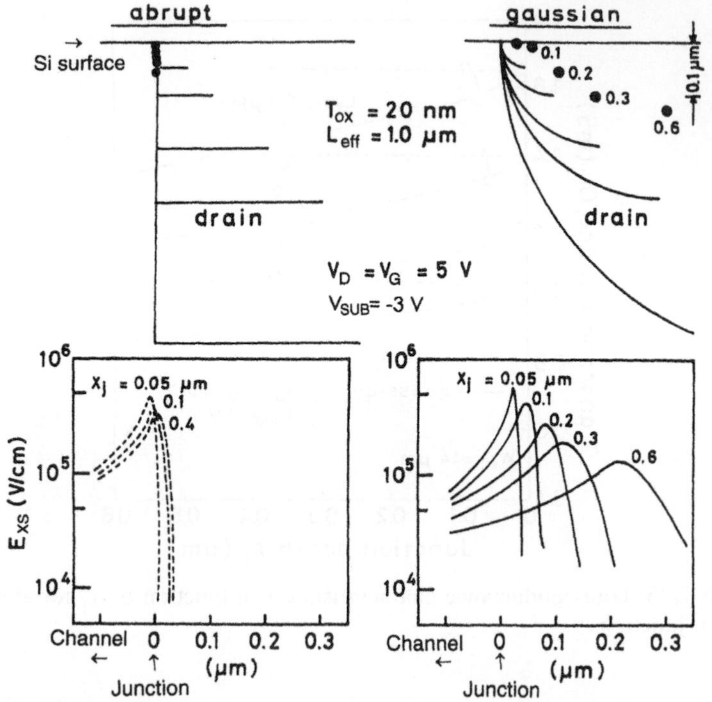

FIGURE 7.26 Position of maximum vertical electric field and the horizontal electric field distribution parallel to the Si surface for both abrupt and Gaussian junctions.

mental results cited here are considered to reflect the above simulation results on vertical offset effects [Hsu *et al.*, 1985b; Ko *et al.*, 1986; Izawa and Takeda, 1987].

7.8 Summary

The device characteristics of As–P (n^+–n^-) double diffused drain and LDD structures have been investigated experimentally and analytically as basic submicrometer MOS device structures. Furthermore, three kinds of device characteristics which are specific to LDD were discussed. "Drain engineering," such as DDD or LDD structures, can improve the hot-carrier sustaining voltage and are indispensable for submicrometer devices. However when $L_{\text{eff}} < 0.5$ μm and the supply voltage is less than 3.3 V, short-channel effects such as punchthrough, in addition to hot-carrier

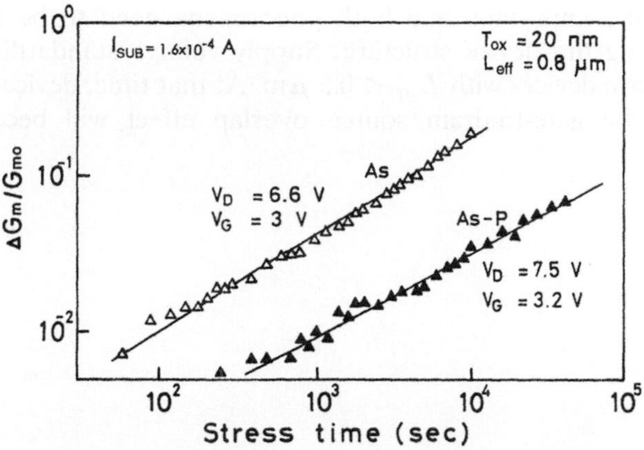

FIGURE 7.27 Stress-time variation in transconductance degradation for both As and As–P drain MOSFETS with $L_{\text{eff}} = 0.8\ \mu$m and $x_j = 0.3\ \mu$m.

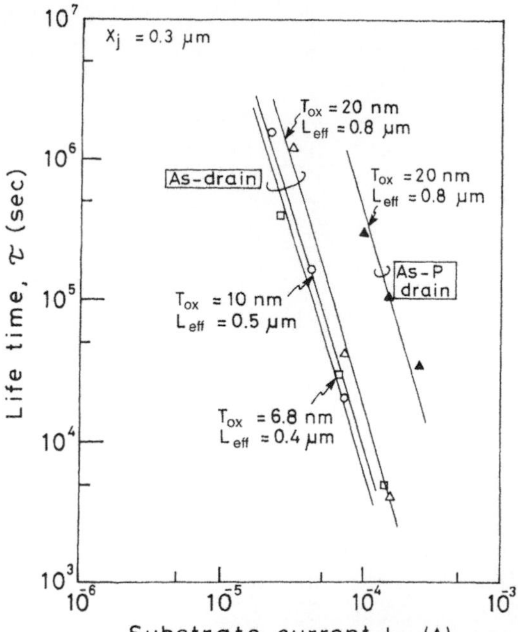

FIGURE 7.28 Lifetime experimentally obtained for both As and As–P drain MOSFETs as a function of substrate current.

effects, become important, and both phenomena need to be considered when designing the device structure. Supply voltage standardization will further improve devices with $L_{eff} < 0.5\ \mu m$. At that time, device structures that utilize the gate-to-drain/source overlap effect will become more important.

CHAPTER

8

Gate-to-Drain Overlapped Devices (GOLD)

8.1 Introduction

With scaling down of device dimensions to under 0.5 μm, problems such as reduced reliability and lowered punchthrough voltage require that power supply voltages be reviewed and/or the device structure rebuilt. "Drain-engineering" methods such as the double diffused drain (DDD) [Takeda *et al.*, 1982c] and the lightly doped drain (LDD) [Ogura *et al.*, 1980] have been used to improve reliability. However, it is difficult to achieve high reliability and high performance at 5 V with these methods because there is a trade-off between transconductance and device breakdown voltage (hot-carrier, drain sustaining). This is because drain-engineering methods optimize only n^- length L_n and n^- dose N_d. However, the gate–drain/source overlap length Γ is another important parameter in the control of device characteristics [Izawa *et al.*, 1988].

This chapter describes a new device, the gate–drain overlapped device (GOLD) [Izawa *et al.*, 1988], to achieve high reliability, high performance, and suppressed punchthrough, all without design trade-offs. The GOLD concept makes use of the strong gate–drain overlap effect, which is in contrast to the weak gate–drain overlap effect in the LDD methods [Ko *et al.*, 1986; Hamamoto *et al.*, 1986]. Another device structure, inverse-T

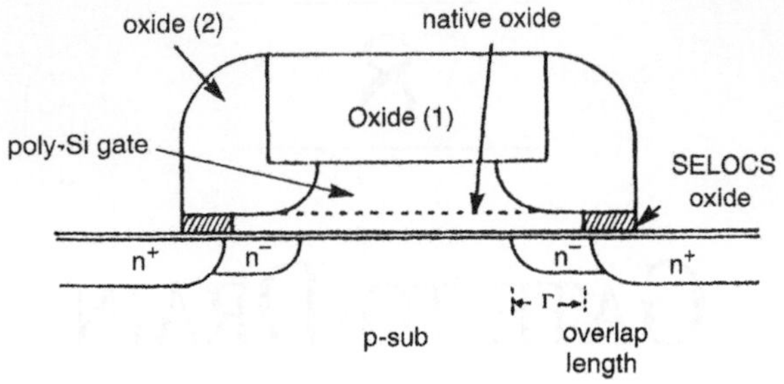

FIGURE 8.1 Schematic cross section of GOLD. GOLD takes advantage of the gate–drain overlap structure. Controllability of thickness and length of overlapped gate are key points in GOLD in order to optimize the gate–drain overlap effect. (Reprinted with permission from R. Izawa, T. Kure, and E. Takeda, "Impact of the Gate-Drain Overlapped Devices (GOLD) for Deep Submicrometer VLSI," *IEEE Trans. Electron Devices* **35**, 2088–2093 (© 1988 IEEE).)

gate LDD (ITLDD) [Huang *et al.*, 1986c], was designed to reduce the structure-enhanced hot-carrier degradation inherent in the LDD structure. However, the increased reliability of GOLD over other LDD devices is made possible by two factors: electric field reduction and suppression of hot-carrier injection, which results from a vertical field induced by an overlapped gate with an optimized gate–drain/source overlap length.

Furthermore, GOLD has a fine overlapped gate structure that also helps optimize the gate–drain/source overlap length. This chapter presents a new method for fabricating the GOLD structure precisely.

8.2 GOLD Structure and Its Fabrication Process

A schematic cross section of GOLD is shown in Figure 8.1. The structure consists of two main parts: (1) a double-layer gate with a thin etch-stop intermediate layer and (2) selective oxide coating of silicon gate (SELOCS) [Sunami *et al.*, 1980] sidewalls to control the overlap length. The GOLD process is shown in Figure 8.2. The process flow up to gate oxidation is conventional. A native oxide film (5–10 Å), indicated by broken lines, is grown by air-curing the wafer after depositing the first 50-nm-thick polysilicon layer. The second polysilicon layer and CVD oxide layer are then deposited on the native oxide (Figure 8.2a). In this GOLD process, gate etching is the key process. The first thinner polysilicon layer, which acts as the overlapped gate, remains after precisely patterning the second polysilicon layer using highly selective dry etching. The selectivity is about 100 times higher. Therefore, the polysilicon etch is stopped by the

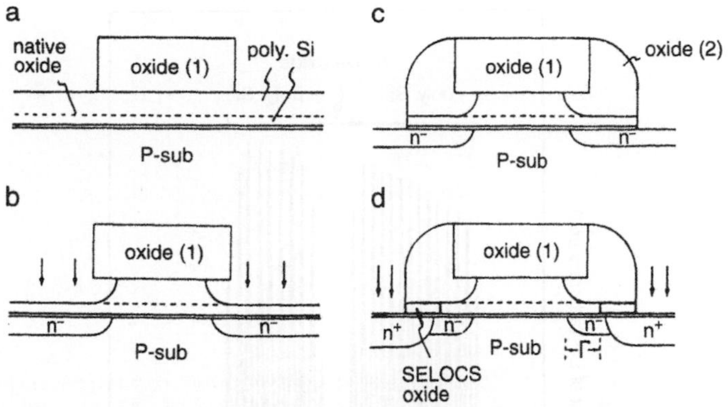

FIGURE 8.2 Process sequence for GOLD. There are four key processes. (a) Double-layer gate with a thin etch-stop native oxide layer to control the first polysilicon layer thickness. (b) The gate–drain (source) overlap structure is formed in self-alignment. Furthermore, the n^- length L_n and the overlap length Γ are independently optimized by the sidewall oxide (2) and SELOCS oxide, as shown in (c) and (d). (Reprinted with permission from R. Izawa, T. Kure, and E. Takeda, "Impact of the Gate-Drain Overlapped Devices (GOLD) for Deep Submicrometer VLSI," *IEEE Trans. Electron Devices* **35**, 2088–2093 (© 1988 IEEE).)

native oxide film, as shown in Figure 8.3. The signal of SiF reveals the etching of polysilicon.

Subsequently, by implanting phosphorus (1×10^{13} cm^{-2}) with higher energy (80 keV) through the thinner polysilicon layer, the n^- layer is formed in self-alignment with the gate (Figure 8.2b). Oxide sidewall spacers are then formed to determine the n^- layer length. The remaining first polysilicon layer outside the sidewall spacer is etched by anisotropic dry etching (Figure 8.2c). For final control of the overlap length, which strongly affects the device characteristics, a SELOCS oxide film is formed. SELOCS oxide can be grown thicker at low temperatures (~ 800° C) by applying wet oxidation to the phosphorus-doped polysilicon gate, while keeping the source–drain impurity profile unchanged and reducing oxidation of the substrate. The optimized overlap length is about 0.2 μm. Finally, the drain and the source n^+ areas are formed by As implantation outside the sidewall spacer (Figure 8.2d).

8.3 Device Characteristics

8.3.1 G_m IMPROVEMENT

GOLD has 1.25 times higher transconductance (G_m) and 1.15 times larger channel currents (I_{DS}) than conventional LDD, with $L_{eff} = 0.5$ μm

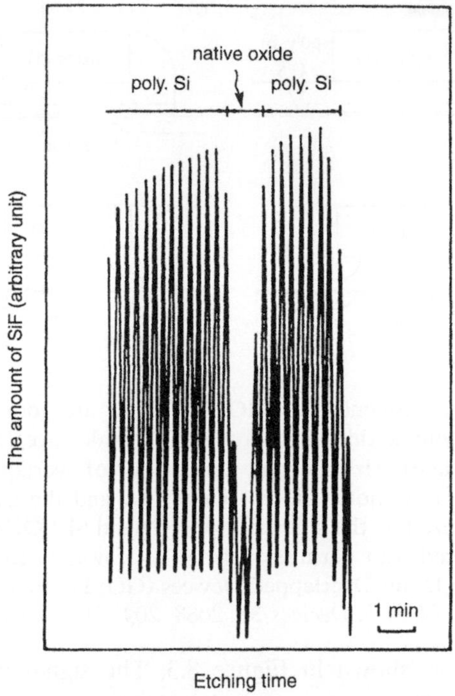

FIGURE 8.3 The characteristics of highly selective dry etching. The signals of SiF are detected during the polysilicon etch. The period in which these signals disappeared demonstrates the etching stop at the native oxide film. (Reprinted with permission from R. Izawa, T. Kure, and E. Takeda, "Impact of the Gate-Drain Overlapped Devices (GOLD) for Deep Submicrometer VLSI," *IEEE Trans. Electron Devices* **35**, 2088–2093 (© 1988 IEEE).)

and $V_D = 5$ V, as shown in Figure 8.4a–b. If $V_D = 3$ V is assumed for LDD, I_{DS} for GOLD is two times larger. Furthermore, this G_m improvement becomes greater with decreasing L_{eff}. L_{eff}, which is determined by using the maximum small-signal transconductance at $V_D = 0.1$ V with $V_G = 0.8$ V, is the effective channel length between n^- regions.

These GOLD characteristics are due to the reduced n^- resistance caused by the vertical field induced by the overlapped gate.

8.3.2 SUPPRESSED AVALANCHE-INDUCED BREAKDOWN

GOLD has higher drain sustaining voltage (BV_{DS}) than conventional LDD, as shown in Figure 8.5. The BV_{DS} in GOLD increases up to 10 V with $L_{eff} = 0.5$ μm. This BV_{DS} improvement is due to not only the reduced n^- resistance but also the reduced lateral electric field. As a

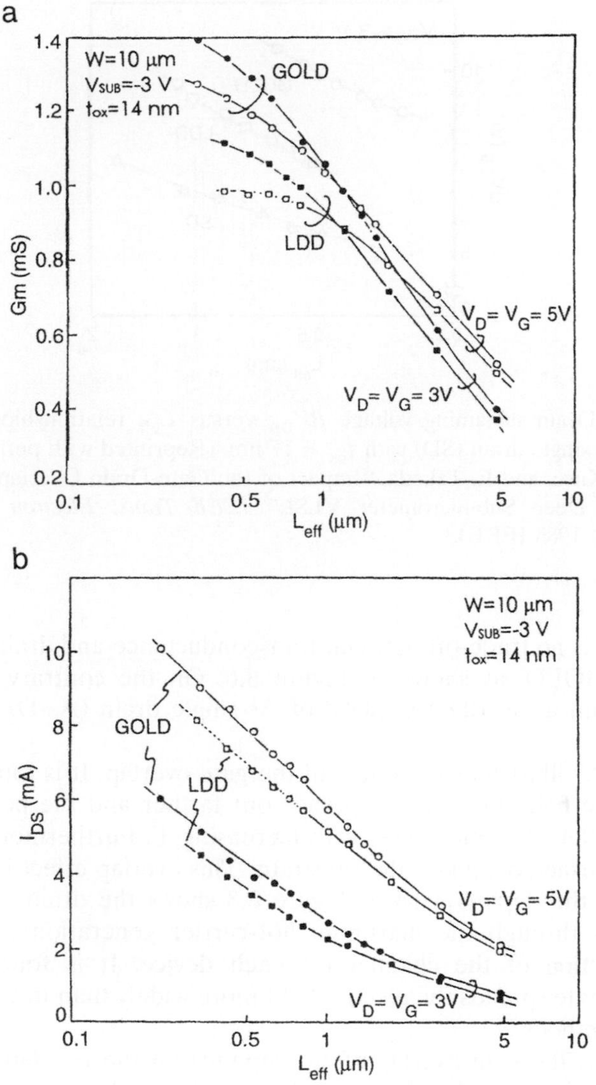

FIGURE 8.4 (a) G_m and (b) I_{DS} versus L_{eff} relationships. V_{th}–L_{eff} characteristics are the same in GOLD and LDD. (Reprinted with permission from R. Izawa, T. Kure, and E. Takeda, "Impact of the Gate-Drain Overlapped Devices (GOLD) for Deep Submicrometer VLSI," *IEEE Trans. Electron Devices* **35**, 2088–2093 (© 1988 IEEE).)

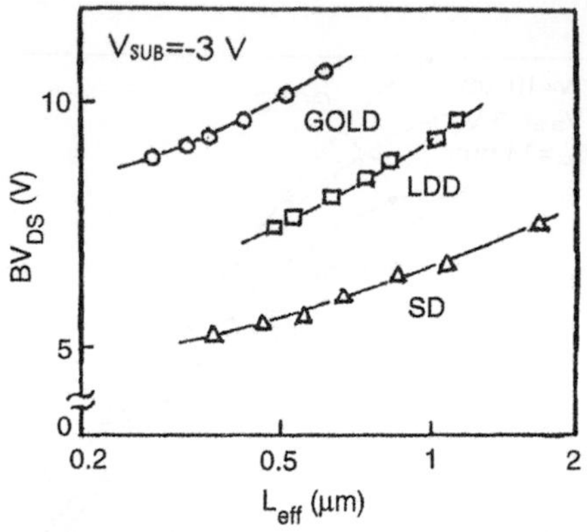

FIGURE 8.5 Drain sustaining voltage BV_{DS} versus L_{eff} relationships in GOLD, LDD, and As single drain (SD) with $t_{ox} = 17$ nm. (Reprinted with permission from R. Izawa, T. Kure, and E. Takeda, "Impact of the Gate-Drain Overlapped Devices (GOLD) for Deep Submicrometer VLSI," *IEEE Trans. Electron Devices* **35**, 2088–2093 (© 1988 IEEE).)

result, there is no trade-off between transconductance and drain sustaining voltage in GOLD, as shown in Figure 8.6. On the contrary, there is a trade-off from using LDD in place of As-single drain (As-D) to improve reliability.

Figure 8.7 illustrates the effect of the gate overlap. It is found that the drain electric field distribution spreads out farther and the peak value of the electric field becomes lower with increasing Γ. Furthermore, the peak position is pushed deep into the substrate. This overlap effect is due to the widening of the depletion layer. Figure 8.8 shows the drain electric field distributions through the maximum hot-carrier generation point in the lateral direction of the channel for each device. It is found that the overlapped gate spreads the electric field more widely than in the drain-engineering methods.

However, the gate overlap effect depends on the overlap length and the n^- dose, as shown in Figure 6.5. The maximum drain electric field $\mathscr{E}_m$ and the maximum hot-carrier generation rate GR_m decrease with increasing Γ until $\Gamma = 0.2$ μm. After that, they level off. The length corresponding to the leveling-off point equals the depletion width in the n^- region. The reason for this is that the vertical field induced by the overlapped gate has a strong influence on the depletion layer in the n^- region. Consequently, this overlap length (0.2 μm) is optimal, and $\mathscr{E}_m$ and GR_m are not changed by increasing the gate–drain overlap capacitance. Further, a lower n^- dose (5×10^{12} cm^{-2}) is preferred.

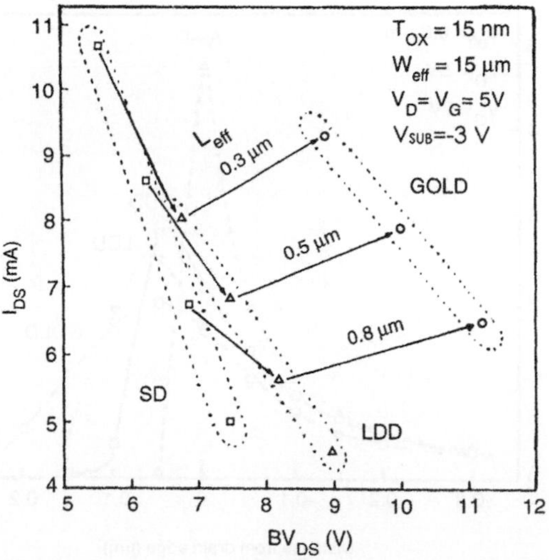

FIGURE 8.6 Relationships between I_{DS} and BV_{DS} in GOLD, LDD, and SD. In LDD, I_{DS} decreases as a result of the improved breakdown voltage. However, I_{DS} and BV_{DS} increase together in GOLD, as compared with LDD. There is no trade-off between transconductance and breakdown voltage in GOLD. (Reprinted with permission from R. Izawa, T. Kure, and E. Takeda, "Impact of the Gate-Drain Overlapped Devices (GOLD) for Deep Submicrometer VLSI," *IEEE Trans. Electron Devices* **35**, 2088–2093 (© 1988 IEEE).)

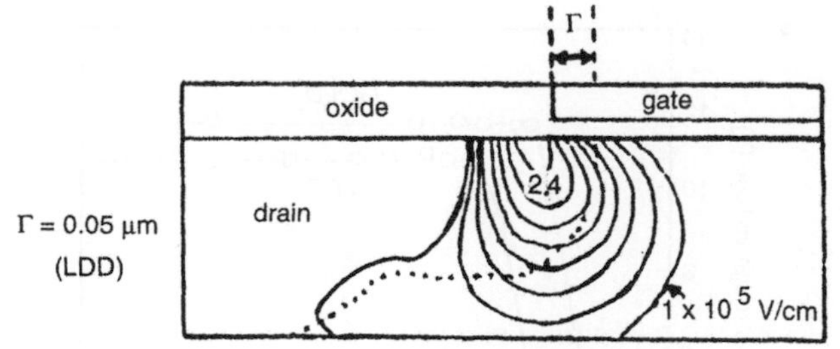

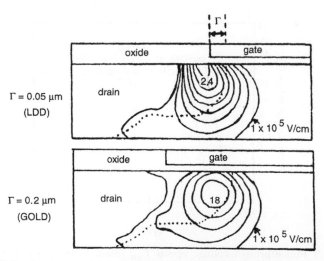

FIGURE 8.7 Drain electric field distribution in GOLD and LDD, where L_{eff} = 0.8 μm, V_D = 5 V and V_G = 2 V. The drain electric field distribution spreads out further due to the overlapped gate. (Reprinted with permission from R. Izawa, T. Kure, and E. Takeda, "Impact of the Gate-Drain Overlapped Devices (GOLD) for Deep Submicrometer VLSI," *IEEE Trans. Electron Devices* **35**, 2088–2093 (© 1988 IEEE).)

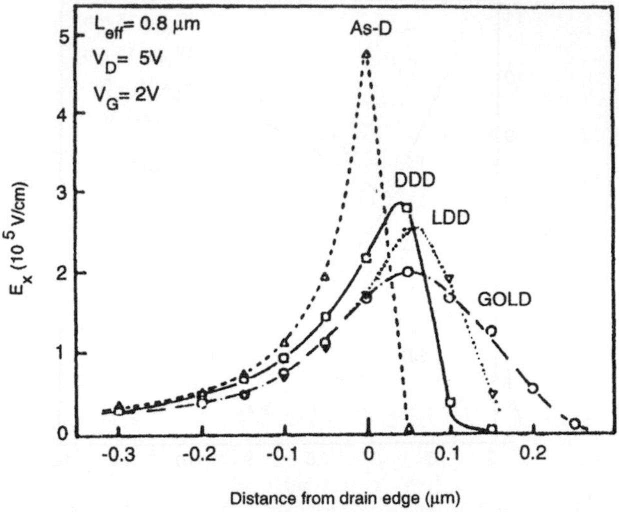

FIGURE 8.8 Drain electric field through the maximum hot-carrier generation point in the lateral direction of the channel for each device structure. The overlapped gate spreads the electric field more widely than in the drain engineering methods. (Reprinted with permission from R. Izawa, T. Kure, and E. Takeda, "Impact of the Gate-Drain Overlapped Devices (GOLD) for Deep Submicrometer VLSI," *IEEE Trans. Electron Devices* **35**, 2088–2093 (© 1988 IEEE).)

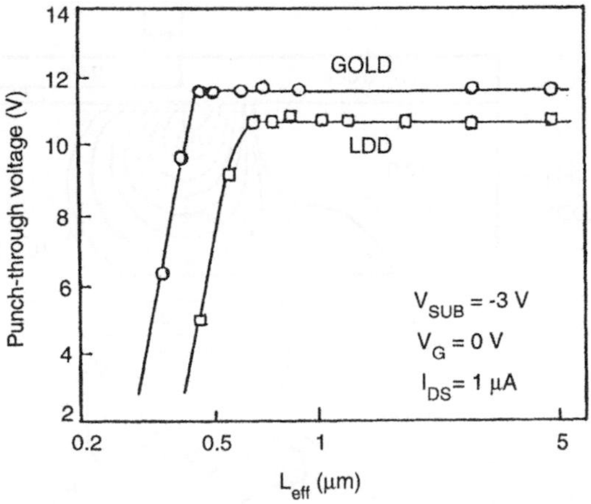

FIGURE 8.9 Punchthrough voltage versus L_{eff} relationships in GOLD and LDD. (Reprinted with permission from R. Izawa, T. Kure, and E. Takeda, "Impact of the Gate-Drain Overlapped Devices (GOLD) for Deep Submicrometer VLSI," *IEEE Trans. Electron Devices* **35**, 2088–2093 (© 1988 IEEE).)

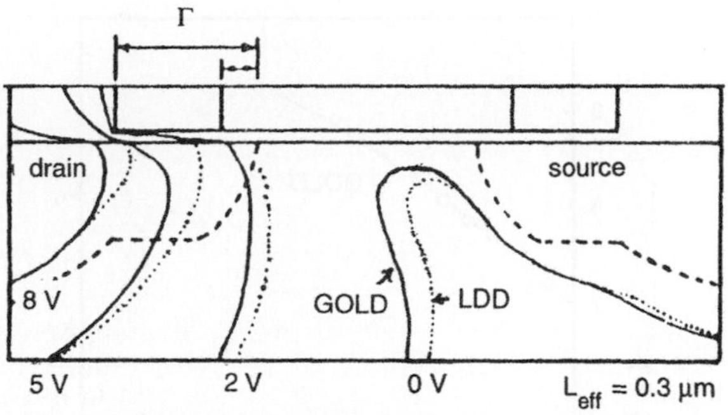

FIGURE 8.10 Potential distribution between source and drain in GOLD and LDD, where $L_{eff} = 0.3$ μm, $V_D = 8$ V, and $V_G = 0$ V. The solid lines show the potential contours of GOLD and the dotted lines show those of LDD. (Reprinted with permission from R. Izawa, T. Kure, and E. Takeda, "Impact of the Gate-Drain Overlapped Devices (GOLD) for Deep Submicrometer VLSI," *IEEE Trans. Electron Devices* **35**, 2088–2093 (© 1988 IEEE).)

GOLD is also excellent with regard to short-channel effects. It is found that GOLD suppresses punchthrough even with $L_{eff} < 0.5$ μm, as shown in Figure 8.9. The reason for this is that the equipotential contours are moved away from the source by the vertical field induced by the overlapped gate. Figure 8.10 shows the equipotential contours between the source and drain in GOLD and LDD. The potential under the gate in GOLD is higher because spreading of the drain field into the channel region is suppressed.

The punchthrough voltage of GOLD in a long channel is also larger than that of LDD in a long channel. This is partly because of the suppressed weak-avalanche breakdown caused by the lateral field reduction.

8.4 Reduced Hot-Carrier Degradation

GOLD is a surprisingly hot-carrier-resistant structure. Hot-carrier breakdown voltage (BV_{HC}, same as BV_{DC})($\Delta G_m/G_{m0} = 0.1$; 10 years) is about 8 V even for $L_{eff} = 0.5$ μm, while in conventional LDD, $BV_{HC} < 5$ V, as shown in Figure 8.11. Figure 8.12 shows the mechanism of reduced hot-carrier degradation. It is found that the electron–hole pair generation rate value (GR_m) in GOLD is one order of magnitude smaller than that in LDD because of the reduced lateral field. Consequently, GOLD reduces not only the trapped electrons in the sidewall spacers specific to LDD but also the avalanche-induced hot-carrier generation. The dependence of

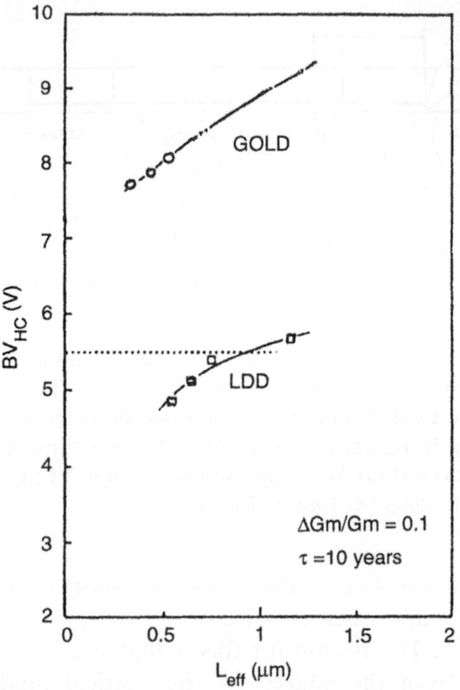

FIGURE 8.11 Hot-carrier breakdown voltage BV_{HC} versus L_{eff} relationships in GOLD and LDD. BV_{HC} is defined as the drain voltage at which the G_m degradation becomes 10% after 10 years. (Reprinted with permission from R. Izawa, T. Kure, and E. Takeda, "Impact of the Gate-Drain Overlapped Devices (GOLD) for Deep Submicrometer VLSI," *IEEE Trans. Electron Devices* **35**, 2088–2093 (© 1988 IEEE).)

stress time on G_m degradation for both GOLD and LDD is shown in Figure 8.13. In this figure, improvement (I) shows how the degradation is improved by eliminating electron trapping in the sidewall spacer. The other improvement (II) is caused by the reduced hot-carrier generation.

8.5 Summary

A new concept that takes advantage of the gate–drain overlap effect provides a guideline for more reliable, higher-speed MOSFET design in the deep submicrometer region. A novel device, GOLD, can easily be

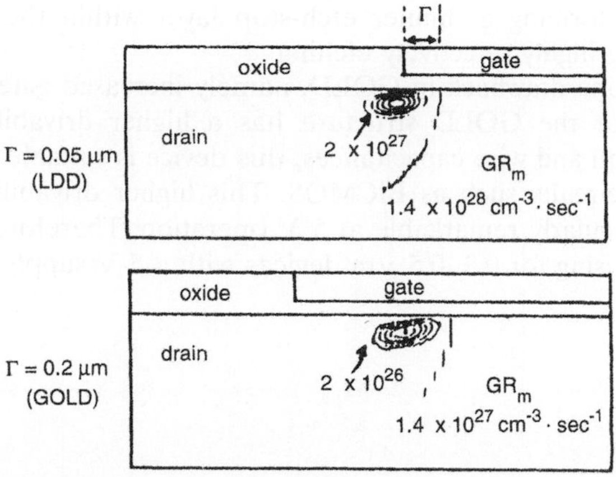

FIGURE 8.12 Electron–hole pair generation rate distributions in the drain with GOLD and LDD, where $L_{\text{eff}} = 0.8\ \mu\text{m}$, $V_D = 5$ V, and $V_G = 2$ V. (Reprinted with permission from R. Izawa, T. Kure, and E. Takeda, "Impact of the Gate-Drain Overlapped Devices (GOLD) for Deep Submicrometer VLSI," *IEEE Trans. Electron Devices* **35**, 2088–2093 (© 1988 IEEE).)

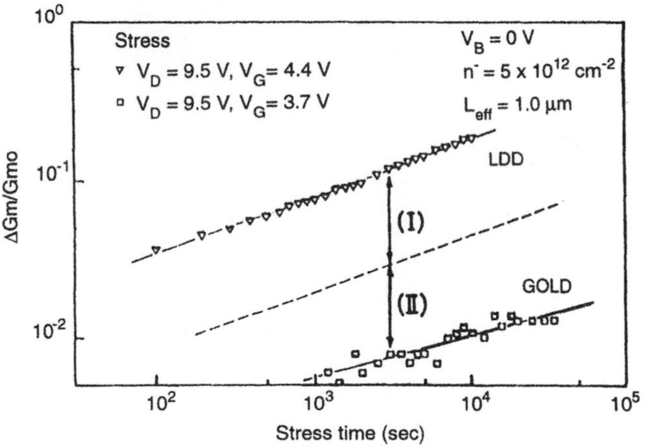

FIGURE 8.13 Dependence of G_m degradation on stress time in GOLD and LDD. The improved reliability of GOLD is caused by mechanisms (I) and (II). (I) is due to the suppressed hot-carrier injection into the sidewall oxide and (II) is caused by the reduced electric field. (Reprinted with permission from R. Izawa, T. Kure, and E. Takeda, "Impact of the Gate-Drain Overlapped Devices (GOLD) for Deep Submicrometer VLSI," *IEEE Trans. Electron Devices* **35**, 2088–2093 (© 1988 IEEE).)

fabricated by forming a thinner etch-stop layer within the double-layer gate and using highly selectively etching.

There is one drawback in GOLD, namely increased gate capacitance. However, since the GOLD structure has a higher drivability (G_m, I_{DS}) against junction and wire capacitances, this device is suitable for nonheavy gate loading circuits such as BiCMOS. This higher drivability advantage becomes particularly remarkable at 5 V operation. Therefore, GOLD will be most promising for 0.3–0.5 μm devices with a 5-V supply voltage.

REFERENCES

Abbas, S. A. and R. C. Dockerty, "Hot-Carrier Instability in IGFETs," *Appl. Phys. Lett.* **27**, 147–148 (1975).

Abeles, J. H., W. K. Chan, E. Colas, and A. Kastalsky, "Junction Field-Effect Transistor Simple Quantum Well Optical Waveguide Modulator Employing the Two-Dimensional Moss–Burstein Effect," *Appl. Phys. Lett.* **54**, 2177–2179 (1989).

Acovic, A. and M. Ilegems, "Characterization of Degradation Due to Hot-Carriers in MOSFETs by the Drain–Substrate Gate Controlled Tunnel Diode," *Proc. Eur. Solid State Devices Res. Conf., 17th*, 265–268 (1988).

Acovic, A., M. Dutoit, and M. Ilegems, "Novel Method for Characterizing Degradation of MOSFETs Caused by Hot-Carriers," *Proc. Workshop on Low Temp. Semicond. Electron.*, Burlington, VT, 118–122 (1989).

Acovic, A., M. Dutoit, and M. Ilegems, "Study of the Increased Effects of Hot-Carrier Stress on *n*-MOSFETs at Low Temperature," *IEEE Trans. Electron Devices* **36**, 2603 (1989).

Acovic, A., M. Dutoit, and M. Ilegems, "Characterization of Hot-Electron-Stressed MOSFETs by Low-Temperature Measurements of the Drain Tunnel Current," *IEEE Trans. Electron Devices* **37**, 1467–1476 (1990).

Agostinelli, V. M., Jr., T. J. Bordelon, X. L. Wang, C. F. Yeap, A. F. Tasch, and C. M. Maziar, "A Two-Dimensional Model for Predicting Substrate Current in Submicrometer MOSFETs," *50th Annual Device Res. Conf.*, IIA-6 (1992).

Agusta, B. and T. Harroun, "Conceptual Design of an Eight Megabyte High Performance Charge-Coupled Storage Device," *Proc. Appl. Conf. Charge Coupled Devices*, Nav. Electron. Lab. Cent., San Diego, 55 (1973).

Ahmad, A., "A Method to Characterize Traps in Silicon," *Semicond. Int.* May, 154–156 (1991).

Ahn, J., W. Ting, and D.-M. Kwong, "Furnace Nitridation of Thermal SiO_2 in Pure N_2O Ambient for ULSI MOS Application," *IEEE Electron Device Lett.* **13**, 117–119 (1992).

Ahn, S. T., S. Hayashida, K. Iguchi, J. Takagi, T. Watanabe, and K. Sakiyama, "Hot-Carrier Degradation of Single-Drain pMOSFETs with Differing Sidewall Spacer Thicknesses," *IEEE Electron Device Lett.* **13**, 211–213 (1992).

Aitken, J. M., D. R. Young, and K. Pan, "Electron Trapping in Electron-Beam Irradiated SiO_2," *J. Appl. Phys.* **49**, 3386–3391 (1978).

Aitken, J. M., "Electron and Hole Trapping in Irradiated SiO_2," *Proc. 3rd Bienni. Univ./Ind. Gov. Microelectron. Symp.*, Lubbock, TX, 50 (1979).

Aitken, J. M., "Radiation-Induced Trapping Centres in Thin Silicon Dioxide Films," *J. Non-Cryst. Solids* **40**, 31–47 (1980).

Akamatsu, S., T. Hori, G. Fuse, and K. Tsuji, "Hot-Carrier Effects in 0.5-μm-Wide Trench-Isolated MOSFETs," *Trans. Inst. Electron. Inf. Commun. Eng. JPN* **72C-11**, 463–468 (1989).

Akers, L. A., M. A. Holly, and C. Lund, "Hot Carriers in Small Geometry CMOS," *Tech. Dig.—Int. Electron Devices Meet.*, 80–83 (1984).

Akers, L. A. and M. Walker, "Hot Carrier Effects in Submicron CMOS," *Physica B + C (Amsterdam)*, **134**, 116–119 (1985).

Alcorn, G. E., H. N. Kotecha, and J. G. Prosser, "Improved Process for Marking Ion Implanted Junctions for MOSFET Structures with High Resistivity Substrate," *IBM Tech. Disclosure Bull.* **21**, 2398–2400 (1978).

Allen, F. G. and G. W. Gobeli, "Work Function, Photoelectric Threshold and Surface States of Atomically Clean Silicon," *Phys. Rev.* **127**, 150–158 (1962).

Allen, S. J., D. C. Tsui, F. DeRosa, K. K. Thornber, and B. A. Wilson, "Direct Measurement of Velocity Overshoot by Hot-Electron–Submillimeter Wave Conductivity in Si Inversion Layers," *J. Phys., Colloq. (Orsay, Fr.)* **42**, Suppl. C7, 369–374 (1981).

Alpern, Y. and J. Shappir, "Hot Electron Injection from Undoped Poly Silicon Floating Gate in MOS Devices," *Proc. IEEE Conv.*, Israel, 2.4.4.1–2.4.4.3 (1983).

Alvarez, A. R., J. Teplik, D. W. Schucker, T. Hulseweh, H. B. Liang, M. Dydyk, and I. Rahim, "Second Generation BI-CMOS Gate Array Technology," *Proc. Bipolar Circuits & Tech. Meet.*, Minneapolis, MN, 113–117 (1987).

Amelio, G. F., M. F. Tompsett, and G. E. Smith, "Experimental Verification of the Charge Coupled Diode Concept," *Bell Syst. Tech. J.* **49**, 593–600 (1970).

Ancona, M. G., N. S. Saks, and D. McCarthy, "Lateral Distribution of Hot-Carrier-Induced Interface Traps in MOSFET's," *IEEE Trans. Electron Devices* **35**, 2221–2228 (1988).

Andhare, P. N., R. K. Nahar, N. M. Devashrayee, S. Chandra, and W. S. Khokle, "Development and Performance Characterization of the Lightly Doped Drain MOS Transistor," *Microelectron. Reliab.* **30**, 681–690 (1990).

Andronov, A. A., Y. N. Nozdrin, and V. N. Shastin, "Tunable Far-IR Lasers Based on Hot Holes in Semiconductors," *Izv. Akad. Nauk SSSR, Ser. Fiz.* **50**, 1103–1110 (1986).

Andronov, A. A., Y. N. Nozdrin, and V. N. Shastin, "Tunable FIR Lasers in Semiconductors Using Hot Hole," *Infrared Phys.* **27**, 31–38 (1987).

Annunziata, R., G. Dalla Libera, E. Ghio, and A. Maggis, "Annealing of Hot Carrier Damaged Double Metal MOSFET," *Proc. Eur. Solid State Devices Res. Conf., 19th*, 715–718 (1989).

Antoniadis, D. A. and J. E. Chung, "Physics and Technology of Ultra Short Channel MOSFET Devices," *Int. Electron Devices Meet., 1991*, 21–24 (1991).

Aoki, M., K. Yano, T. Masuhara, S. Ikeda, and S. Meguro, "Optimum Crystallographic Orientation of Submicron CMOS Devices," *Tech. Dig.—Inst. Electron Devices Meet.*, 577–580 (1985).

Aoki, M., S. Hanamura, T. Masuhara, and K. Yano, "Performance and Hot-Carrier Effects of Small CRYO-CMOS Devices," *IEEE Trans. Electron Devices* **34**, 8–18 (1987).

Aoki, M., S. Hanamura, T. Masuhara, S. Ikeda, and S. Meguro, "Optimum Crystallographic Orientation of Submicrometer CMOS Devices Operated at Low Temperatures," *IEEE Trans. Electron Devices* **34**, 52–57 (1987).

Aoki, M., K. Yano, T. Masuhara, and K. Komiyaji, "Hot-Carrier Effects under Pulsed Stress in CMOS Devices," *Tech. Dig. VLSI Tech. Symp.*, 49–50 (1987).

Aoki, M., K. Shimohigashi, K. Yano, and T. Masuhara, "Improvement of Hot-Carrier Degradation in Cooled CMOS," *Tech. Dig. VLSI Tech. Symp.*, 81–82 (1989).

Aoki, T. and A. Yoshii, "Analysis of Latchup-Induced Photoemission," *Tech. Dig.—Electron Devices Meet.*, 281–284 (1989).

Aoki, T. and A. Yoshii, "Analysis of Latchup-Induced Photon Emissions," *IEEE Trans. Electron Devices* **37**, 2080–2083 (1990).

Arimoto, K., T. Yamagata, H. Miyamoto, K. Mashiko, M. Yamada, S. Sato, and H. Shibata, "An Effect of Hiller-Induced-Stress to DRAM Sense Amplifier," *Tech. Dig. VLSI Tech. Symp.* 92–93 (1985).

Arnold, D., K. Hess, and G. J. Iafrate, "Electron Transport in Heterostructure Hot-Electron Diodes," *Appl. Phys. Lett.* **53**, 373–375 (1988).

Arnold, D. and M. S. Sharma, "Modeling the Anomalous Threshold Voltage Behavior of Submicrometer MOSFETs," *IEEE Electron Devices Lett.* **13**, 92–94 (1992).

Arora, N. D. and M. S. Sharma, "MOSFET Substrate Current Model for Circuit Simulation," *IEEE Trans. Electron Devices* **38**, 1392–1398 (1991).

Artaki, M., "Hot-Electron Flow in an Inhomogeneous Field," *Appl. Phys. Lett.* **52**, 141–143 (1988).

Asai, S., "Semiconductor Memory Trends," *Proc. IEEE* **74**, 1623–1635 (1986).

Aschenbach, B., "Galactic Supernova Remnants," *Space Sci. Rev.* **40**, 447–465 (1985).

Asenov, A., M. Bollu, F. Koch, and J. Scholz, "On the Nature and Energy of Defect States Caused by Hot Electrons in Si," *Appl. Surf. Sci.* **30**, 319–324 (1987).

Asenov, A., J. Berger, W. Weber, M. Bollu, and F. Koch, "Hot-Carrier Degradation Monitoring in LDD *n*-MOSFETs Using Drain Gated-Diode Measurements," *Microelectron. Eng.* **15**, 445–448 (1991).

Asenov, A., J. Berger, P. Speckbacher, F. Koch, and W. Weber, "Spatially-Resolved Measurements of Hot-Carrier Generated Defects at the Si–SiO_2 Interface," *Proc. 7th Bienn. Eur. Conf.*, 247–250 (1991).

Assaderaghi, F., J. Chen, R. Solomon, T.-Y. Chan, P.-K. Ko, and C. Hu, "Transient Behavior of Subthreshold Characteristics of Fully Depleted SOI MOSFETs," *IEEE Electron Devices Lett.* **12**, 518–520 (1991).

Atalla, M. M., M. Tannenbaum, and E. J. Scheibner, "Stabilization of Silicon Surface by Thermally Grown Oxide," *Bell Syst. Tech. J.* **38**, 749–783 (1959).

Aur, S., P. Yang, P. Pattnaik, and P. K. Chatterjee, "Modeling of Hot Carrier Effects for LDD MOSFETs," *Tech. Dig. VLSI Tech. Symp.*, 112–113 (1985).

Aur, S., D. E. Hocevar, and P. Yang, "Circuit Hot Electron Effect Simulator," *Tech. Dig.—Int. Electron Devices Meet.*, 498–501 (1987).

Aur, S., D. Hocevar, E. Dale, and P. Yang, "Hotron—A Circuit Hot Electron Effect Simulator," *Tech. Dig. IEEE ICCAD*, 256–259 (1987).

Aur, S., A. Chatterjee, and T. Polgreen, "Hot-Electron Reliability and ESD Latent Damage," *IEEE Trans. Electron Devices* **35**, 2189–2193 (1988).

Aur, S., "Kinetics of Hot Carrier Effects for Circuit Simulation," *Proc. Int. Reliab. Phys. Symp.*, 88–91 (1989).

Avni, E. and J. Shappir, "Oxide Trapping under Spatially Variable Oxide Electric Field in the Metal-Oxide-Silicon Structure," *Appl. Phys. Lett.* **51**, 463–467 (1987).

Avni, E. and J. Shappir, "Effects of Temperature Annealing on Charge-Injection-Induced Trapping in Gate Oxides of Metal-Oxide-Silicon Transistors," *J. Appl. Phys.* **63**, 1563–1568 (1988).

Avni, E. and J. Shappir, "Modeling of Charge-Injection Effects in Metal-Oxide-Semiconductor Structures," *J. Appl. Phys.* **64**, 734–742 (1988).

Av-Ron, M., M. Shatzkes, T. H. DiStefano, and I. B. Cadoff, "The Nature of Electron Tunneling in SiO_2," In *The Physics of SiO_2 and Its Interfaces*, S. T. Pantelider, ed., Pergamon, New York, 46 (1978).

Azechi, H., H. Fujita, S. Ido, Y. Izawa, Y. Kato, Y. Kawamura, M. Matoba, K. Mima, T. Mochizuki, M. Monma, S. Nakai, K. Nishihara, H. Nishimura, T. Norimatsu, T. Sasaki, H. Takabe, T. Ueda, T. Yabe, K. Yoshida, T. Yamanaka, and C. Yamanaka, "Laser Fusion Research at Institute of Laser Engineering—Osaka," *Nucl. Fusion, Suppl.* **2**, 13–23 (1981).

Baba, S., A. Kita, and J. Ueda, "Mechanism of Hot Carrier Induced Degradation in MOSFETs," *Tech. Dig.—Int. Electron Devices Meet.*, 734–737 (1986).

Bach, D. R., D. E. Casperson, D. W. Forslund, S. J. Gitomer, P. D. Goldstone, A. Hauer, J. F. Kephart, J. M. Kindel, R. Kristal, G. A. Kyrala, K.B. Mitchell, D. B. Van Hulsteyn, and A. H. Williams, "Intensity-Dependent Absorption in 10.6 μ Laser-Illuminated Spheres," *Phys. Rev. Lett.* **50**, 2082–2085 (1983).

Baglee, D. A. and C. Duvvury, "Reduced Hot-Electron Effects in MOSFET's with an Optimized LDD Structure," *IEEE Electron Devices Lett.* **5**, 389–391 (1984).

Baker, R. L., C. N. Duckworth, and M. C. Syrett, "Preliminary Examination of Hot Carriers in 2 μ and 1.25 μ CMOS," *IEE Colloq. Dig.*, 7.1–7.2 (1987).

Balasinski, A., C. Wenliang, and T.-P. Ma, "Effects of Combined X-Ray Irradiation and Hot-Electron Injection on NMOS Transistors," *J. Electron. Mater.* **21**, 737–743 (1992).

Balasubramanyam, K., M. J. Hargrove, H. I. Hanafi, M. S. Lin, D. Hoyniak, L. Larue, and D. R. Thomas, "Characterization of As–P Double Diffused Drain Structure," *Tech. Dig.—Int. Electron Devices Meet.*, 782–785 (1984).

Baliga, B. J., T. Syau, and P. Venkatraman, "The Accumulation-Mode Field-Effect Transistor: A New Ultralow On-Resistance MOSFET," *IEEE Electron Devices Lett.* **13**, 427–429 (1992).

Balk, P., "Low Temperature Annealing in the Al–SiO_2–Si System," *Ext. Abstr./Electron. Div.* **14**, 29–32 (1965).

Balk, P., "Effects of Hydrogen Annealing on Silicon Surfaces," *Ext. Abstr./Electron. Div.* **14**, 237–240 (1965).

Balk, P., P. G. Burkhardt, and L. V. Gregor, "Orientation Dependence of Built-in Surface Charge on Thermally Oxidized Silicon," *Proc. IEEE* **53**, 2133–2134 (1965).

Balk, P., "Hot Carrier Injection in Oxides and the Effect on MOSFET Reliability," *Conf. Ser.—Inst. Phys.*, 63–82 (1984).

Balland, B., C. Plossu, and C. Choquet, "Nonuniformity of the Surface State Density in the Channel of CMOS Transistors," *Rev. Phys. Appl.* **23**, 1837–1845 (1988).

Ballay, N. and B. Baylac, "CAD MOSFET Model for EPROM Cells," *J. Phys., Colloq. (Orsay, Fr.)* **49**, 681–685 (1988).

Ballet, J., J. F. Luciani, and P. Mora, "Suprathermal Ionization in Evaporating Clouds. Non-local Electron Distribution Function," *Astron. Astrophys.* **218**, 292–298 (1989).

Bampi, S. and J. D. Plummer, "Modified LDD Device Structures for VLSI," *Tech. Dig.—Int. Electron Devices Meet.*, 234–237 (1985).

Banerjee, S., R. Sundaresan, H. Shichijo, and S. Malhi, "Hot-Electron Degradation of *n*-Channel Polysilicon MOSFETs," *IEEE Trans. Electron Devices* **35**, 152–157 (1988).

Baranskii, P. I., I. S. Buda, and V. V. Savyak, "Mechanism Responsible for Anisotropy of Thermal EMF in Semiconductor," *Phys. Status Solidi A* **112**, 11–27 (1989).

Barbe, D. F., "Imaging Devices Using the Charge-Coupled Concept," *Proc. IEEE* **63**, 38–67 (1975).

Bardeen, J., "On the Surface States Associated with a Periodic Potential," *Phys. Rev.* **56**, 317–323 (1939).

Bardeen, J., "Surface States and Rectification at a Metal Semiconductor Contact," *Phys. Rev.* **71**, 717–727 (1947).

Bardeen, J. and W. H. Brattain, "The Transistor, a Semiconductor Triode," *Phys. Rev.* **74**, 230–231 (1948).

Bardeen, J., "Three-Electrode Circuit Element Utilizing Semiconductive Materials," *Solid State Technol.*, 68–71 (1987).

Barnes, J. and W. Fitchner, "Next Generation VLSI Power Supply Standards," *Tech. Dig.—Int. Electron Devices Meet.*, 545 (1984).

Barron, M. B., "Low Level Currents in Insulated Gate Field Effect Transistors," *IEEE Trans. Electron Devices* **15**, 293 (1972).

Bartelink, D. J., "Limits of Applicability of the Depletion Approximation and Its Recent Augmentation," *Appl. Phys. Lett.* **38**, 461–463 (1981).

Bassous, E. and T. H. Ning, "Self-Aligned Polysilicon Gate MOSFETs with Tailored Source and Drain Profiles," *IBM Tech. Disclosure Bull.* **22**, 5146–5147 (1980).

Bastiaens, J. J. J. and W. C. H. Gubbels, "256-Kbit SRAM: An Important Step on the Way to Submicron IC Technology," *Philips Tech. Rev.* **44**, 33–42 (1988).

Basu, P. K., "Monte Carlo Calculation of Hot Electron Drift Velocity in Silicon (100)-Inversion Layer by Including Three Subbands," *Solid State Commun.* **27**, 657–660 (1978).

Basu, P. K. and J. B. Roy, "Warm Electron Coefficient of Two-Dimensional Electron Gas in Silicon Inversion Layer at Low Temperature," *Phys. Status Solidi B* **121**, 743–748 (1984).

Batchelor, D. B., "Comparison of Electron Cyclotron Heating Theory and Experiment in EBT," *Proc. Int. Symp. Heating Toroidal Plasmas, 4th*, Rome, Italy, 2, 779–794 (1984).

Bauer, F. and P. Balk, "Relationship between Short Channel Behavior and Long Term Stability of *n*-Channel Enhancement and Depletion MOSFETs," *Solid-State Electron.* **29**, 797–806 (1986).

Bauer, F., S. C. Jain, J. Korec, V. Lauer, M. Offenberg, and P. Balk, "Incompatibility of Requirements for Optimizing Short Channel Behavior and Long Term Stability in MOSFETs," *Solid-State Electron.* **31**, 27–33 (1988).

Bellens, R., P. Heremans, G. Groeseneken, and H. Maes, "Analysis of Hot Carrier Degradation in AC Stressed *n*-Channel MOS Transistors Using the Charge Pumping Technique," *J. Phys., Colloq. (Orsay, Fr.)* **49**, 651–655 (1988).

Bellens, R., P. Heremans, G. Groeseneken, and H. E. Maes, "Analysis of Mechanism for the Enhanced Degradation during AC Hot Carrier Stress of MOSFETs," *Tech. Dig.—Int. Electron Devices Meet.*, 212–215 (1988).

Bellens, R., P. Heremans, G. Groeseneken, and H. Maes, "Characterization and Analysis of Hot-Carrier Degradation in *p*-Channel Transistors using Constant Current Stress Experiment," *Proc. Eur. Solid State Devices Res. Conf., 17th*, 261–264 (1988).

Bellens, R., P. Heremans, G. Groeseneken, and H. Maes, "Hot-Carrier Effects in *n*-Channel MOS Transistors under Alternating Stress Conditions," *IEEE Electron Devices Lett.* **9**, 232–234 (1988).

Bellens, R., P. Heremans, G. Groeseneken, and H. Maes, "New Procedure for Lifetime Prediction of *n*-Channel MOS-Transistor Using the Charge Pumping Technique," *Proc. Int. Reliab. Phys. Symp.*, 8–14 (1988).

Bellens, R., P. Heremans, G. Groeseneken, and H. E. Maes, "On the Channel-Length Dependence of the Hot-Carrier Degradation of *n*-Channel MOSFETs," *IEEE Electron Devices Lett.* **10**, 553–555 (1989).

Bellens, R., P. Heremans, G. Groeseneken, H. Maes, and W. Weber, "The Influence of the Measurement Setup on Enhanced AC Hot Carrier Degradation of MOSFETs," *IEEE Trans. Electron Devices* **37**, 310–313 (1990).

Bellens, R., P. Heremans, G. Groeseneken, and H. E. Maes, "AC Hot Carrier Degradation Behavior in *n*- and *p*-Channel MOSFETs and in CMOS Invertors," *Proc. 7th Bienn. Eur. Conf.*, 271–274 (1991).

Bellens, R., E. deSchrijver, G. Groeseneken, P. Heremans, and H. E. Maes, "Influence of Post-Stress Effects on the Dynamic Hot-Carrier Degradation Behavior of Passivated *n*-Channel MOSFETs," *IEEE Electron Devices Lett.* **13**, 357–359 (1992).

Bellens, R., G. Groeseneken, P. Heremans, and H. E. Maes, "On the Different Time Dependence of Interface Trap Generation and Charge Trapping during Hot Carrier Degradation in CMOS," *Microelectron. Eng.* **19**, 465–468 (1992).

Bellone, S., G. Busatto, and C. M. Ransom, "Recombination Measurement of *n*-Type Heavily Doped Layer in High/Low Silicon Junctions," *IEEE Trans. Electron Devices* **38**, 532–537 (1991).

Bennett, H. S., M. Gaitan, P. Roitman, T. J. Russell, and J. S. Suehle, "Modeling MOS Capacitors to Extract Si–SiO_2 Interface Trap Densities in the Presence of Arbitrary Doping Profiles," *IEEE Trans. Electron Devices* **33**, 759–765 (1986).

Berenga, F., "VLSI Technologies," *Elettron. Oggi*, 101–112 (1980).

Berglund, C. N., "Surface States at Steam-Grown Silicon–Silicon Dioxide Interface," *IEEE Trans. Electron Devices* **13**, 701–705 (1966).

Berglund, C. N. and R. J. Powell, "Photoinjection into SiO_2, Electron Scattering in the Image Force Potential Well," *J. Appl. Phys.* **42**, 573–579 (1971).

Bergonzoni, C., R. Benecchi, and P. Caprara, "Hot Carrier Stress Induced Changes in MOST Transconductance Structure," *J. Phys., Colloq. (Orsay, Fr.)* **49**, 779–782 (1988).

Bergonzoni, C. and G. Dalla Libera, "A Physical Characterization of Dynamically Stressed CMOS Transistors," *Microelectron. Eng.* **15**, 453–456 (1991).

Bergonzoni, C. and G. Dalla Libera, "Physical Characterization of Hot-Electron-Induced MOSFET Degradation through an Improved Approach to the Charge-Pumping Technique," *IEEE Trans. Electron Devices* **39** 1895–1901 (1992).

Bergonzoni, C., G. Dalla Libera, R. Benecchi, and A. Nannini, "Dynamic Hot Carrier Degradation Effects in CMOS Submicron Transistors," *Microelectron. Reliab.* **32**, 1515–1519 (1992).

Berz, F. and J. A. Morice, "Accuracy of the Transmission Coefficient across Parabolic Barriers as Obtained from a Generalized WKB Approach," *Phys. Status Solidi* **139**, 573–582 (1987).

Bhattacharyya, A., "Hot-Electron Trapping in SiO_2 and Its Effects on the Reliability of VLSI," *Proc. Int. Conf. Phys. Semicond. Devices*, New Delhi, India, 222–230 (1982).

Bhattacharyya, A. and S. N. Shabde, "Degradation of Short-Channel MOSFETs under Constant Current Stress across Gate and Drain," *IEEE Trans. Electron Devices* **33**, 1329–1333 (1986).

Bhattacharyya, A. and S. N. Shabde, "Generation of Interface States during the Electrical Stressing of MOS Transistors," *Philips J. Res.* **42**, 583–592 (1987).

Bhattacharyya, A. and S. N. Shabde, "Mechanism Degradation of LDD MOSFETs Due to Hot-Electron Stress," *IEEE Trans. Electron Devices* **35**, 1156–1158 (1988).

Bhattacharyya, S., R. Kovelamudi, S. Batra, S. Banerjee, B.-Y. Nguyen, and P. P. Tobin, "Parallel Hot-Carrier-Induced Degradation Mechanisms in Hydrogen-Passivated Polysilicon-on-Insulator LDD *p*-MOSFETs, *IEEE Electron Devices Lett.* **13**, 491–493 (1992).

Bibyk, S. B., H. Wang, and P. Borton, "Analyzing Hot-Carrier Effects on Cold CMOS Devices," *IEEE Trans. Electron Devices* **34**, 83–88 (1987).

Biermans, P. T. J., T. Poorter, and H. J. H. Merks-Eppingbroek, "The Impact of Different Hot-Carrier-Degradation Components on the Optimization of Submicron *n*-Channel LDD Transistors," *J. Phys., Colloq.* (*Orsay, Fr.*) **49**, 787–790 (1988).

Bimberg, D., "Pulse Dispersion and Preheating Effects in Ultrafast Photoconductive Detectors: $In_{0.53}/Ga_{0.47}/As$ as Example," *Appl. Phys. Lett.* **41**, 368–370 (1982).

Blatter, G. and F. Greeuter, "Electrical Breakdown at Semiconductor Grain Boundaries," *Phys. Rev. B* **34**, 8555–8572 (1986).

Blewer, R. S., "Progress in LPCVD Tungsten for Advance Microelectronics Applications," *Solid State Technol.* 117–126 (1986).

Boesch, H. E., Jr., F. B. McLean, J. M. Benedetto, J. M. McGarrity, and W. E. Bailey, "Saturation of Threshold Voltage Shift in MOSFETs at High Total Dose," *IEEE Trans. Nucl. Sci.* **33**, 1191–1197 (1986).

Bogert, H. Z., C. T. Sah, and D. A. Tremere, "Applications of the Surface Potential Controlled Transistor Tetrodes," *IEEE Proc. Int. Conf. Solid State Circuits*, 34–35 (1962).

Bojarczuk, N. A., "Process Conditions Affecting Hot Electron Trapping in DC Magnetron Sputtered MOS Devices," *J. Vac. Sci. Technol.* **18**, 890–894 (1981).

Bojarczuk, N. A. and M. Petroski, "Electron Trapping Characteristics of MOS Capacitors Produced by Ion Beam Sputter Deposition," *J. Vac. Sci. Technol.* **20**, 876–877 (1982).

Bold, B. S. and M. P. Brassington, "Critical Evaluation of the Performance of CMOS Circuits Incorporating Double-Implanted *n*-MOS Source and Drain Regions," *IEEE Electron Devices Lett.* **7**, 211–213 (1986).

Bollu, M., F. Koch, A. Madenach, and J. Scholz, "Electrical Switching and Noise Spectrum of $Si-SiO_2$ Interface Defects Generated by Hot Electrons," *Appl. Surf. Sci.* **30**, 142–147 (1987).

Booth, R., S. Yoon, M. White, and D. Young, "Comparison of Symmetrical and Asymmetrical Hot-Electron Injection in MOS Transistors," *Solid-State Electron.* **34**, 599–604 (1991).

Borchert, B., K. R. Hofmann, and G. Dorda, "Positive and Negative Charge Generation by Hot Carriers in *n*-MOSFETs" *Electron. Lett.* **19**, 746–747 (1983).

Borchert, B. and G. Dorda, "Experimental Evidence for Different Saturation Velocities of Electrons in Silicon," *Proc. Eur. Solid State Devices Res. Conf., 17th*, 111–113 (1988a).

Borchert, B. and G. Dorda, "Hot-Electron Effects on Short-Channel MOSFET's Determined by the Piezoresistance Effect," *IEEE Trans. Electron Devices* **35**, 483–488 (1988b).

Bordelon, T. J., X.-L. Wang, C. M. Maziar, and A. F. Tasch, "An Evaluation of Energy Transport Models for Silicon Device Simulation," *Solid-State Electron.* **34**, 617–628 (1991).

Bordone, P., C. Jacoboni, P. Lugli, and L. Reggiani, "Effect of a Perturbed Acoustic-Phonon Distribution on Hot-Electron Transport: A Monte Carlo Analysis," *J. Appl. Phys.* **61**, 1460–1468 (1987).

Bottom, V. E., "Invention of the Solid-State Amplifier," *Phys. Today* **17**, 24–26 (1964).

Boukriss, B., H. Haddara, S. Cristoloveanu, and A. Chovet, "Modeling of the $1/f$ Noise Overshoot in Short-Channel MOSFETs Locally Degraded by Hot-Carrier Injection," *IEEE Electron Devices Lett.* **10**, 433–436 (1989).

Bourcerie, M., B. S. Doyle, J.-C. Marchetaux, A. Boudou, and H. Mingam, "Hot-Carrier Stressing Damage in Wide and Narrow LDD *n*-MOS Transistors," *IEEE Electron Devices Lett.* **10**, 132–134 (1989).

Bourcerie, M., B. S. Doyle, J.-C. Marchetaux, J.-C. Soret, and A. Boudou, "Relaxable Damage in Hot-Carrier Stressing of *n*-MOS Transistors—Oxide Traps in the Near Interfacial Region of the Gate Oxide," *IEEE Trans. Electron Devices* **37**, 708–717 (1990).

Bower, R. W., H. G. Dill, K. G. Aubuchon, and S. A. Thompson, "MOSFET Formed by Gate Masked Ion Implantation," *IEEE Trans. Electron Devices* **15**, 757–761 (1968).

Boyle, W. S. and G. E. Smith, "Charge-Coupled Semiconductor Devices," *Bell Syst. Tech. J.* **49**, 587 (1970).

Boyle, W. S. and G. E. Smith, "Charge-Coupled Devices—A New Approach to MIS Device Structures," *IEEE Spectrum* **8**, 18–27 (1971).

Bracchitta, J. A., T. L. Honan, and R. L. Anderson, "Hot-Electron-Induced Degradation in MOSFET's at 77 K," *IEEE Trans. Electron Devices* **32**, 1850–1857 (1985).

Braithwaite, N. S. J., A. Montes, and L. M. Wickens, "Some Comments on Thermal Flux Inhibition in Laser Produced Plasmas," *Plasma Phys.* **23**, 713–723 (1981).

Brassington, M. P. and R. R. Razouk, "Relationship between Gate Bias and Hot-Carrier Induced Instabilities in *p*-Channel MOSFETs," *Tech. Dig. VLSI Tech. Symp.*, 55–56 (1987).

Brassington, M. P., M. El-Diwany, P. Tuntasood, and R. R. Razouk, "Advanced Submicron BiCMOS Technology for VLSI Applications," *Tech. Dig. VLSI Tech. Symp.*, 89–90 (1988).

Brassington, M. P. and R. R. Razouk, "Relationship between Gate Bias and Hot-Carrier-Induced Instabilities in Buried- and Surface-Channel MOSFETs" *IEEE Trans. Electron Devies* **35**, 320–321 (1988).

Brassington, M. P., M. W. Poulter, and M. El-Diwany, "Suppression of Hot-Carrier Effects in Submicrometer Surface-Channel *p*-MOSFET's" *IEEE Trans. Electron Devices* **35**, 1149–1151 (1988).

Bravaix, A. and D. Vuillaume, "A Simple Charge-Pumping Method to Measure the Logarithmic-Time Dependence of Trapped Oxide Charge in *p*-MOSFET's," *Microelectron. Eng.* **19**, 469–472 (1992).

Brennan, K. and K. Hess, "A Theory of Enhanced Impact Ionization Due to the Gate Field and Mobility Degradation in the Inversion Layer of MOSFETs," *IEEE Electron Devices Lett.* **7**, 86–88 (1986).

Brews, J. R., "Correcting Interface-State Errors in MOS Doping Profile Determinations," *J. Appl. Phys.* **44**, 3228–3231 (1973).

Brews, J. R., "A Simplified High-Frequency MOS Capacitance Formula," *Solid-State Electron.* **20**, 607 (1977).

Brews, J. R., "A Charge-Sheet Model of the MOSFET," *Solid-State Electron.* **21**, 345 (1978).

Brews, J. R., "Subthreshold Behavior of Uniformly and Nonuniformly Doped Long-Channel MOSFET," *IEEE Trans. Electron Devices* **26**, 1282–1291 (1979).

Brews, J. R., "Threshold Shifts Due to Nonuniform Doping Profiles in Surface Channel MOSFETs," *IEEE Trans. Electron Devices* **26**, 1696–1710 (1979).

Brews, J. R., W. Fichtner, E. H. Nicollian, and S. M. Sze, "Generalized Guide for MOSFET Miniaturization," *IEEE Electron Devices Lett.* **1**, 2–4 (1980).

Brews, J. R., "Physics of the MOS Transistor," *Appl. Solid State Sci., Suppl.* **2A** 1–120 (1981).

Brox, M. and W. Weber, "Proof and Quantification of Dynamic Effects in Hot-Carrier-Induced Degradation under Pulsed Operation Conditions," *29th Annu. Proc. Reliab. Phys.*, Las Vegas, NV, 133–141 (1991).

Brox, M. and W. Weber, "The Role of the Two-Step Process in Hot-Carrier Degradation," *Proc. 7th Bienn. Eur. Conf.*, 267–270 (1991).

Brox, M., E. Wohlrab, and W. Weber, "A Physical Lifetime Prediction Method for Hot-Carrier-Stressed *p*-MOS Transistors," *Tech. Dig.—Int. Electron Devices Meet., 1991*, 525–528 (1991).

Brozek, T. and B. Pesic, "The Impact of High-Field Stressing on *C–V* Characteristics of Irradiated Gate Oxides," *Appl. Surf. Sci.* **63**, 295–300 (1993).

Bryant, A., T. Furukawa, J. Mandelman, S. Mittl, W. Noble, and E. Nowak, "Angled Implant Fully Overlapped LDD (AI-FOLD) *n*-FETs for Performance and Reliability," *Proc. Int. Reliab. Phys. Symp.*, 152–157 (1989).

Bryant, A., B. El-Kareh, T. Furukawa, W. Noble, E. Nowak, and W. Tonti, "A Fundamental Performance Limit of Optimized 3.3 V Subquarter Micron Fully Overlapped LDD MOSFETs," *Tech. Digest VLSI Technol. Symp.* 45–46 (1990).

Bryant, F. J. and A. Krier, "Direct Current Electroluminescence in Rare-Earth-Doped Zinc Sulphide," *Phys. Status Solidi* **81**, 681–686 (1984).

Budde, W. and W. H. Lamfried, "A Charge-Sheet Capacitance Model Based on Drain Current Modeling," *IEEE Trans. Electron Devices* **37**, 1678–1687 (1990).

Budinsky, "The MOS Technological Processes for VLSI," *Slaboproudy Obz.* **46**, 431–439 (1985).

Budinsky, "The Development of Very Fast MODFET Transistors," *Slaboproudy Obz.* **48**, 89–92 (1987).

Buehler, M. G., N. Zamani, and J. A. Zoutendyk, "CMOS-ASIC Life—Predictions from Test-Coupon Data," *Proc. Int. Conf. Microelectron. Test Struct., 1992*, 6–11 (1992).

Bulucea, C., "Avalanche Injection into the Oxide in Silicon Gate Controlled Devices. I. Theory," *Solid-State Electron.* **18**, 363–374 (1974).

Bunyan, R. J. T., M. J. Uren, N. J. Thomas, and J. R. Davis, "Degradation in Thin-Film SOI MOSFET's Caused by Single-Transistor Latch," *IEEE Electron Devices Lett.* **11**, 359–361 (1990).

Burnett, J. D. and C. Hu, "Hot-Carrier Degradation in Bipolar Transistors at 300 and 110 K—Effect on BiCMOS Inverter Performance," *IEEE Trans. Electron Devices* **37**, 1171–1173 (1990).

Burt, D. J., "Basic Operation of the Charge-Coupled Device," *Int. Conf. Tech. Appl. Charge Coupled Devices* [*Proc.*], Edinburgh, *1974*, 1 (1974).

Buti, T. N., S. Ogura, N. Rovedo, K. Tobimatsu, and C. F. Codella, "Asymmetrical Halo Source GOLD Drain (HS-GOLD) Deep Sub-Half Micron *n*-MOSFET Design for Reliability and Performance," *Tech. Dig.—Electron Devices Meet.*, 617–620 (1989).

Buti, T. N., S. Ogura, N. Rovedo, and K. Tobimatsu, "A New Asymmetrical Halo Source GOLD Drain (HS-GOLD) Deep Sub-Half-Micrometer *n*-MOSFET Design for Reliability and Performance," *IEEE Trans. Electron Devices* **38**, 1757–1764 (1991).

Butler, A. L., J. N. Ellis, and N. F. Stogdale, "Improved Device Performance as a Result of Shallow Phosphorous Source–Drains by Rapid Thermal Annealing," *Proc. 1st Int. ULSI Sci. Tech. Symp.*, Philadelphia, 624–631 (1987).

Cable, J. S. and J. C. S. Woo, "Hot-Carrier-Induced Interface State Generation in Submicrometer Reoxidized Nitrided Oxide Transistors Stressed at 77 K," *IEEE Trans. Electron Devices* **38**, 2612–2618 (1991).

Cabon-Till, B. and G. Ghibaudo, "Modeling of Mobility Degradation in Submicron MOSFETs after Electrical Stressing," *Proc. Int. Conf. High-Speed Electron: Basic Phys. Phenom. Devices Princ.*, 108–111 (1986).

Cacharelis, P., E. Fong, E. Torgerson, M. J. Converse, and P. Denham, "Single Transistor Electrically Alterable Cell," *IEEE Electron Devices Lett.* **6**, 519–521 (1985).

Cadene, M., Y. N'Goran, I. Youm, G. W. Cohen-Solal, and D. Laplaze, "Germinate Recombination Process in CdZnS-$(CH)_x$ Heterojunctions," *Proc. Int. Photovoltaic Sol. Energy Conf.*, Sevilla, Spain, 589–593 (1987).

Campos, V. B., S. Das Sarma, and M. A. Stroscio, "Phonon-Confinement Effect on Electron Energy Loss in One-Dimensional Quantum Wires," *Phys. Rev. B* **46**, 3849–3853 (1992).

Canali, C., F. Nava, and L. Reggiani, "Drift Velocity and Diffusion Coefficients from Time-of-Flight Measurements," *Top. Appl. Phys.* **58**, 87–112 (1985).

Caplan, P. J., E. H. Poindexter, P. K. Vasudev, and R. C. Henderson, "Process-Induced Point Defects in Oxidized Silicon Structures," *Proc. Mater. Res. Soc. Symp.* **76**, 241–245 (1987).

Card, H. C. and A. G. Worrall, "Reversible Floating-Gate Memory," *J. Appl. Phys.* **44**, 2326–2330 (1973).

Card, H. C. and M. I. Elmasry, "Functional Modeling of Nonvolatile MOS Memory Devices," *Solid-State Electron.* **19**, 863 (1976).

Card, H. C. and E. L. Heassell, "Modeling of Channel Enhancement Effects on the Write Characteristics of FAMOS Devices," *Solid-State Electron.* **19**, 965–968 (1976).

Carley, L. R., "Trimming Analog Circuits Using Floating-Gate Analog MOS Memory," *Tech. Dig. IEEE Int. Solid-State Circuits Conf.* **32**, 202–203 (1989).

Carnes, J. E., W. F. Kosonocky, and E. G. Ramberg, "Free Charge Transfer in Charge-Coupled Devices," *IEEE Trans. Electron Devices* **19**, 798 (1972).

Carrillo, J. L., G. Leon-Acosta, J. Arriaga, and M. A. Rodriguez, "Influence of Recombination Centers on the Relaxation Process of a 2D Photoexcited Hot Electron Plasma," *Solid State Commun.* **63**, 773–778 (1987).

Cartier, E. and F. R. McFeely, "Hot-Electron Dynamics in SiO_2 Studied by Soft-X-Ray–Induced Core-Level Photoemission," *Phys. Rev. B* **44**, 10689–10705 (1991).

Cassi, D. and B. Ricco, "An Analytical Model of the Energy Distribution of Hot Electrons," *IEEE Trans. Electron Devices* **37**, 1514–1521 (1990).

Castagne, R., "Physics and Modeling of Hot Electron Effects in Submicron Devices," *Physica B + C (Amsterdam)* **134**, 1–3 (1985).

Cernea, R.-A., G. Samachisa, S.-S. Su, H.-F. Tsai, Y.-S. Kao, C.-Y. M. Wang, Y.-S. Chen, A. Renninger, T. Wong, J. Brennan, Jr., and J. Haines, "1 Mb Flash EEPROM," *Tech. Dig. IEEE Int. Solid-State Circuits Conf.* **32**, 138–139 (1989).

Cham, K. M. and R. G. Wheeler, "Electron–Phonon Interactions in *n*-Type Silicon Inversion Layers at Low Temperature," *Surf. Sci.* **98**, 210 (1980).

Cham, K. M., S.-Y. Oh, D. Chin, and J. L. Moll, *Computer-Aided Design and VLSI Device Development*, Kluwer Academic Publishers, Dordrecht, Netherlands (1986).

Cham, K. M., J. Hui, P. Vande Voorde, and H. S. Fu, "Self-Limiting Behavior of Hot Carrier Degradation and Its Implication on the Validity of Lifetime Extraction by Accelerated Stress," *Proc. Int. Reliab. Phys. Symp.*, 191–194 (1987).

Chan, T. Y., P.-K. Ko, and C. Hu, "Dependence of Channel Electric Field on Device Scaling," *IEEE Electron Devices Lett.* **6**, 551–553 (1980).

Chan, T. Y., P.-K. Ko, and C. Hu, "A Simple Method to Characterize Substrate Current in MOSFET's," *IEEE Electron Devices Lett.* **5**, 505–507 (1984).

Chan, T. Y., P.-K. Ko, and C. Hu, "Dependence of Channel Electric Field on Device Scaling," *IEEE Electron Devices Lett.* **6**, 551–553 (1985).

Chan, T. Y., A. T. Wu, P.-K. Ko, C. Hu, and R. Razouk, "Asymmetrical Characteristics in LDD and Minimum-Overlap MOSFETs," *IEEE Electron Devices Lett.* **7**, 16–19 (1986).

Chan, T. Y., A. T. Wu, P.-K. Ko, and C. Hu, "Effects of the Gate-to-Drain/Source Overlap on MOSFET Characteristics," *IEEE Electron Devices Lett.* **8**, 326–328 (1987).

Chan, T. Y., C. L. Chiang, and H. Gaw, "New Insight into Hot-Electron-Induced Degradation of *n*-MOSFET's," *Tech. Dig.—Int. Electron Devices Meet.*, 196–199 (1988).

Chan, T. Y. and H. Gaw, "Performance and Hot-Carrier Reliability of Deep-Submicrometer CMOS," *Tech. Dig.—Int. Electron Devices Meet.*, 71–74 (1989).

Chan, T.-Y., S.-W. Lee, and H. Gaw, "Experimental Characterization and Modeling of Electron Saturation Velocity in MOSFET's Inversion Layer from 90 to 350 K." *IEEE Electron Devices Lett.* **11**, 466–468 (1990).

Chang, C., M. S. Liang, C. Hu, and R. W. Broderson, "Carrier Tunneling Related Phenomena in Thin Oxide MOSFET's, *Tech. Dig.—Int. Electron Devices Meet.*, **131**, 153–162 (1984).

Chang, C., C. Hu, and R. W. Brodersen, "Quantum Yield of Electron Impact Ionization in Silicon," *J. Appl. Phys.* **57**, 302–309 (1985).

Chang, C., S. Haddad, B. Swaminathan, and J. Lien, "Drain-Avalanche and Hole-Trapping Induced Gate Leakage in Thin-Oxide MOS Devices," *IEEE Electron Devices Lett.* **9**, 588–590 (1988).

Chang, C. C. and W. C. Johnson, "Frequency and Temperature Tests for Lateral Nonuniformities in MIS Capacitors," *IEEE Trans. Electron Devices* **24**, 1249–1255 (1977).

Chang, C. Y., F. C. Tzeng, C. T. Chen, and Y. W. Mao, "Hot-Carrier Memory Effect in an Al/SiN/SiO_2/Si *n*-MOS Diode Due to Electrical Stress," *IEEE Electron Devices Lett.* **6**, 448–449 (1985).

Chang, J. J., "Nonvolatile Semiconductor Memory Devices," *Proc. IEEE* **64**, 1039–1059 (1976).

Chang, L. L. and H. N. Yu, "The Germanium Insulated-Gate Field-Effect Transistor (FET)," *Proc. IEEE* **53** 316–317 (1965).

Chang, S.-T. and K. Chiu, "Effects of Titanium Silicide Process on Oxide Charge Trapping and Hot-Electron Reliability in Submicron MOS Devices for VLSI," *Tech. Dig. VLSI Tech. Symp.*, 65–66 (1987).

Chang, S.-T. and K. Chiu, "Reduced Oxide Charge Trapping and Improved Hot-Electron Reliability in Submicron MOS Devices Fabricated by Titamium Salicide Process," *IEEE Electron Devices Lett.* **9**, 244–246 (1988).

Chang-Liao, J.-S. and J.-G. Hwu, "Improvement of Hot-Carrier Resistance and Radiation Hardness of *n*MOSFETs by Irradiation-Then-Anneal Treatments," *Solid-State Electron.* **34**, 761–764 (1991).

Chao-Song, H., Q. Li-Jian, and R. Zhao-Xing, "Stabilizing Effects of Hot Electron on Low Frequency Plasma Drift Waves," *Acta Phys. Sin.* **37**, 1284–1290 (1988).

Chapman, R. A., C. C. Wei, D. A. Bell, S. Aur, G. A. Brown, and R. A. Haken, "0.5 Micron CMOS for High Performance at 3.3 V," *Tech. Dig.—Int. Electron Devices Meet.*, 52–55 (1988).

Chatterjee, A., S. Aur, T. Niuya, P. Yang, and J. A. Seitchik, "Failure in CMOS Circuits Induced by Hot Carriers in Multi-Gate Transistors," *Proc. Int. Reliab. Phys. Symp.*, 26–29 (1988).

Chatterjee, P. K., G. W. Taylor, R. L. Easley, H.-S. Fu, and A. F. Tasch, "A Survey of High-Density Dynamic RAM Cell Concepts," *IEEE Trans. Electron Devices* **26**, 827–839 (1979).

Chatterjee, P. K., "VLSI Dynamic *n*-MOS Design Constraints Due to Drain Induced Primary and Secondary Impact Ionization," *Tech. Dig.—Int. Electron Devices Meet.*, **14** (1979).

Chatterjee, P. K., W. R. Hunter, T. C. Holloway, and Y. T. Lin, "Impact of Scaling Laws on the Choice of *n*-Channel or *p*-Channel for MOS VLSI," *IEEE Electron Devices Lett.* **1**, 220–223 (1980).

Chattopadhyay, D., "Mobility in *n*-(Al, Ga)/As/GaAs Heterojunctions at Moderate Electric Fields," *Electron. Lett.* **20**, 466–468 (1984).

Chen, H.-S. and S. S. Li, "Determination of Interface State Density in Small-Geometry MOSFET's by High–Low-Frequency Transconductance Method," *IEEE Electron Devices Lett.* **12**, 13–15 (1991).

Chen, I.-C., S. E. Hoagland, and C. Hu, "Electrical Breakdown in Thin Gate and Tunneling Oxides," *IEEE Trans. Electron Devices* **32**, 413–422 (1985).

Chen, I.-C., S. Holland, and C. Hu, "Electron Trap Generation by Recombination of Electrons and Holes in SiO_2," *J. Appl. Phys.* **61**, 4544–4548 (1987).

Chen, I.-C., J. Y. Choi, T.-Y. Chan, and C. Hu, "The Effect of Channel Hot-Carrier Stressing on Gate-Oxide Integrity in MOSFET's," *IEEE Trans. Electron Devices* **35**, 2253–2258 (1988).

Chen, I.-C., D. J. Coleman, and C. W. Teng, "Gate Current Injection Initiated by Electron Band-to-Band Tunneling in MOS Devices," *IEEE Electron Devices Lett.* **10**, 297–300 (1989).

Chen, I.-C., J. P. Lin, and C. W. Teng, "A Highly Reliable 0.3 μ *n*-Channel MOSFET Using Poly Spacers," *Tech. Digest VLSI Technol. Symp.* 39–40 (1990).

Chen, I.-C., C. C. Wei, and C. W. Teng, "Simple Gate-to-Drain Overlapped MOSFET's Using Poly Spacers for High Immunity to Channel Hot-Electron Degradation," *IEEE Electron Devices Lett.* **11**, 78–81 (1990).

Chen, I.-C., R. A. Chapman, and C. W. Teng, "A Sub-Half Micron Partially Gate-to-Drain Overlapped MOSFET Optimized for High Performance and Reliability," *Int. Electron Devices Meet., 1991*, 545–548 (1991).

Chen, J., T.-Y. Chan, P.-K. Ko, and C. Hu, "Gate Current in OFF-State MOSFET," *IEEE Electron Devices Lett.*, 203–205 (1989).

Chen, J., P. Fang, P.-K. Ko, C. Hu, R. Solomon, T.-Y. Chan, and C. G. Sodini, "Noise Overshoot at Drain Current Kink in SOI MOSFET," *1990 IEEE SOS/SOI Tech. Conf.*, 40–41 (1990).

Chen, J., K. N. Quader, R. Solomon, T.-Y. Chan, P.-K. Ko, and C. Hu, "Hot Electron Gate Current and Degradation in P-Channel SOI MOSFET's," *1991 IEEE Int. SOI Conf.*, Vail Valley, CO (1991).

Chen, J. V., "CMOS—The Emerging VLSI Technology," *Proc. IEEE Int. Conf. Comput. Des.*, 130–141 (1985).

Chen, J. V., "CMOS—The Emerging VLSI Technology," *IEEE Circuits Devices Mag.* **2**, 16–31 (1986).

Chen, K.-L., S. A. Saller, I. A. Groves, and D. B. Scott, "Reliability Effects on MOS Transistors Due to Hot-Carrier Injection," *IEEE J. Solid-State Circuits* **20**, 306–313 (1985a).

Chen, K.-L., S. A. Saller, and R. Shah, "Some Methods to Reduce Hot-Carrier Effects," *Tech. Dig. VLSI Tech. Symp.*, 102–103 (1985b).

Chen, K.-L., S. Saller, and R. Shah, "The Case of AC Stress in the Hot-Carrier Effect," *IEEE Trans. Electron Devices* **33**, 424–426 (1986).

Chen, L. I., K. A. Pickar, and S. M. Sze, "Carrier Transport and Storage Effects in Au Ion Implanted SiO_2 Structures," *Solid-State Electron.* **15**, 979 (1972).

Chen, M.-J., "New Observation of Gate Current in Off-State MOSFET," *IEEE Trans. Electron Devices* **38**, 2118–2120 (1991).

Chen, M.-L., C.-W. Leung, W. T. Cochran, S. Jain, H. P. W. Hey, H. Chew, and C. Dziuba, "Hot Carrier Aging in Two Level Metal Processing," *Tech. Dig.—Int. Electron Devices Meet.*, 55–58 (1987).

Chen, M.-L., "CMOS Hot Carrier Protection with LDD," *Semicond. Int.* **11**, 78–81 (1988).

Chen, M.-L., W. T. Cochran, T. S. Yang, C. Dziuba, C.-W. Leung, W. Lin, and W. Juengling, "Constraints in *p*-Channel Device Engineering for Submicron OS Technologies," *Tech. Dig.—Int. Electron Devices Meet.*, 390–393 (1988).

Chen, M.-L., C.-W. Leung, W. T. Cochran, W. Juengling, C. Dziuba, and T. Yang, "Suppression of Hot-Carrier Effects in Submicrometer CMOS Technology," *IEEE Trans. Electron Devices* **35**, 2210–2220 (1988).

Chen, S. L., J. Gong, and S. H. Yang, "Dynamic Device Degradation during Hot-Carrier Stress in DDD *n*-MOSFETs," *Solid-State Electron.* **36**, 501–511 (1993).

Chen, W. and T.-P. Ma, "A New Technique for Measuring Lateral Distribution of Oxide Charge and Interface Traps near MOSFET Junctions," *IEEE Electron Devices Lett.* **12**, 393–395 (1991).

Chen, W. and T.-P. Ma, "Channel Hot-Carrier Induced Oxide Charge Trapping in NMOSFET'S," *Int. Electron Devices Meet., 1991.* 731–734 (1991).

Chen, W., A. Balasinski, B. Zhang, and T.-P. Ma, "Hot-Carrier Effects on Interface-Trap Capture Cross Section in MOSFET's as Studied by Charge Pumping," *IEEE Electron Devices Lett.* **13**, 201–202 (1992).

Chen, Y.-Z. and T.-W. Tang, "Numerical Simulation of Avalanche Hot-Carrier Injection in Short-Channel MOSFET's," *IEEE Trans. Electron Devices* **35**, 2180–2188 (1988).

Chen, Y.-Z. and T.-W. Tang, "Computer Simulation of Hot-Carrier Effects in Asymmetric LDD and LDS MOSFET Device," *IEEE Trans. Electron Devices* **36**, 2492–2498 (1989).

Cheney, G. T., R. M. Jacobs, H. W. Korb, H. E. Nigh, and J. Stack, "Al_2O_3—SiO_2 IGFET Integrated Circuits," *Tech. Dig.—Int. Electron Devices Meet.*, 18–21 (1967).

Ching-Ho Cheng and C. Surya, "The Effect of Hot-Electron Injection on the Properties of Flicker Noise in N-Channel MOSFETs," *Solid-State Electron.* **36**, 475–479 (1993).

Cheng, S. K. and P. Manos, "Effects of Operating Temperature on Electrical Parameters in an Analog Process," *IEEE Circuits Devices Mag.* **5**, 31–38 (1989).

Cheng, Y. C. and E. A. Sullivan, "Effect of Coulombic Scattering on Silicon Surface Mobility," *J. Appl. Phys.* **45**, 187–192 (1974).

Chevrier, J., D. Delagebeaudeuf, C. Rumehard, and P. Briere, "III–V Heterojunction Devices—Classification and State of the Art," *Rev. Tech. Thomson-CSF* **19**, 473–495 (1987).

Childs, P. A., W. Eccleston, and R. A. Stuart, "Alternative Mechanism for Substrate Minority Carrier Injection in MOS Devices Operating in Low Level Avalanche," *Electron. Lett.* **17**, 281–282 (1981).

Chiu, K. Y., J. L. Moll, and J. Manoliu, "A Bird's Beak Free Local Oxidation Technology Feasible for VLSI Circuits Fabrication," *IEEE Trans. Electron Devices* **29**, 536–540 (1982).

Choi, J.-Y., P.-K. Ko, and C. Hu, "Effect of Oxide Field on Hot-Carrier-Induced Degradation of Metal-Oxide-Semiconductor Field-Effect Transistor," *Appl. Phys. Lett.* **50**, 1188–1190 (1987a).

Choi, J-Y., P.-K. Ko, and C. Hu, "Hot-Carrier-Induced MOSFET Degradation: AC versus DC Stressing," *Tech. Dig. VLSI Tech. Symp.*, 45–46 (1987b).

Choi, J.-Y., P.-K. Ko, and C. Hu, "Hot Carrier Induced MOSFET Degradation under AC Stress," *IEEE Electron Devices Lett.* **8**, 333–335 (1987c).

Choi, J.-Y., P.-K. Ko, C. Hu, and W. F. Scott, "Hot-Carrier-Induced Degradation of Metal-Oxide-Semiconductor Field-Effect Transistors: Oxide Charge versus Interface Traps," *J. Appl. Phys.* **65**, 354–360 (1989).

Choi, J.-Y., J. G. Fossum, and R. Sundaresan, "SOI Design for Submicron CMOS," *IEEE SOS/SOI Tech. Conf.*, 23–24 (1989).

Choi, J.-Y., R. Sundaresan, and J. G. Fossum, "Monitoring Hot-Electron-Induced Degration of Floating-Body SOI MOSFET's, *IEEE Electron Devices Lett.* **11**, 156–158 (1990).

Choon, E. C., K. Thornewell, P. Tsai, T. Gukerberger, J. Sylvestri, and J. Orro, "Hot Electron Induced Retention Time Degradation in MOS Dynamic RAMs," *Proc. Int. Reliab. Phys. Symp.*, 195–198 (1986).

Christensen, M. B., "Complex Semiconductors Operating at Low Temperatures: Comparative Life Test Evaluation of Three Variants of 256 K $\times$ 1*n*-MOS DRAM and Three Variants of 8 K $\times$ 8 CMOS SRAM Operating at Low Temperatures," *Elektronikcentralen Rep.* **201**, 33 (1987).

Chu, S. T., J. Dikken, C. D. Hartgring, F. J. List, J. G. Raemarkers, S. A. Bell, B. Walsh, and R. H. W. Salters, "25-ns Low-Power Full-CMOS 1-Mbit (128 K Multiplied by 8) SRAM," *IEEE J. Solid-State Circuits* **SC-23**, 1078–1084 (1988).

Chu, T. L., J. R. Szedon, and C. H. Lee, "The Preparation and *C–V* Characteristics of Si–Si_3N_4 and Si–SiO_2–Si_3N_4 Structure," *Solid-State Electron.* **10**, 625 (1967).

Chung, J., M.-C. Jeng, G. May, P.-K. Ko, and C. Hu, "Hot-Electron Currents in Deep-Submicrometer MOSFETs," *Tech. Dig.—Int. Electron Dig. Meet.*, 200–203 (1988).

Chung, J., M.-C. Jeng, G. May, P.-K. Ko, and C. Hu, "Low-Voltage Hot-Electron Currents and Degradation in Deep-Submicrometer MOSFETs," *Proc. Int. Reliab. Phys. Symp.*, 92–97 (1989).

Chung, J. E. and R. S. Muller, "Development and Application of a Si–SiO_2 Interface-Trap Measurement System Based on the Staircase Charge-Pumping Technique," *Solid-State Electron.* **32**, 867–882 (1989).

Chung, J. E., M.-C. Jeng, J. E. Moon, P.-K. Ko, and C. Hu, "Low-Voltage Hot-Electron Currents and Degradation in Deep-Submicrometer MOSFET's," *IEEE Trans. Electron Devices* **37**, 1651–1657 (1990).

Chung, J. E., M.-C. Jeng, J. E. Moon, P.-K. Ko, and C. Hu, "Performance and Reliability Design Issues for Deep-Submicrometer MOSFETs," *IEEE Trans. Electron Devices* **38**, 545–554 (1991).

Chung, J. E., J. Chen, P.-K. Ko, C. Hu, and M. Levi, "The Effects of Low-Angle Off-Axis Substrate Orientation on MOSFET Performance and Reliability," *IEEE Trans. Electron Devices* **38**, 627–633 (1991).

Chung, J. E., P.-K. Ko, and C. Hu, "A Model for Hot-Electron-Induced MOSFET Linear-Current Degradation Based on Mobility Reduction Due to Interface-State Generation," *IEEE Tras. Electron Devices* **38**, 1362–1370 (1991).

Chung, S. S.-S., "A Complete Model of the *I–V* Characteristics for Narrow-Gate MOSFET's," *IEEE Trans. Electron Devices* **37**, 1020–1030 (1990).

Chung, S. S.-S., J.-S. Lee, Y.-G. Chen, and P.-C. Hsu, "UNI-MOS: A Unified SPICE Built-in MOSFET Model for Circuit Simulation and Lifetime Evaluation," *Proc. 3rd Annu. IEEE ASIC Semin. Exhibit*, Rochester, NY P5/6.1–4 (1990).

Chynoweth, A. G. and K. G. McKay, "Photon Emission from Avalanche Breakdown in Silicon," *Phys. Rev.* **102**, 369–376 (1956).

Cirit, M. A., "Switch Level Random Pattern Testability Analysis," *Proc. IEEE Des. Autom. Conf.*, 587–590 (1988).

Cirit, M. A., "Hot Carrier Effects on CMOS Circuit Performance," *Proc. IEEE Custom Conf.*, 26.5.1–26.5.4 (1989).

Clark, W. F., B. El-Kareh, R. G. Pires, S. L. Titcomb, and R. L. Anderson, "Low Temperature CMOS—A Brief Review," *IEEE Trans. Components, Hybrids, Manuf. Tech.* **CHMT-15**, 397–404 (1992).

Cochran, B. "Tradeoffs in Submicron CMOS Process Design," *Semicond. Int.* **14**, 146–148 (1991).

Codella, C. and S. Ogura, "Halo Doping Effects in Submicron DI-LDD Device Design," *Tech. Dig.—Int. Electron Devices Meet.*, 230–233 (1985).

Coe, D. J., "Changes in Effective Channel Length Due to Hot-Electron Trapping in Short-Channel MOSTS," *IEE J. Solid-State Electron Devices* **2**, 57–61 (1978).

Coen, R. W. and R. S. Muller, "Velocity of Surface Carriers in Inversion Layers on Silicon," *Solid-State Electron.* **23**, 35 (1980).

Colinge, J.-P., "Hot-Electron Effects in Silicon-on-Insulator *n*-Channel MOSFET's," *IEEE Trans. Electron Devices* **34**, 2173–2177 (1987).

Colinge, J.-P., "Some Properties of Thin-Film SOI MOSFETs," *IEEE Circuits Devices Mag.* **3**, 16–20 (1987).

Collins, D. R. and C. T. Sah, "Effects of X-Ray Irradiation on the Characteristics of MOS Structures," *Appl. Phys. Lett.* **8**, 124 (1966).

Collins, H. W. and B. Pelly, "HEXFET, a New Power Technology, Cuts On-Resistance, Boosts Rating," *Electron. Des.* **17**, 36 (1979).

Colman, D., R. T. Bate, and J. P. Mize, "Mobility Anisotropy and Piezoresistance in Silicon *p*-Type Inversion Layers," *J. Appl. Phys.*, **39**, 1923–1931 (1968).

Comeau, A., "Comments, with Reply, on 'Modeling of the $1/f$ Noise Overshoot in Short-Channel MOSFETs Locally Degraded by Hot-Carrier Injection' by B. Boukriss *et al.*," *IEEE Electron Devices Lett.* **11**, 129–130 (1990).

Conley, J. F. and P. M. Lenahan, "Room Temperature Reactions Involving Silicon Dangling Bond Centers and Molecular Hydrogen in Amorphous SiO_2 Thin Films on Silicon," *IEEE Trans. Nucl. Sci.* **NS-39**, 2186–2191 (1992).

Conn, T., J. Eachus, J. Klema, R. Pyle, and B. Schwiesow, "Update of Reliability Concerns on MOS Memory Integrated Circuits," *IEEE Reg. 5 Conf.*, 183–188 (1986).

Conwell, E., "High Field Transport in Semiconductors," *Solid State Phys., Suppl.* **9** (1967).

Cooper, J. A. and D. F. Nelson, "High Field Drift Velocity of Electrons at the $Si–SiO_2$ Interface as Determined by a Time-of-Flight Technique," *J. Appl. Phys.* **54**, 1445–1456 (1983a).

Cooper, J. A. and D. F. Nelson, "Measurement of the High-Field Drift Velocity of Electron in Inversion Layer in Silicon," *IEEE Electron Devices Lett.*, **EDL-2**(7), 169–173 (1983b).

Cottrell, P. E. and R. R. Troutman, "Characterization of Electronic Gate Current in IGFETs Operating in the Linear and Saturation Regions," *IEEE Trans. Electron Devices* **24**, 1200 (1977).

Cottrell, P. E., R. R. Troutman, and T. H. Ning, "Hot-Electron Emission in *n*-Channel IGFETs," *IEEE Trans. Electron Devices* **26**, 520–533 (1979).

Cristoloveanu, S., "Interface and Oxide Engineering for High Quality SIMOX Devices," *Proc. 7th Bienn. Eur. Conf.*, 53–63 (1991).

Cristoloveanu, S., S. M. Gulwadi, D. E. Ioannou, G. J. Campisi, and H. L. Hughes, "Hot-Electron-Induced Degradation of Front and Back Channels in Partially and Fully Depleted SIMOX MOSFETs," *IEEE Electron Devices Lett.* **13**, 603–605 (1992).

Crowe, T. W. and R. J. Mattauch, "Analysis and Optimization of Millimeter- and Submillimeter-Wavelength Mixer Diodes," *IEEE Trans. Microwave Theory Tech.* **MTT-35**, 159–168 (1987).

Cuevas, P., "Simple Explanation for the Apparent Relaxation Effect Associated with Hot-Carrier Phenomenon in MOSFET's," *IEEE Electron Devices Lett.* **9**, 627–629 (1988).

Dacey, G. C. and I. M. Ross, "Unipolar FET," *Proc. IRE* **41**, 970–979 (1953).

Dalton, J. V. and J. Drobek, "Structure and Sodium Migration in Silicon Nitride Films," *J. Electrochem. Soc.* **115**, 865 (1968).

Dang, R. L. M. and M. Konaka, "Two-Dimensional Computer Analysis of Triode-Like Characteristics of Short-Channel MOSFETs," *IEEE Trans. Electron Devices* **27**, 1533–1539 (1980).

Dang, R. L. M., M. Nakamura, and T. Wada, "Computer-Simulated Optimization of a LDD-MOSFET of Submicron Channel Length," *Proc. Int. Symp. Appl. Simul. Model.*, Santa Barbara, CA, 155–159 (1987).

Dan'ko, D. B., R. D. Fedorovich, A. V. Gaidar, and V. N. Poroshin, "Electron and Photon Emission from Discontinuous Carbon Films," *Int. J. Electron.* **73**, 1005–1008 (1992).

Das, N. C., W. S. Khokle, and S. Mohanty, "Hot Carrier Effects in Depletion-Mode MOSFETs," *Int. J. Electron.* **60**, 495–503 (1986).

Das, N. C., P. W. C. Duggan, and V. Nathan, "Annealing Characteristics of Hot Carrier Induced Damage in *n*-Channel MOSFETs," *Int. J. Electron.* **73**, 1201–1213 (1992).

Das, N. C., P. W. C. Duggan, and V. Nathan, "Transconductance Technique for Measurement of Interface State Density and Oxide Charge in LDD-MOSFETs," *Phys. Status Solidi A* **133**, 167–177 (1992).

Das, N. C., V. Nathan, S. Dacus, and J. Cable, "Combined Effects of Hot-Carrier Stressing and Ionizing Radiation in SiO_2, NO, and ONO MOSFETs," *IEEE Electron Devices Lett.* **14**, 40–42 (1993).

Da Silva, E. F., Jr., Y. Nishioka, and T.-P. Ma, "Effects of Trichloroethane during Oxide Growth on Radiation-Induced Interface Traps in Metal/SiO_2/Si Capacitor," *Appl. Phys. Lett.* **51**, 1262–1264 (1987).

Da Silva, E. F., Jr., Y. Nishioka, and T.-P. Ma, "Radiation and Hot-Electron Hardened MOS Structures," *Tech. Dig.—Int. Electron Devices Meet.*, 848–849 (1987).

Da Silva, E. F., Jr., "Reliability of MOSFETs as Affected by the Interface Trap Transformation Process," *IEEE Electron Devices Lett.* **10**, 537–539 (1989).

Davis, J., Jr., "Degradation Behaviour of *n*-Channel MOSFETs Operated at 77 K," *IEE Proc., Part I* **127**, 183–187 (1980).

Davis, M. and R. Lahri, "Gate Oxide Charge-to-Breakdown Correlation to MOSFET Hot-Electron Degradation," *IEEE Electron Devices Lett.* **9**, 183–185 (1988).

Davis, M. and F. Haas, "In Line Wafer Level Reliability Monitors," *Solid State Technol* **32**, 107–110 (1989).

Deal, B. E., E. H. Snow, and C. A. Mead, "Barrier Energies in Metal–Silicon Dioxide–Silicon Structures," *J. Phys. Chem. Solids* **27**, 1873 (1966).

Deal, B. E., M. Sklar, A. S. Grove, and E. H. Snow, "Characteristics of the Surface-State Charge (Q_{ss}) of Thermally Oxidized Silicon," *J. Electrochem. Soc.* **114**, 266 (1967).

Deal, B. E., "Standardized Terminology for Oxide Charges Associated with Thermally Oxidized Silicon," *IEEE Trans. Electron Devices* **27**, 606–608 (1980).

Dean, C. C. and M. Pepper, "One-Dimensional Electron Localization and Conduction of Electron–Electron Scattering in Narrow Silicon MOSFETs," *J. Phys. C* **17**, 5663–5676 (1984).

Declerck, G., "Trends in VLSI Processing," *Nucl. Instrum. Methods Phys. Res. Sect. A* **275**, 475–478 (1989).

Decoste, R. and B. H. Ripin, "High Energy Ions from a Nd-Laser Plasma," *IEEE Int. Conf. Plasma Sci.*, **59** (1977).

Decoste, R., J.-C. Kieffer, M. Piche, H. Pepin, and P. Lavigne, "Reduction in Hot Electron Transport on Dielectric Targets at High Laser Irradiance," *Phys. Fluids* **25**, 1699–1670 (1982).

Deen, M. J. and J. Wang, "Design Considerations for the Operation of CMOS Inverters at Cryogenic Temperatures," *Proc. Symp. Low Temp. Electron. High Temp. Supercond.*, Honolulu, HI, 108–116 (1988).

Deen, M. J. and J. Wang, "Substrate Currents in Short Buried-Channel PMOS Devices at Cryogenic Temperatures," *Cryogenics* **30**, 1113–1117 (1990).

Delord, J. F., D. G. Hoffman, and G. Stringer, "Use of MOS Capacitors in Determining the Properties of Surface State at the Si–SiO_2 Interface," *Bull. Am. Phys. Soc.* [2], **10**, 546 (1965).

Dennard, R. H., F. H. Gaensslen, H.-N. Yu, V. L. Rideout, E. Bassons, and A. R. LeBlanc, "Design of Ion-Implanted MOSFET's with Very Small Physical Dimensions," *IEEE J. Solid State Circuits* **SC-9**, 256–268 (1974).

Dennard, R. H., "Evolution of the MOSFET Dynamic RAM—A Personal View," *IEEE Trans. Electron Devices* **31**, 1549–1555 (1984).

deSchrijver, E., P. Heremans, R. Bellens, G. Groeseneken, and H. E. Maes, "Analysis of Post-Stress Effects in Passivated MOSFETs after Hot-Carrier Stress," *Microelectron. Eng.* **15**, 437–440 (1991).

deSchrijver, E., P. Heremans, R. Bellens, G. Groeseneken, and H. E. Maes, "Post-Stress Interface Trap Generation: A New Hot-Carrier Induced Degradation Phenomenon in Passivated *n*-Channel MOSFET's," *30th Annu. Proc. Reliab. Phys.*, 112–115 (1992).

Deyhimy, I., R. C. Eden, R. J. Anderson, and I. S. Harris, "A 500-MHz GaAs Charge-Coupled Device," *Appl. Phys. Lett.* **36**, 151–153 (1980).

Dickinger, P., G. Nanz, and S. Selberherr, "Measurement and Simulation of Degradation Effects in High Voltage DMOS Devices," *20th Eur. Solid State Devices Res. Conf.*, 369–372 (1990).

Dienys, V. and A. Dargys, "Microwave Experiments Including Avalanche," *J. Phys., Colloq. (Orsay, Fr.)* **42**, 33–49 (1981).

Dierickx, B., E. Simoen, and G. Declerck, "Transient Response of Silicon Devices at 42 K. I. Theory," *Semicond. Sci. Technol.* **6**, 896–904 (1991).

DiMaria, D. J., "Determination of Insulator Bulk Trapped Charge Densities and Centroids from Photocurrent–Voltage Characteristics of MOS Structure," *J. Appl. Phys.* **47**, 4073–4077 (1976).

DiMaria, D. J., K. M. DeMeyer, and D. W. Dong, "Electrically-Alterable Memory Using a Dual Electron Injection Structure," *IEEE Electron Devices Lett.* **1**, 179–181 (1980).

DiMaria, D. J. and M. V. Fischetti, "Electron Transport and Heating in Silicon Dioxide Films," *Annu. Rep—Electr. Insul. Dielectr. Phenom. Conf.*, 169–179 (1987).

DiMaria, D. J. and M. V. Fischetti, "Hot Electrons in Silicon Dioxide: Ballistic to Steady-State Transport," *Appl. Surf. Sci.* **33**, 278–297 (1987).

DiMaria, D. J. and M. V. Fischetti, "Vacuum Emission of Hot Electrons from Silicon Dioxide at Low Temperatures," *J. Appl. Phys.* **64**, 4683–4691 (1988).

DiMaria, D. J. and J. W. Stasiak, "Trap Creation in Silicon Dioxide Produced by Hot Electrons," *J. Appl. Phys.* **65**, 2342–2356 (1989).

DiMaria, D. J., "The Relationship of Trapping and Trap Creation in Silicon Dioxide Films to Hot Carrier Degradation of Si MOSFETs," *Proc. 7th Bienn. Eur. Conf.*, 65–72 (1991).

Ditali, A., P. Fazan, and I. Khan, "Hot-Carrier Reliability in Double-Implanted Lightly Doped Drain Devices for Advanced DRAMs," *Electron. Lett.* **28**, 19–21 (1992).

Ditali, A., V. Mathews, and P. Fazan, "Hot-Carrier-Induced Degradation of Gate Dielectrics Grown in Nitrous Oxide under Accelerated Aging," *IEEE Electron Devices Lett.* **13**, 538–540 (1992).

Djahli, F., C. Plossu, and B. Balland, "Micronic *n*-Channel MOSFET Degradation under Strong and Short-Time Hot-Carrier Stress," *Mater. Sci. Eng., B* **B15**, 164–168 (1992).

Doany, F. E. and D. Grischkowsky, "Measurement of Ultrafast Hot-Carrier Relaxation in Silicon by Thin-Film-Enhanced, Time-Resolved Reflectivity," *Appl. Phys. Lett.* **52**, 36–38 (1988).

Dorda, G., "Piezoresistance in Quantized Conduction Bands in Silicon Inversion Layers," *J. Appl. Phys.*, **42**, 2053–2060 (1971).

Dori, L., J. Sun, M. Arienzo, S. Basavaiah, Y. Taur, and D. Zichermann, "Very Thin Nitride/Oxide Composite Gate Insulator for VLSI CMOS," *Tech. Dig. VLSI Tech. Symp.*, 25–26 (1987).

Dorosti, J. and C. R. Viswanathan, "Photo-Injection Studies of Traps in HCl/H_2O Oxides," *Proc. Int. Top. Conf. Phys.* SiO_2 *Interfaces*, Yorktown Heights, NY, 184–188 (1978).

Downer, M. C., D. H. Reitze, and G. Focht, "Ultrafash Laser Probe of Interband Absorption Edges in 3D and 2D Semiconductors," *Proc. SPIE—Int. Soc. Opt. Eng.* **1282**, 121–131 (1990).

Doyle, B. S. M. Bourcerie, J.-C. Marchetaux, and A. Boudou, "Relaxation Effects in *n*-MOS Transistors after Hot-Carrier Stressing," *IEEE Electron Devices Lett.* **8**, 234–236 (1987).

Doyle, B. S., M. Bourcerie, J.-C. Marchetaux, and A. Boudou, "Dynamic Channel Hot-Carrier Degradation of *n*-MOS Transistors by Enhanced Electron-Hole Injection into the Oxide," *IEEE Electron Devices Lett.* **8**, 237–239 (1987).

Doyle, B. S., M. Bourcerie, P. Leclaire, A. Boudou, and P. Dars, "Hot Electron Degradation in the Source of Asymmetrical LDD Structures," *Electron. Lett.* **23**, 1356–1357 (1987).

Doyle, B. S., M. Bourcerie, J.-C. Marchetaux, and A. Boudou, "Surface State Creation by Dynamic Aging," *IEE Colloq. Dig.*, 11.1–11.4 (1987).

Doyle, B. S., M. Bourcerie, J.-C. Marchetaux, and A. Boudou, "The Voltage Dependence of Degradation in *n*-MOS Transistors," *Proc. Eur. Solid State Devices Res. Conf., 17th*, 257–260 (1988).

Doyle, B. S., M. Bourcerie, J.-C. Marchetaux, and A. Boudou, "Effect of Substrate Bias on Hot-Carrier Damage in *n*-MOS Devices," *IEEE Electron Devices Lett.* **10**, 11–13 (1989).

Doyle, B. S., and D. Lau, "The Properties and Annealing of Gate Oxide Damage of Oxynitride-Passivated CMOS Transistors Arising from Mechanical Stresses during Packaging," *IEDM Tech. Dig.—Int. Electron Devices Meet.*, 91–94 (1989).

Doyle, B. S., M. Bourcerie, J.-C. Marchetaux, and A. Boudou, "Interface State Creation and Charge Trapping in the Medium-to-High Gate Voltage Range ($V_d/2 \geq V_g \geq V_d$) during Hot-Carrier Stressing of *n*-MOS Transistors," *IEEE Trans. Electron Devices* **37**, 744–754 (1990a).

Doyle, B. S. and K. R. Mistry, "Lifetime Prediction Method for Hot-Carrier Degradation in Surface-Channel *p*-MOS Devices," *IEEE Trans. Electron Devices* **37**, 1301–1307 (1990).

Doyle, B. S., M. Bourcerie, C. Bergonzoni, R. Benecchi, A. Bravis, K. R. Mistry, and A. Boudou, "The Generation and Characterization of Electron and Hole Traps Created by Hole Injection during Low Gate Voltage Hot-Carrier Stressing of *n*-MOS Transistors," *IEEE Trans. Electron Devices* **37**, 1869–1876 (1990b).

Doyle, B. S. and K. R. Mistry, "A General Gate-Current *p*-MOS Lifetime Predication Method Applicable to Different Channel Structures," *IEEE Electron Devices Lett.* **11**, 547–548 (1990).

Doyle, B. S. and G. J. Dunn, "Dynamic Hot-Carrier Stressing of Reoxidized Nitrided Oxide," *IEEE Electron Devices Lett.* **12**, 63–65 (1991).

Doyle, B. S. and K. R. Mistry, "A Lifetime Prediction Method for Oxide Electron Trap Damage Created during Hot-Electron Stressing of *n*-MOS Transistors," *IEEE Electron Devices Lett.* **12**, 178–180 (1991).

Doyle, B. S., K. R. Mistry, and G. J. Dunn, "Rexodizied Nitrided Oxides (RNO) for Latent ESD-Resistant MOSFET Dielectrics," *IEEE Electron Devices Lett.* **12**, 184–186 (1991).

Doyle, B. S., C. Bergonzoni, and A. Boudou, "The Influence of Gate Edge Shape on the Degradation in Hot-Carrier Stressing of *n*-Channel Transistors," *IEEE Electron Devices Lett.* **12**, 363–365 (1991).

Doyle, B. S. and R. S. O'Conner, K. R. Mistry, and G. J. Grula, "Comparison of Shallow Trench and LOCOS Isolation for Hot-Carrier Resistance," *IEEE Electron Devices Lett.* **12**, 673–675 (1991).

Doyle, B. S., B. J. Fishbein, and K. R. Mistry, "NBTI-Enhanced Hot Carrier Damage in *p*-Channel MOSFETs," *Int. Electron Devices Meet., 1991*, 529–532A (1991).

Doyle, B. S. and G. J. Dunn, "Recovery of Hot-Carrier Damage in Reoxidized Nitrided Oxide MOSFET's," *IEEE Electron Devices Lett.* **13**, 38–40 (1992).

Doyle, B. S., K. R. Mistry, and D. B. Jackson, "Examination of Gradual-Junction *p*-MOS Structures for Hot Carrier Control Using a New Lifetime Extraction Method," *IEEE Trans. Electron Devices* **39**, 2290–2297 (1992).

Doyle, B. S., J. Faricelli, K. R. Mistry, and D. Vuillaume, "Characterization of Oxide Trap and Interface Trap Creation during Hot-Carrier Stressing of *n*-MOS Transistors Using the Floating-Gate Technique," *IEEE Electron Devices Lett.* **14**, 63–65 (1993).

Doyle, B. S. and K. R. Mistry, "The Characterization of Hot Carrier Damage in *p*-Channel Transistors," *IEEE Trans. Electron Devices* **40**, 152–156 (1993).

Dozier, C. M., "Ionizing Radiation in Microelectronics Processing," *Solid State Technol.* **29**, 105–108 (1986).

Dressendorfer, P. V. and R. C. Barker, "Photoemission Measurements of Interface Barrier Energies for Tunnel Oxides on Silicon," *Appl. Phys. Lett.* **36**, 933–935 (1980).

Dunn, G. J. and S. A. Scott, "Channel Hot-Carrier Stressing of Reoxidized Nitrided Silicon Dioxide," *IEEE Trans. Electron Devices* **37**, 1719–1726 (1990).

Dunn, G. J. and J. T. Krick, "Channel Hot-Carrier Stressing of Reoxidized Nitrided Oxide *p*-MOSFETs," *IEEE Trans. Electron Devices* **38**, 901–906 (1991).

Dunn, G. J., "Effect of Synchrotron X-Ray Radiation on the Channel Hot-Carrier Reliability of Reoxidized Nitrided Silicon Dioxide," *IEEE Electron Devices Lett.* **12**, 8–9 (1991).

Dunn, G. J. and J. T. Krick, "Effects of Radiative Processing Steps on Inversion Layer Mobility and Channel Hot Carrier Damage in Reoxidized Nitrided Oxide MOSFETs," *J. Electron. Mater.* **21**, 677–681 (1992).

Duvvury, C., D. Baglee, M. Smayling, and M. P. Duane, "Series Resistance Modeling for Optimum Design of LDD Transistors," *Tech. Dig.—Int. Electron Devices Meet.* 388–341 (1983).

Duvvury, C., D. Baglee, M. P. Duane, A. Hyslop, M. Smayling, and M. Maekawa, "Analytical Method for Determining Intrinsic Drain/Source Resistance of Lightly Doped Drain (LDD) Devices," *Solid-State Electron.* **27**, 89–96 (1984).

Duvvury, C., D. Redwine, H. Kitagawa, R. Haas, Y. Chuang, C. Beydler, and A. Hyslop, "Impact of Hot Carriers on DRAM Circuits," *Proc. Int. Reliab. Phys. Symp.*, 201–206 (1987).

Duvvury, C., D. Redwine, and H. J. Stiegler, "Leakage Current Degradation in *n*-MOSFETs Due to Hot-Electron Stress," *IEEE Electron Devices Lett.* **9**, 579–581 (1988).

Duvvury, C., R. N. Rountree, H. J. Stiegler, T. Polgreen, and D. Corum, "ESD Phenomena in Graded Junction Devices," *Proc. Int. Reliab. Phys. Symp.* 71–76 (1989).

Eastman, L. F., "Experimental Studies of Ballistic Transport in Semiconductors (FETs)," *J. Phys., Colloq.* (*Orsay, Fr.*) **42**, 263–269 (1981).

Eccleston, W. and M. Uren, eds., "Insulating Films on Semiconductors 1991," Proceedings from the 7th Biennial European Conference, Including Satellite Workshops on Silicon on Insulator, Materials and Device Technology and the Physics of Hot Electron Degradation in Si MOSFETs, Liverpool, UK, (1991).

Edwards, D. G., "Testing for MOS Integrated Circuit Failure Modes," *Dig. IEEE Conf Test 80s*, 407–416 (1980).

Edwards, D. G., "Reliability of MOS Technologies Including MOS Power Transistors," *Proc. EUROCON Reliab. Electr. Electron. Component Syst., 5th*, Copenhagen, Denmark, 254–258 (1982).

Edwards, D. G., "Testing for MOS IC Failure Modes," *IEEE Trans. Reliab.* **R-31**, 9–18 (1982).

Eimori, T., H. Ozaki, H. Oda, S. Ohsaki, J. Mitsuhashi, S. Satoh, and T. Matsukawa, "Improvement of LDD MOSFET's Characteristics by the Oblique-Rotating Ion Implantation," *Ext. Abs. Conf. Solid State Devices and Materials*, 27–30 (1987).

Einspruch, N. G. and G. S. Gildenblat, *Advanced MOS Device Physics*, Academic Press, San Diego (1989).

Eitan, B., D. Frohman-Bentchkowsky, and J. Shappir, "Electron Trapping in SiO_2—An Injection Mode Dependent Phenomenon," *Tech. Dig.—Int. Electron Devices Meet.*, 604–607 (1981a).

Eitan, B. and D. Frohman-Bentchkowsky, "Hot-Electron Injection into the Oxide in *n*-Channel MOS Devices," *IEEE Trans. Electron Devices* **28**, 328–340 (1981).

Eitan, B., D. Frohman-Bentchkowsky, and J. Shappir, "Holding Time Degradation in Dynamic MOS RAM by Injection-Induced Electron Currents," *IEEE Trans. Electron Devices* **28**, 1515–1519 (1981b).

Eitan, B., D. Frohman-Bentchkowsky, and J. Shappir, "Impact Ionization at Very Low Voltage in Silicon," *J. Appl. Phys.* **53**, 1244–1247 (1982).

Eklund, R., R. Chapman, C. Wei, C. Blanton, T. Holloway, M. Rodder, J. Graham, H. Terazawa, V. Rao, H. Tran, T. Suzuki, R. Havemann, R. Sundaresan, D. Scott, and R. Haken, "A 0.5-m BiCMOS Technology for Logic and 4 Mbit-Class SRAM's," *Tech. Dev.—Int. Electron Devices Meet.*, 425–428 (1989).

El-Banna, M. and M. El-Nokali, "Quasi-Two-Dimensional Model for Small-Geometry MOSFETs Including the Source/Drain Resistances," *Int. J. Electron.* **66**, 585–595 (1989).

El-Banna, M. and M. A. El-Nokali, "A Simple Analytical Model for Hot-Carrier MOSFET's," *IEEE Trans. Electron Devices* **36**, 979–986 (1989).

El-Hennawy, A., J. Borel, and D. Barbier, "Hot Carrier Effects in MOSFET Devices: Injection and Surface Trapping," *11th Eur. Solid State Devices Res. Conf. 6th SSSDT—Europhys. Conf. Abst.* **5F**, 51–55 (1981).

El-Hennawy, A., M. H. El Siad, J. Borel, and G. Kamarinos, "Modeling MOSFETs at Strong Narrow Pulses for VLSI Applications," *Solid-State Electron.* **30**, 519–526 (1987).

El-Hennawy, A., "Design and Simulation of a High Reliability Non-Volatile CMOS EEPROM Memory Cell Compatible with Scaling-Down Trends," *Int. J. Electron.* **72**, 73–87 (1992).

El-Kareh, B., "High Density Field-Effect Transistor Memory Cell," *IBM Tech. Disclosure Bull.* **26**, 4411–4412 (1984).

El-Kareh, B., T. B. Hook, M. E. Johnson, J. J. Lajza, and R. W. McLaughlin, "Field-Induced Instabilities in Polyimide Passivated Lateral *pnp* Transistors," *IEEE Trans. Components, Hybrids, Manuf. Technol.* **CHMT-13**, 623–628 (1990).

El-Kareh, B., W. W. Abadeer, and W. R. Tonti, "Design of Submicron PMOSFETs for DRAM Array Applications," *Tech. Dig.—Int. Electron Devices Meet.*, 379–384 (1991).

Elliot, A. B. M., "The Use of Charge Pumping Currents Surface State Densities in MOS Transistors," *Solid-State Electron.* **19**, 241–247 (1976).

Ellis, J. N. and A. L. Butler, "Hot Electron Effect Reduction Using Phosphorous Source Drains," *IEE Colloq. 'Hot Carrier Degradation Short Channel MOS'*, 9.1–9.19 (1987).

El-Mansy, Y. A. and D. M. Caughey, "Modeling Weak Avalanche Multiplication Currents in IGFETs and SOS Transistors for CAD," *Tech. Dig.—Int. Electron Devices Meet.*, **31** (1975).

El-Mansy, Y. A. and A. R. Boothroyd, "A Simple-Two-Dimensional Model for IGFET Operation in the Saturation Region," *IEEE Trans. Electron Devices* **24**, 254–262 (1977).

El-Mansy, Y. A., "Limits to Scaling MOS Devices," *Tech. Dig. VLSI Tech. Symp.*, 16–17 (1981).

El-Mansy, Y. A., "MOS Device and Technology Constraints in VLSI," *IEEE Trans. Electron Devices* **29**, 567–573 (1982).

Elsaid, M. H., S. G. Chamberlain, and L. A. K. Watt, "Computer Model and Charge Transport Studies in Short Gate Charge-Coupled Devices," *Solid-State Electron.* **20**, 61 (1977).

Enomoto, T., K. Nagasawa, N. Tsubouchi, and H. Matsumoto, "Production Technology for High-Density Semiconductor Memories," *Mitsubishi Denki Giho* **53**, 495–498 (1979).

Eom, J. and C. B. Su, "Observation of Positive and Negative Nonlinear Gain in Semiconductor Diode Lasers Using Optical Modulation," *IEEE Conf. Laser Electr. Opt.*, 220–222 (1989).

Estabrook, K., "Hot Electrons Due to Laser Absorption by Ion Acoustic Turbulence," *Phys. Rev. Lett.* **47**, 1396–1399 (1981).

Estreich, D. B., A. Ochoa, and R. W. Dutton, *Tech. Dig.—Int. Electron Devices Meet.*, 230–234 (1978).

Fair, R. B. and R. C. Sun, "Threshold-Voltage Instability in MOSFETs Due to Channel Hot-Hole Emission," *IEEE Trans. Electron Devices* **28**, 83–94 (1981).

Fang, F. F. and B. B. Golberg, "Nonlinear Behavior of Magnetoconductance in Two-Dimensional Electron Gas," *Surf. Sci.* **170**, 187–192 (1986).

Fang, P., K. K. Hung, P.-K. Ko, and C. Hu, "Characterizing a Single Hot-Electron-Induced Trap in Submicron MOSFET Using Random Telegraph Noise," *1990 Symp. VLSI Technol., Dig. Tech. Pap.*, 37–38 (1990).

Fang, P., K. K. Hung, P.-K. Ko, and C. Hu, "Hot-Electron-Induced Traps Studied through the Random Telegraph Noise," *IEEE Electron Devices Lett.* **12**, 273–275 (1991).

Fang, P., J. T. Yue, and D. Wollessen, "A Method to Project Hot Carrier Induced Punch Through Voltage Reduction for Deep Submicron LDD PMOS FETs at Room and Elevated Temperatures," *Proc. Int. Reliab. Phys. Symp.*, 131–135 (1992).

Fang, Z. H., S. Cristoloveanu, and A. Chovet, "Analysis of Hot-Carrier-Induced Aging from $1/f$ Noise in Short-Channel MOSFET's," *IEEE Electron Devices Lett.* **7**, 371–373 (1986).

Fang, Z. H., "Low-Frequency Pseudogeneration–Recombination Noise of MOSFET's Stressed by Channel Hot Electrons in Weak Inversion," *IEEE Trans. Electron Devices* **33**, 516–519 (1986).

Fantini, F., "Reliability Problems with VLSI," *Microelectron. Reliab.* **24**, 275–296 (1984).

Fazan, P., M. Dutoit, C. Martin, and M. Ilegems, "Charge Generation in Thin SiO_2 Polysilicon-Gate MOS Capacitor," *Solid-State Electron.* **30**, 829–834 (1987).

Feldbaumer, D. W. and D. K. Schroder, "MOSFET Doping Profiling," *IEEE Trans. Electron Devices* **38**, 135–140 (1991).

Ferry, D. K. and P. Das, "Microwave Hot Electron Effects in Semiconductor Quantized Inversion Layer," *Solid-State Electron.* **20**, 355–359 (1977).

Ferry, D. K. and P. Lugli, "Carrier Interactions in Semiconductor Plasma," *IEEE Int. Conf. Plasma Sci.*, 4 (1986).

Ferry, D. K. and L. A. Akers, "Hot Carriers in Semiconductors—Proceedings of the 6th International Conference," *Solid-State Electron.* **32**, 1051–1923 (1989).

Fichtner, W. and H. W. Poetzl, "MOS Modeling by Analytical Approximations. I. Subthreshold Current and Threshold Voltage," *J. Int. Electron.* **46**, 33 (1979).

Fichtner, W., "Scaling Calculation for MOSFET," *IEEE Solid State Circuits Tech. Workshop on Scaling Microlithog.*, New York (1980).

Fiegna, C., E. Sangiorgi, F. Venturi, A. Abramo, and B. Ricco, "Modeling of High Energy Electrons in *n*-MOSFETs," *Tech. Dig.—Int. Electron Devices Meet.*, 119–122 (1991).

Fischer, W., "Equivalent Circuit and Gain of MOSFET," *Solid-State Electron.* **9**, 71 (1966).

Fischetti, M. V., Gastaldi, F. Maggioni, and A. Modelli, "Hot Electrons Induced Defects at the Si/SiO_2 Interface," *11th Eur. Solid State Devices Res. Conf. & 6th SSDT—Europhys. Conf. Abstr.* **5F**, 56–57 (1981).

Fischetti, M. V., R. Gastaldi, F. Maggioni, and A. Modelli, "Slow and Fast States Induced by Hot Electrons at Si–SiO_2 Interface," *J. Appl. Phys.* **53**, 3136–3144 (1982).

Fischetti, M. V., "Importance of the Anode Field in Controlling the Generation Rate of the Donor States at the Si–SiO_2 Interface," *J. Appl. Phys.* **56**, 575–577 (1984).

Fischetti, M. V., Z. A. Weinberg, and J. A. Calise, "The Effect of Gate Metal and SiO_2 Thickness on the Generation of Donor States at the Si–SiO_2 Interface," *J. Appl. Phys.* **57**, 418–425 (1985).

Fischetti, M. V., "High Field Electron Transport in SiO_2 and Generation of Positive Charge at the Si–SiO_2 Interface," *Proc. Int. Conf. Insul. Films Semicond.*, Toulouse, France, 181–189 (1986).

Fischetti, M. V., S. E. Laux, and W. Lee, "Monte Carlo Simulation of Hot-Carrier Transport in Real Semiconductor Devices," *Solid-State Electron.* **32**, 1723–1729 (1989).

Fischetti, M. V., S. E. Laux, and D. J. DiMaria, "Physics of Hot-Electron Degradation of Si MOSFET's—Can We Understand It?" *Appl. Surf. Sci.* **39**, 578–596 (1989).

Fischetti, M. V., "Monte Carlo Simulation of Transport in Technological Significant Semiconductors of the Diamond and Zinc-Blende Structures. Part I: Homogeneous Transport," *IEEE Trans. Electron Devices* **38**, 634–649 (1991).

Fischetti, M. V. and S. E. Laux, "Monte Carlo Simulation of Transport in Technological Significant Semiconductors of the Diamond and Zinc-Blende Structures. Part II: Submicrometer MOSFET's," *IEEE Trans. Electron Devices* **38**, 650–660 (1991).

Fishbein, B. J., J. T. Watt, and J. D. Plummer, "Characterization of Cesium Oxide Implants for Use in MOS Devices Operated at 77 K," *Proc. Symp. Low Temp. Electron. High Temp. Supercond.*, Honolulu, HI, 102–107 (1988).

Fitting, H. J., "Electron and Ion Emission from MOS-Field Cathodes," *Exp. Tech. Phys.* **24**, 459–566 (1976).

Ford, J. M. and D. K. Stemple, "Effects of Arsenic Drain Profile on Submicrometer Salicide MOSFET's," *IEEE Trans. Electron Devices* **35**, 302–308 (1988).

Forslund, D. W., J. M. Kindel, K. Lee, and E. L. Lindman, "Stability at High Laser Intensity," *IEEE Int. Conf. Plasma Sci.*, 148 (1977).

Fossum, J. G., "Modeling and Design Issues for Submicron SOI CMOS Technology," *Proc. Electrochem. Soc.* **90**, 491–498 (1990).

Fossum, J. G., J.-Y. Choi, and R. Sundaresan, "SOI Design for Competitive CMOS VLSI," *IEEE Trans. Electron Devices* **37**, 724–729 (1990).

Fraenkel, A., E. Finkman, G. Bahir, A. Brandel, G. Livescu, and M. T. Asom, "Vertical Transport, Transmission Coefficients, and Dwell Time in Asymmetric Quantum Well Structures," *Superlattices Microstruct.* **12**, 557–560 (1992).

Frankl, D. R., "Some Effects of Material Parameters on the Design of Surface Space-Charge Varactors," *Solid-State Electron.* **2**, 71 (1961).

Frenkel, J., "On Pre-Breakdown Phenomena in Insulators and Electron Semiconductors," *Phys. Rev.* **54**, 647–648 (1938).

Frenkel, J., "On the Theory of Electric Breakdown of Dielectrics and Electronic Semiconductors," *Tech. Phys. USSR* **5**, 685 (1938).

Frese, K. W., Jr., and C. Chen, "Theoretical Models of Hot Carrier Effects at Metal–Semiconductor Electrodes," *J. Electrochem. Soc.* **139**, 3234–3243 (1992).

Frey, J., "Where Do Hot Electrons Come From? (MOSFETs)," *IEEE Circuits Devices Mag.* **7**, 31–34 (1991).

Friedrich, D., H. Bernt, A. Kaatz, F. Naumann, and L. Schmidt, "Comparison between Surface Channel PMOS Transistors Processed with Optical and X-Ray Lithography with Regard to X-Ray Damage," *Microelectron. Eng.* **11**, 259–262 (1990).

Frohman-Bentchkowsky, D., "The Metal-Nitride-Oxide-Silicon (MNOS) Transistor Characteristics and Applications," *Proc. IEEE* **58**, 1207–1219 (1970).

Frohman-Bentchkowsky, D., "A Fully Decoded 2048-Bit Electrically-Programmable MOS-ROM," *IEEE J. Solid-State Circuits* **SC-6**, 301–306 (1971).

Frohman-Bentchkowsky, D., "Memory Behavior in a Floating-Gate Avalanche-Injection MOS (FAMOS) Structure," *Appl. Phys. Lett.* **18**, 332–334 (1971).

Frohman-Bentchkowsky, D., "FAMOS—A New Semiconductor Charge Storage Device," *Solid-State Electron.* **17**, 517–529 (1974).

Frosch, C. J. and L. Derrick, "Surface Protection and Selective Masking during Diffusion in Silicon," *J. Electrochem. Soc.* **104**, 547–552 (1957).

Fu, K.-Y. and K.-W. Teng, "Comparison of Channel Current and Drain Avalanche Current–Induced Hot-Carrier Effects in MOSFET's," *IEEE Electron Devices Lett.* **8**, 132–134 (1987).

Fujita, S., Y. Uemoto, and A. Sasaki, "Trap Generation in Gate Oxide Layer of MOS Structures Encapsulated by Silicon Nitride," *Tech. Dig.—Int. Electron Devices Meet.*, 64–67 (1985).

Fukuda, H., M. Yasuda, T. Iwabuchi, and S. Ohno, "Novel N_2O-Oxynitridation Technology for Forming Highly Reliable EEPROM Tunnel Oxide Films," *IEEE Electron Devices Lett.* **12**, 587–589 (1991).

Fukuma, M. and M. Matsumura, "A Simple Model for Short Channel MOSFETs," *Proc. IEEE* **65**, 1212–1213 (1977).

Fukuma, M. and Uebbing R. H., "Two-Dimensional MOSFET Simulation with Energy Transport Phenomena," *Tech. Dig.—Int. Electron. Devices Meet.*, 621–624 (1984).

Fukuta, M. and Y. Hirachi, "Hemt's and New Devices for High-Speed Applications," *Proc. Euro. Microwave Conf.*, Rome, Italy, 95–101 (1987).

Fukuta, M., "Microwave Device HEMT 'FHR 01FH'," *J. Jpn. Soc. Precis. Eng.* **54**, 1445–1446 (1988).

Fuller, C. S. and J. A. Ditzenberg, "The Diffusion of Boron and Phosphorus into Silicon," *J. Appl. Phys.* **25**, 1439–1440 (1954).

Fuller, R. T., W. R. Richards, Y. Nissan-Cohen, J. C. Tsang, and P. M. Sandow, "Effects of Nitride Layers on Surface State Density and the Hot Electron Lifetime of Advanced CMOS Circuits," *Proc. Custom Int. Circuits Conf.*, Portland, OR, 337-340 (1987).

Furukawa, S., "What Are New Concept Devices?" *J. Inst. Electron. Inf. Commun. Eng. Jpn.* **75**, 366–371 (1992).

Furuyama, T., T. Ohsawa, Y. Watanabe, H. Ishiuchi, T. Tanaka, K. Ohuchi, H. Tango, K. Natori, and O. Ozawa, "Experimental 4 MB CMOS DRAM," *Tech. Dig. IEEE Int. Solid-State Circuits Conf.* **29**, 272–372, 370 (1986).

Furuyama, T., T. Ohsawa, Y. Watanabe, H. Ishiuchi, T. Watanabe, T. Tanaka, K. Natori, and O. Ozawa, "Experimental 4-MBIT CMOS DRAM," *IEEE J. Solid-State Circuits* **SC-21**, 605–611 (1986).

Gaensslen, F., V. L. Rideout, E. J. Walker, and J. J. Walker, "Very Small MOSFET's for Low-Temperature Operation," *IEEE Trans. Electron Devices* **24**, 218–229 (1977).

Gaensslen, F. H. and J. M. Aitken, "Sensitive Technique for Measuring Small MOS Gate Currents," *IEEE Electron Devices Lett.* **1**, 231–233 (1980).

Gaensslen, F. H. and D. E. Nelsen, "Solid-State Devices—Low Temperature Device Operation," *Tech. Dig.—Int. Electron Devices Meet.*, 379–408 (1987).

Gaitan, M., I. D. Mayergoyz, and C. E. Korman, "Investigation of the Threshold Voltage of MOSFET's with Position- and Potential-Dependent Interface Trap Distributions Using a Fixed-Point Interaction Method," *IEEE Trans. Electron Devices* **37**, 1031–1038 (1990).

Gallagher, R. C. and W. S. Corak, "An MOS Hall Element," *Solid-State Electron.* **9**, 571 (1966).

Galloway, K. F., C. L. Wilson, and L. C. Witte, "MOSFET Electrical Parameter Extraction from Charge-Sheet Model Fitting," *Proc. 6th Bienn. Univ./Gov./Ind. Microelectron. Symp.*, Auburn, AL, 77–81 (1985).

Galup, C. and P. Gentil, "analysis of Short *n*-Channel Most Degradation under Hot Electron Effect," *11th Eur. Solid State Devices Res. Conf. & 6th SSSDT —Durophys. Conf. Abstr.* **5F**, 47–48 (1981).

Garone, P. M., V. Venkataraman, and J. C. Stuem, "Hole Confinement in MOS-Gated Ge_xSi_{1-x}/Si Heterostructures," *IEEE Electron Devices Lett.* **12**, 230–232 (1991).

Garrett, C. B. B. and W. H. Brattain, "Physical Theory of Semiconductor Surfaces," *Phys. Rev.* **99**, 376–387 (1955).

Garrigues, M. and Y. Hellouin, "Emission Probability of Hot Electrons for Highly Doped Silicon-on-Sapphire IGFET," *IEEE Trans. Electron Devices* **28**, 928–936 (1981).

Garrigues, M. and A. Pavlin, "New Technique for Measuring Small MOS Gate Currents," *Electron. Lett.* **21**, 16–17 (1985).

Garrigues, M. and B. Balland, "Hot Carrier Injection into SiO_2," In *Instabilities in Si Devices—Si Passivation and Related Instability*, G. Barbottin and A. Vapaille, eds., North-Holland Publ., Amsterdam, Vol. 1, 441–502 (1986).

Garrigues, M. and P. Rojo, "Two-Dimensional Computer Simulation of Hot Carrier Degradation in *n*-MOSFETs," *J. Phys., Colloq. (Orsay, Fr.)* **49**, 673–676 (1988).

Gasgnier, M. and M.-O. Ruault, "Microstructural Observation of $RBa_2Cu_3O_7$ (R = Gd, Y, and Tm) Superconductor Thin Crystals between 15 and 100 K," *J. Appl. Phys.* **62**, 4935–4938 (1987).

Gasner, J., A. Ito, and R. Nowak, "An Investigation of Device Instabilities Arising from the Encapsulation Material and Composition," *Proc. Int. IEEE VLSI Multilevel Interconnect. Conf., 7th* Santa Clara, CA, 354–356 (1990).

Gdula, R. A., "Effects of Processing on Hot Electron Trapping in SiO_2," *J. Electrochem. Soc.* **123**, 42–47 (1976).

Gdula, R. A. and P. C. Li, "Electron Trapping in CVD PSG Films," *Proc. Electrochem. Soc. Fall Meet.*, Las Vegas, NV, 634–636 (1976).

Ge, D. Y., N. Hwang, and L. Forbes, "Composite *n*-MOSFET for Submicrometre Circuits," *Electron. Lett.* **29**, 623–625 (1993).

Geis, M. W., D. C. Flanders, D. A. Antoniadis, and H. I. Smith, "Crystalline Silicon on Insulators by Graphoepitaxy," *Tech. Dig.—Int. Electron Devices Meet.*, 210 (1979).

Geis, M. W., J. A. Gregory, and B. B. Pate, "Capacitance–Voltage Measurements on Metal–SiO_2–Diamond Structures Fabricated with (100)- and (111)-Oriented Substructures," *IEEE Trans. Electron Devices* **38**, 619–626 (1991).

Gelmont, B. L., "Hot Electrons in Narrow Gap Semiconductor," *Proc. Int. Summer Sch. Narrow Gap Semicond.—Phys. Appl.*, Nimes, France, 371–387 (1980).

Gesch, H., J. P. Leburton, and G. E. Dorda, "Localization of Hot Electron Injection into the Oxide of MOSFETs," *11th Eur. Solid State Devices Res. Conf. & 6th SSSDT—Europhys. Conf. Abstr.*, 45–46 (1981).

Gesch, H., J. P. Leburton, and G. E. Dorda, "Generation of Interface States by Hot-Hole Injection in MOSFETs," *IEEE Trans. Electron Devices* **29**, 913–918 (1982).

Gharabagi, R. and M. A. El-Nokali, "A Charge-Based Model for Short-Channel MOS Transistor Capacitances," *IEEE Trans. Electron Devices* **37**, 1064–1073 (1990).

Ghibaudo, G. and B. Cabon, "Influence of Hot-Electron-Induced Aging on the Dynamic Conductance of Short-Channel MOSFETs," *Solid-State Electron.* **30**, 1049–1052 (1987).

Ghibaudo, G., "A Simple Model for the Drain Saturation Voltage Dependence with Gate Voltage for Short-Channel MOSFETs," *Phys. Status Solidi A* **99**, K149–K153 (1987).

Gibbons, J. F., "Ion Implantation in Semiconductor. Part I. Range Distribution Theory and Experiment," *Proc. IEEE* **56**, 295–319 (1968).

Gibbons, J. J. and K. F. Lee, "One-Gate-Wide CMOS Inverter on Laser-Recrystallized Polysilicon," *IEEE Electron Devices Lett.* **1**, 117–118 (1980).

Giebel, T. and K. Goser, "Hot Carrier Degradation of *n*-Channel MOSFET's Characterized by a Gated-Diode Measurement Technique," *IEEE Electron Devices Lett.* **10**, 76–78 (1989).

Gildenblat, G. S., C.-L. Huang, and S. A. Grot, "Temperature Dependence of Electron Trapping in Metal-Oxide-Semiconductor Devices as a Function of the Injection Mode," *J. Appl. Phys.* **64**, 2150–2152 (1988).

Glaccum, A. E., I. P. Thomas, I. Hogg, and T. Hull, "Effect of 2-D Diffusion Profile on the Device Characteristics Predicted by MINIMOS," *IEE Colloq. Dig.*, 7.1–7.3 (1985).

Gnudi, A., D. Ventura, G. Baccarani, and F. Odeh, "Two-Dimensional MOSFET Simulation by Means of a Multidimensional Spherical Expansion of the Boltzman Transport Equation," *Solid-State Electron.* **36**, 575–581 (1993).

Goetzberger, A., "Ideal MOS Curves for Silicon," *Bell Syst. Tech. J.* **45**, 1097–1122 (1966).

Goetzberger, A., "Behavior of MOS Inversion Layers at Low Temperature," *IEEE Trans. Electron Devices* **14**, 787–792 (1967).

Goetzberger, A. and E. H. Nicollian, "Temperature Dependence of Inversion Layer Frequency Response in Silicon," *Bell Syst. Tech. J.* **46**, 513–522 (1967).

Goldsman, N. and J. Frey, "Prediction of Hot-Electron-Induced Gate Currents," *IEEE Trans. Electron Devices* **33**, 1861 (1986).

Goldsman, N. and J. Frey, "Electron Energy Distribution for Calculation of Gate Leakage Current in MOSFETs," *Solid-State Electron.* **31**, 1089–1092 (1988).

Goldsman, N., L. Hendrickson, and J. Frey, "Reconciliation of a Hot-Electron Distribution Function with the Lucky Electron–Exponential Model in Silicon," *IEEE Electron Devices Lett.* **11**, 472–474 (1990).

Gomi, T., H. Miwa, H. Sasaki, H. Yamamoto, M. Nakamura, and A. Kayanuma, "A Sub-30psec Si Bipolar LSI Technology," *Tech. Dig.—Int. Electron Devices Meet.*, 744–747 (1988).

Goodman, J. W., F. I. Leonberger, S. Y. Kung, and R. A. Athale, "Optical Interconnections for VLSI Systems," *Proc. IEEE* **72**, 850–866 (1984).

Gornik, E. and D. C. Tsui, "Far Infrared Emission from Hot Electrons in Si-Inversion Layers," *Solid-State Electron.*, 139–142 (1978).

Gosney, W. M., "Subthreshold Drain Leakage Current in MOSFET," *IEEE Trans. Electron Devices* **19**, 213–219 (1972).

Gosney, W. M., "DIFMOS—A Floating-Gate Electrically Erasable Nonvolatile Semiconductor Memory Technology," *IEEE Trans. Electron Devices* **24**, 495–599 (1977).

Gourrier, S., P. Friedel, P. Dimitriou, and J. B. Theeten, "Growth of Dielectric Films on Semiconductors and Metals Using a Multipole Plasma," *Thin Solid Films* **84**, 379–388 (1981).

Graffeuil, J., J.-F. Sautereau, G. Blasquez, and P. Rossel, "Noise Temperature in GAAS Epi-Layer for FETs," *IEEE Trans. Electron Devices* **25**, 596–599 (1978).

Gray, P. V. and D. M. Brown, "Density of SiO_2–Si Interface States," *Appl. Phys. Lett.* **8**, 31–33 (1966).

Green, C. W., "p-MOS Dynamic RAM Reliability—A Case Study," *Proc. Int. Reliab. Phys. Symp.*, 213–219 (1979).

Greenwood, C. J., "Construction and Characterization of an ASIC EEPROM," *Microelectron. Conf., 1991, Inst. Eng. Aust.*, 58–63 (1991).

Grinberg, A. A., A. Kastalsky, and S. Luryi, "Theory of Hot-Electron Injection in CHINT/NERFET Device," *IEEE Trans. Electron Devices* **34**, 409–419 (1987).

Grinolds, H. R., M. Kinugawa, and M. Kakumu, "Reliability and Performance of Submicron LDD *n*-MOSFET's with Buried-As n^- Impurity Profiles," *Tech. Dig. —Int. Electron Devices Meet.*, 246–249 (1985).

Groeseneken, G., H. E. Maes, N. Beltran, and R. F. de Keersmaecker, "A Reliable Approach to Charge-Pumping Measurements in MOS Transistors," *IEEE Trans. Electron Devices* **31**, 42–53 (1984).

Gross, B. J., K. S. Krisch, and C. G. Sodini, "An Optimized 850° C Low-Pressure-Furnace Reoxidized Nitrided Oxide (ROXNOX) Process," *IEEE Trans. Electron Devices* **38**, 2036–2041 (1991).

Grosvalet, J. and C. Jund, "Influence of Illumination on MIS Capacitance in the Strong Inversion Region," *IEEE Trans. Electron Devices* **14**, 777–780 (1967).

Grot, S. A., C.-L. Huang, and G. S. Gildenblat, "Temperature Dependence of Charge Trapping and Dielectric Breakdown in MOS Devices," *Proc. Symp. Low Temp. Electron. High Temp. Supercond.*, Honolulu, HI, 142–150 (1988).

Grove, A. S., E. H. Snow, B. E. Deal, and C. T. Sah, "Simple Physical Model for the Space-Charge Capacitance of MOS Structures," *J. Appl. Phys.* **35**, 2458–2460 (1964).

Grove, A. S., E. H. Snow, B. E. Deal, and C. T. Sah, "Investigation of Thermally Oxidized Silicon Surfaces Using MOS Structures," *Solid-State Electron.* **8**, 145 (1965).

Grove, A. S. and D. J. Fitzgerald, "Surface Effects on *p–n* Junctions: Characteristics of Surface Space-Charge Regions under Nonequilibrium Conditions," *Solid-State Electron.* **9**, 783 (1966).

Grove, A. S., *Physics and Technology of Semiconductor Devices*, Wiley, New York, 311–1315 (1967).

Guebels, P. P. and F. Vand De Wiele, "A Small Geometry MOSFET Model for CAD Application," *Solid-State Electron.* **26**, 267–273 (1983).

Guenther, B. D. and P. W. Kruse, "Submillimeter Wave Detector Workshop," *Int. J. Infrared Millimeter Waves* **7**, 1091–1109 (1986).

Gupta, A., K. Raol, S. Pradhan, and K. P. Roenker, "Comparison of the Inversion Layer Mobilities on *n*-Channel MOSFETs with Oxide and Nitrided Oxide Gate Dielectrics," *Proc. 7th Bienn. Univ./Gov./Ind. Microelectron. Symp.*, 39–43 (1987).

Gupta, A., S. Pradhan, and K. P. Roenker, "Hot-Carrier-Induced Degradation in Nitrided Oxide MOSFETs," *IEEE Trans. Electron Devices* **36**, 577–588 (1989).

Gurevich, Yu. G., G. N. Logvinov, and V. B. Yurchenko, "Inverted Energy Distribution of Carriers in a Current Flowing along a Submicron Semiconductor Layer," *Fiz. Tverd. Tela* **34**, 1666–1670 (1992).

Guterman, D. C., I. H. Rimawi, R. D. Halvorson, D. J. McElroy, and W. W. Chan, "Electrically Alterable Hot-Electron Injection Floating Gate MOS Memory Cell with Series Enhancement," *Tech. Dig.—Int. Electron Devices Meet.*, 340–343 (1978).

Hadara, H. and S. Cristoloveanu, "Profiling of Stress Induced Interface States in Short Channel MOSFETs Using a Composite Charge Pumping Technique," *Solid-State Electron.* **29**, 767–772 (1986).

Hadara, H. and S. Cristoloveanu, "Parameter Extraction and Two-Dimensional Modeling of Locally Damaged MOSFET," *IEE Colloq. 'Hot Carrier Degradation Short Channel MOS Dig.*, 3.1–3.4 (1987).

Hadara, H. and S. Cristoloveanu, "Two-Dimensional Modeling of Locally Damaged Short-Channel MOSFET's Operating in the Linear Region," *IEEE Trans. Electron Devices* **34**, 378–385 (1987).

Haddad, H., L. Forbes, P. Burke, and W. Richling, "Carbon Doping Effects on Hot Electron Trapping," *Proc. Int. Reliab. Phys. Symp.*, 288–289 (1990).

Haddad, S., C. Chang, A. Wang, J. Bustillo, J. Lien, T. Montalvo, and M. Van Buskirk, "An Investigation of Erase-Mode Dependent Hole Trapping in Flash EEPROM Memory Cell," *IEEE Electron Devices Lett.* **11**, 514–516 (1990).

Haddara, H. S. and S. Cristoloveanu, "Parameter Extraction Method for Inhomogeneous MOSFETs Locally Damaged by Hot Carrier Injection," *Solid-State Electron.* **31**, 1573–1581 (1988).

Haddara, H. S., S. Bassiouni, and H. F. Ragaie, "Analysis and Simulation of the Frequency Response of MOSFETs with Uniform and/or Local Oxide Degradation," *Solid-State Electron.* **36**, 741–748 (1993).

Haecker, W., "Infrared Radiation from Breakdown Plasmas in Si, GaSb, and Ge: Evidence for Direct Free Hole Radiation," *Phys. Status Solidi A* **25**, 301–310 (1974).

Haensch, W., W. Werner, and S. Selberherr, "Hot Carrier Analysis Utilizing MINIMOS 3.0," *Tech. Dig. VLSI Tech. Symp.*, 63–64 (1986).

Haines, M. G., "Magnetic-Field Generation in Laser Fusion and Hot-Electron Transport," *Can. J. Phys.* **64**, 912–919 (1986).

Haken, R. A., "Application of the Self-Aligned Titanium Silicide Process to Very Large-Scale Integrated *n*-Metal-Oxide-Semiconductor and Complementary Metal-Oxide-Semiconductor Technologies," *J. Vac. Sci. Technol.* **3**, 1657–1663 (1985).

Hall, R. N. and W. C. Dunlap, "P–N Junctions Prepared by Impurity Diffusion," *Phys. Rev.* **80**, 467–468 (1950).

Hamada, A., Y. Igura, R. Izawa, and E. Takeda, "N^- Source/Drain Compensation Effects in Submicrometer LDD MOS Devices," *IEEE Electron Devices Lett.* **8**, 398–400 (1987).

Hamada, A., T. Toyabe, and E. Takeda, "Development of Hot-Carrier Simulator H^2-CAST and Its Application to LDD," *Electron. Commun. Jpn.* **71**, 75–82 (1988).

Hamada, A. and E. Takeda, *Ext. Abstr. Jpn. Soc. Appl. Phys.*, 630 (1988).

Hamada, A., T. Furusawa, and E. Takeda, "A New Aspect on Mechanical Stress Effect in Scaled MOS Devices," *Tech. Dig. VLSI Tech. Symp.*, 113–114 (1990).

Hamada, A., T. Furusawa, N. Saito, and E. Takeda, "A New Aspect of Mechanical Stress Effects in Scaled MOS Devices," *IEEE Trans. Electron Devices* **38**, 895–900 (1991).

Hamada, A. and E. Takeda, "Structure Dependence of Hot-Carrier Degraded Region in Deep Submicron MOS Devices," *Tech. Dig. VLSI Tech. Symp.* 21–22 (1991).

Hamada, A. and E. Takeda, "AC Hot-Carrier Effect under Mechanical Stress," *Tech. Dig. LSI Tech. Symp.*, 98–99 (1992).

Hamaguchi, C., "Physics of Microdevices," *Solid State Phys.* **19**, 511–518 (1984).

Hamaguchi, C., "Physics of Microdevices I," *Solid State Phys.* **19**, 459–466 (1984).

Hamaguchi, C., T. Mori, T. Wada, K. Terashima, K. Taniguchi, K. Miyatsuji, and H. Hihara, "Physics of Nanometer Structure Devices," *Microelectron. Eng.* (*Jpn.*) **2**, 34–43 (1984).

Hamaguchi, C., "Hot Electron Transport in Very Short Semiconductors," *Phys. B + C (Amsterdam)* **134**, 87–96 (1985).

Hamaguchi, C., A. Hasegawa, and M. Shirahata, "Hot Electron Phenomena in Very Small Semiconductor Devices," *Oyo Butsuri* **55**, 772–781 (1986).

Hamamoto, T., Y. Oowaki, K. Hieda, and K. Ohuchi, "Asymmetry of the Substrate Current Characteristics Enhanced by the Gate Bird's Beak," *Tech. Dig. VLSI Tech. Symp.* 67–68 (1986).

Hamel, J. S., D. J. Rouston, and C. R. Selvakumar, "Trade-off Between Emitter Resistance and Current Gain in Polysilicon Emitter Bipolar Transistors with Intentionally Grown Interfacial Oxide Layers," *IEEE Electron Devices Lett.* **13**, 332–334 (1992).

Hamilton, T., P. Harrod, and S. Dehgan, "MC68030 Process and Circuit Technology," *Proc. IEEE Int. Conf. Comput. Des.: VLSI Comput. Processors*, 590–593 (1987).

Han, D., J. I. Pankove, Y. S. Tsuo, C. H. Qui, and Y. Xu, "Improvement of a-$Si_{1-x}C_x$:H/a-Si:Hp/i Interface by Hydrogen-Plasma Flushing Studied by Photoluminescence," *J. Mater. Res.* **6**, 1900–1904 (1991).

Han, Y. P., J. P. Mize, B. T. Moore, J. Pinto, and R. Worley, "Practical Limitations of Gate-Oxide Thickness Minimization in the MOSFET," *IEEE Trans. Electron Devices* **ED-32**, 559 (1985).

Hanada, K., H. Tanaka, M. Iida, S. Ide, T. Minami, M. Nakamura, T. Maekawa, Y. Terumichi, S. Tanaka, M. Yamada, J. Manickam, and R. B. White, "Sawtooth Stabilization by Localized Electron Cyclotron Heating in a Tokamak Plasma," *Phys. Rev. Lett.* **66**, 1974–1977 (1991).

Hanafi, H. I., "Device Advantages of DI-LDD/LDD MOSFET over DD MOSFET," *IEEE Circuits Devices Mag.* **1**, 13–15 (1985).

Hänsch, W. and C. Werner, "A Hot Carrier Analysis Utilizing MINIMOS 3.0," *Tech. Dig. VLSI Tech. Symp.*, 63–64 (1986).

Hänsch, W. and S. Selberherr, *IEEE Trans. Electron Devices*, **34**, 1074–1078 (1987).

Hänsch, W. and C. Werner, "The Hot-Electron Problem in Submicron MOSFET," *J. Phys., Colloq. (Orsay, Fr.)* **49**, 597–606 (1988).

Hänsch, W. and W. Weber, "Effect of Transients on Hot Carriers," *IEEE Electron Devices Lett.* **10**, 252–254 (1989).

Hänsch, W., A. V. Schwerin, and F. Hofmann, "A New Self-Consistent Modeling Approach to Investigating MOSFET Degradation," *IEEE Electron Devices Lett.* **11**, 362–364 (1990).

Hänsch, W., C. Mazure, A. Lill, M. K. Orlowski, "Hot Carrier Hardness Analysis of Submicrometer LDD Devices," *IEEE Trans. Electron Devices* **38**, 512–517 (1991).

Hao, C., M. Charef, J. Zimmermann, R. Fauquembergue, and E. Constant, "A Semi-Classical Model for Simulating Inversion Carrier Transport in Si MOS Devices," *Phys. Status Solidi A* **81**, 569–577 (1984).

Hao, C., M. Charef, J. Zimmermann, R. Fauquembergue, and E. Constant, "Monte Carlo Simulation of Electron Dynamics in MOS Inversion Channels," *Proc. Phys. Submicron Struct.*, 107–113 (1984).

Harada, Y., H. Hayashi, S. Masuda, T. Fukuda, N. Sato, S. Kato, K. Kobayashi, H. Kuroda, and H. Ozaki, "Characterization of Organic Ultra-Thin Films by Penning Ionization Electron Spectroscopy during Layer-by-Layer Preparation: TFTCNQ, TMTSF and Their Charge Transfer Complex," *Surf. Sci.* **242**, 95–101 (1991).

Harame, D. L., E. Ganin, J. H. Comfort, J. Y.-C. Sun, A. Acovic, S. A. Cohen, P. A. Ronsheim, S. Verdonckt-Vandebroek, P. J. Restle, S. Ratanaphanyarat, M. J. Saccamango, J. B. Johnson, and S. A. Furkay, "Low Thermal Budget Antimony/Phosphorus NMOS Technology for CMOS," *Int. Electron Devices Meet., 1991, Tech. Dig.*, 645–648 (1991).

Harari, E., "Dielectric Breakdown in Electrically Stressed Thin Films of Thermal SiO_2," *J. Appl. Phys.* **49**, 2478 (1978).

Harding, W. E., "Semiconductor Manufacturing in IBM, 1957 to the Present: A Perspective," *IBM J. Res. Dev.* **25**, 647–658 (1981).

Harrell, W. R. and J. Frey, "Apparent Positive Carrier Conduction in SiO_2 Films and Implications for MOSFET Scaling," *IEEE Trans. Electron Devices* **39**, 2671–2672 (1992).

Harter, J., W. Pribyl, M. Bahring, A. Lill, H. Mattes, W. Muller, L. Risch, D. Sommer, R. Strunz, W. Weber, and K. Hoffmann, "60 ns Hot Electron Resistant 4 M DRAM with Trench Cell," *Tech. Dig. IEEE Int. Solid-State Circuits Conf.*, 244–245 (1988).

Haruta, T., Y. Ohji, Y. Nishioka, I. Yoshida, K. Mukai, and T. Sufano, "Improvement of Hardness of MOS Capacitors to Electron-Beam Irradiation and Hot-Electron Injection by Ultradry Oxidation of Silicon," *IEEE Electron Device Lett.* **10**, 27–29 (1989).

Hasegawa, H. and H. Ohno, "Unified Disorder Induced Gap State Model for Insulator–Semiconductor and Metal–Semiconductor Interfaces," *J. Vac. Sci. Technol., B* **4**, 1130–1138 (1986).

Hassein-Bey, A. and S. Cristoloveanu, "Modeling and Characterization of Submicron *p*-Channel MOSFETs," *Proc. 7th Bienn. Eur. Conf.*, 251–255 (1991).

Hassein-Bey, A., and S. Cristoloveanu, "Simulation of Interface Coupling Effects in Ultra-Thin Silicon on Insulator MOSFET's," *Int. Conf. Number. Analy. Semicond. Dev. Integrat. Circuits, 8th*, Vienna, Austria (1992).

Hauer, A., R. Goldman, R. Kristal, M. A. Yates, M. Mueller, F. Begay, D. Van Hulsteyn, K. Mitchell, J. Kephart, H. Oona, E. Stover, J. Brackbill, and D. Forslund, "Suprathermal Electron Generation Transport and Deposition in CO_2 Laser Irradiated Targets," *Proc. Laser Interact. Relat. Plasma Phenom.*, Monterey, CA, 479–492 (1984).

Hayashi, H. and R. Dang, "Effects of Lattice Temperature in MOSFET Analysis under Non-Isothermal Conditions," *COMPEL-Int. J. Comput. Math. Electr. Electron. Eng.* **11**, 489–503 (1992).

Hayashi, T., M. Ohno, A. Uchiyama, H. Fukuda, T. Iwabuchi, and S. Ohno, "Effectiveness of N_2O-Nitrided Gate Oxide for High-Performance CMOSFETs," *IEEE Trans. Electron Devices* **38**, 2711 (1991).

Hayashi, T. and A. Uchiyama, "The Optimization of P^--layer Dopant Concentration in Sub-Micrometer Lightly Doped Drain PMOSFETs," *Oki Tech. Rev.* **57**, 65–70 (1991).

Hayashi, T., H. Fukuda, A. Uchiyama, and T. Iwabuchi, "A New Prediction Method for Hot Carrier Degradation of Submicron PMOSFET with Charge Pumping Technique," *Ext. Abstr., 1992 Int. Conf. Solid State Devices Mater.*, 509–511 (1992).

Hayden, J. D., F. X. Baker, S. A. Ernst, R. E. Jones, J. Klein, M. Lein, T. McNelly, T. C. Mele, H. Mendez, B.-Y. Nguyen, L. C. Parrillo, W. Paulson, J. R. Pfiester, F. Pintchovski, T.-C. See, R. D. Sivan, and B. M. Somero, "A High-Performance Sub-Half Micron CMOS Technology for Fast SRAMs," *Tech. Dig.—Int. Electron Devices Meet.*, 417–420 (1989).

Hayden, J. D., F. K. Baker, S. A. Ernst, R. E. Jones, J. Klein, M. Lien, T. F. McNelly, T. C. Mele, H. Mendez, B.-Y. Nguyen, L. C. Parrillo, W. Paulson, J. R. Pfiester, F. Pintchovski, Y.-C. See, R. D. Sivan, B. M. Somero, and E. O. Travis, "A High-Performance Half-Micrometer Generation CMOS Technology for Fast SRAMs," *IEEE Trans. Electron Devices* **38**, 876–886 (1991).

Heiblum, M., "Tunneling Emitter for Hot Electron Transistor," *IBM Tech. Disclosure Bull.* **24**, 4504–4505 (1982).

Heiblum, M., I. M. Anderson, and C. M. Knoedler, "DC Performance of Ballistic Tunnleing Hot-Electron Transfer Amplifiers," *Appl. Phys. Lett.* **49**, 207–209 (1986).

Hellman, E. S., J. S. Harris, C. B. Hanna, and R. B. Laughlin, "One Dimensional Polaron Effects and Current Inhomogeneities in Sequential Phonon Emission," *Physica B + C (Amsterdam)* **134**, 41–46 (1985).

Hellouin, Y., F. Chehade, and M. Garrigues, "Hot-Hole Injection Probabilities into the Insulator of Metal-Insulator-Silicon Devices," *J. Appl. Phys.* **61**, 5342–5345 (1987).

Hellouin, Y. "Hot Carrier Injection into the Gate Insulator of MOS Transistors and Practical Consequences," *Onde Electri.* **72**, 57–62 (1992).

Hemink, G. J., R. C. M. Wijiburg, R. Cuppens, and J. Middelhoek, "High Efficiency Hot Electron Injection for EEPROM Applications Using a Buried Injector," *Ext. Abstr., Solid State Devices Mater.*, 133–136 (1989).

Hemink, G. J., R. C. M. Wijburg, P. B. M. Wolbert, and H. Wallinga, "Modeling of VIPMOS Hot Electron Gate Currents," *Microelectron. Eng.* **15**, 65–68 (1991).

Hendriks, E. A., R. J. J. Zijlstra, and J. Wolter, "Current Noise in *n*-Channel Si-MOSFETs at 4.2 K," *Proc. Int. Conf. 'Noise Phys. Syst.' '1/f Noise'*, Rome, Italy, 175–179 (1986).

Hendriks, E. A. and R. J. J. Zijlstra, "Field-Induced Generation–Recombination Noise in (100) *n*-Channel Si-MOSFETs at $T = 4.2$ K. I. Theory," *Physica B + C (Amsterdam)* **147**, 282–290 (1988).

Hendriks, E. A., R. J. J. Zijlstra, and J. Middlehoek, "Current Noise in (111) *n*-Channel Si-MOSFETs at $T = 4.2$ K," *Physica B + C (Amsterdam)*, 297–304 (1988).

Hendriks, E. A. and R. J. J. Zijlstra, "Diffusion and Inter-Valley Noise in (100) *n*-Channel Si-MOSFETs from $T = 4.2$ to 295 K," *Solid-State Electron.* **31**, 171–180 (1988).

Hendrikson, T. E., "A Simplified Model for Subpinchoff Condition in Depletion-Mode IGFET's," *IEEE Trans. Electron Devices* **25**, 435–441 (1978).

Henning, A. K., N. N. Chan, and J. D. Plummer, "Substrate Current in *n*-Channel and *p*-Channel MOSFETs between 77 K and 300 K: Characterization and Simulation," *Tech. Dig.—Int. Electron Devices Meet.*, 573–576 (1985).

Henning, A. K., J. D. Plummer, and N. N. Chan, "Characterization and Two-Dimensional Simulation on Impact Ionization Current in MOSFET's between 77 and 300 K," *IEEE Trans. Electron Devices* **32**, 2543 (1985).

Henning, A. K., N. N. Chan, J. T. Watt, and J. D. Plummer, "Substrate Current at Cryogenic Temperatures: Measurements and a Two-Dimensional Model for CMOS Technology," *IEEE Trans. Electron Devices* **34**, 64–74 (1987).

Henning, A. K., and J. D. Plummer, "Thermionic Emission Probability for Semiconductor–Insulator Interfaces," *IEEE Trans. Electron Devices* **34**, 2211–2212 (1987).

Henning, A. K., "Low Temperature Gate Current and 'Channel' Hot Carriers in MOS Transistors," *Proc. Workshop Low Temp. Semicond. Electron.*, Burlington, VT, 109–113 (1989).

Her, T.-D., P. S. Liu, D. S. Quon, G. P. Li, R. Kjar, and J. White, "Parasitic Bipolar Transistor Induced Latch and Degradation in SOI MOSFETs," *1991 IEEE Int. SOI Conf.*, Vail Valley, CO (1991).

Heremans, P., H. E. Maes, and N. Saks, "Evaluation of Hot Carrier Degradation of *n*-Channel MOSFETs with the Charge Pumping Technique," *IEEE Electron Device Lett.* **7**, 428–430 (1986).

Heremans, P., G. Groeseneken, and H. E. Maes, "Comparative Study of the Hot Carrier Degradation Phenomena in *n*-MOS and *p*-MOS Short Channel Transistors by Means of the Charge Pumping Technique," *IEE Colloq.*, 2.1–2.5 (1987).

Heremans, P., Y.-C. Sun, G. Groeseneken, and H. E. Maes, "Evaluation of Channel Hot Carrier Effects in n-MOS Transistors at 77 K with the Charge Pumping Technique," *Appl. Surf. Sci.* **30**, 313–318 (1987).

Heremans, P., R. Bellens, G. Groeseneken, and H. E. Maes, "Consistent Model for the Hot-Carrier Degradation in n-Channel and p-Channel MOSFETs," *IEEE Trans. Electron Devices* **35**, 2194–2208 (1988).

Heremans, P., J. Witters, G. Groeseneken, and H. E. Maes, "Analysis of the Charge Pumping Technique and Its Application for the Evaluation of MOSFET Degradation," *IEEE Trans. Electron Devices* **36**, 1318–1335 (1989).

Heremans, P., G. V. D. Bosch, R. Bellens, G. Groeseneken, and H. E. Maes, "The Dependence of Channel Hot-Carrier Degradation on Temperature in the Range 77 K to 300 K," *Proc. Eur. Solid State Devices Res. Conf., 17th*, 727–731 (1989).

Heremans, P., G. V. D. Bosch, R. Bellens, G. Groeseneken, and H. E. Maes, "Understanding of the Temperature Dependence of Channel Hot-Carrier Degradation in the Range 77K to 300K," *Tech. Dig.—Int. Electron Devices Meet.*, 67–70 (1989).

Heremans, P., G. V. D. Bosch, R. Bellens, G. Groeseneken, and H. E. Maes, "Temperature Dependence of the Channel Hot-Carrier Degradation of n-Channel MOSFET's," *IEEE Trans. Electron Devices* **37**, 980–993 (1990).

Herzog, M. and F. Koch, "Hot-Carrier Light Emission from Silicon Metal-Oxide-Semiconductor Devices," *Appl. Phys. Lett.* **53**, 2620–2622 (1988).

Herzog, M., M. Schels, F. Koch, C. Moglestue, and J. Rozenzwerg, "Electromagnetic Radiation from Hot Carriers in FET-Devices," *Solid-State Electron.* **32**, 1765–1769 (1989).

Hess, K., A. Neugroschel, C. C. Shiue, and C. T. Sah, "Non-Ohmic Electron Conduction in Silicon Surface Inversion Layers at Low Temperatures," *J. Appl. Phys.* **46**, 1721–1727 (1975).

Hess, K., "Hot Electron Transport in MOS Structures," *Surf. Sci.* **73**, 135 (1977).

Hess, K., "Review of Experimental Aspects of Hot Electron Transport in MOS Structures," *Solid-State Electron.* **21**, 123–132 (1978).

Hess, K. and H. Shichijo, "Charge-Handling Capacity of Burried-Channel Structures under Hot-Electron Conditions," *IEEE Trans. Electron Devices* **27**, 503–504 (1980).

Hess, K., "Phenomenological Physics of Hot Carriers in Semiconductors," *Phys. Nonlinear Transp. Semicond., Proc. NATO Adv. Study Inst. Phys. Nonlinear Electron Transp.*, 1–42 (1980).

Hess, K. and N. Holonyak, "Hot-Electron and Phonon Effects in Layered Semiconductor Structures and Heterostructures," *Comments Solid State Phys.* **10**, 67–84 (1981).

Hesto, P., J. C. Vaissiere, D. Gasquet, R. Catagne, and J. P. Nougier, "A Suggestion of Noise Experiment for Showing Ballistic Transport (FETs)," *J. Phys., Colloq. (Orsay, Fr.)* **42**, 235–241 (1981).

Hewett, N. P., P. A. Russell, L. J. Challis, F. F. Quail, V. W. Rampton, A. J. Kent, and A. G. Every, "Hot Electron Effects and Phonon Emission from a Two-Dimensional Electron Gas (2 DEG)," *Semicond. Sci. Technol.* **4**, 955–957 (1989).

Heyns, M. and R. F. De Keersmaecker, "Characteristics of Slow Interface Traps Created during Avalanche Injection of Electrons in MOS Devices," *11th Eur. Solid State Devices Res. Conf. & 6th SSDT—Europhys. Conf. Abstr.*, 58–59 (1981).

Higman, J. M., I. C. Kizilyalli, and K. Hess, "Nonlocality of the Electron Ionization Coefficient in *n*-MOSFET's: An Analytical Approach," *IEEE Electron Device Lett.* **9**, 399–401 (1988).

Higman, J. M., K. Hess, C. G. Hwang, and R. W. Dutton, "Coupled Monte Carlo–Drift Diffusion Analysis of Hot-Electron Effects in MOSFET's," *IEEE Trans. Electron Devices* **36**, 930–937 (1989).

Hillmer, H., T. Kuhn, B. Laurich, A. Forchel, and G. Mahler, "Experimental and Theoretical Investigations of Free Exciton Transport in Si," *Phys. Scri.* **35**, 520–523 (1987).

Hilsum, C., "Historical Background of Hot Electron Physics—(A Look over the Shoulder)," *Solid-State Electron.* **21**, 5–8 (1978).

Hirakawa, K. and H. Sakaki, "Hot-Electron Transport in Selectively Doped *n*-type AlGaAs/GaAs Heterojunctions," *J. Appl. Phys.* **63**, 803–808 (1988).

Hirayama, M., H. Miyoshi, N. Tsubouchi, and H. Abe, "High Pressure Oxidation for Thin Gate Insulator Process," *Tech. Dig. VLSI Tech. Symp.* 74–75 (1981).

Hiroki, A. and S. Odanaka, "Physical Modeling of MOSFET Degradation Induced by High-Energy Hot-Carriers," *Ext. Abstr., Solid State Devices Mater.*, 221–224 (1988).

Hiroki, A., S. Odanaka, K. Ohe, and H. Esaki, "Mobility Model for Submicrometer MOSFET Simulations Including Hot-Carrier-Induced Device Degradation," *IEEE Trans. Electron Devices* **35**, 1487–1493 (1988).

Hiroki, A., S. Odanaka, and T. Morii, "A Spill Over Effect of Avalanche Generated Electrons in Buried *p*MOSFETs," *1990 Symp. VLSI Technol., Dig. Tech. Pap.*, 71–72 (1990).

Hiruta, Y., K. Maeguchi, and K. Kanzaki, "Impact of Hot Electron Trapping on Half Micron *p*-MOSFET's with p^+ Poly Si Gate," *Tech. Dig.—Int. Electron Devices Meet.*, 718–721 (1986).

Hiruta, Y., H. Oyamasu, H. S. Momose, H. Iwai, and K. Maeguchi, "Gate Oxide Thickness Dependence of Hot-Carrier Induced Degradation in *p*-MOSFETs," *Proc. Eur. Solid State Devices Res. Conf., 19th*, 732–735 (1989).

Ho, F. Y. and M. J. Kim, "Computer Model for SAMOS Structures," *Proc. 6th Bienn. Univ./Gov./Ind. Microelectron. Symp.*, Auburn, AL, 86–88 (1985).

Ho, J., R. O. Grondin, P. A. Blakey, and W. Porod, "High Frequency Application of Hot Electrons in Superlattices," *Physica B + C (Amsterdam)* **134**, 502–505 (1985).

Ho, S. K. and I. Beinglass, "Formation of Spacers by A Two Steps Etching of Polysilicon," *Ext. Abstr., Electrochem. Soc.* **84-2**, 582–583 (1984).

Hodges, D., *Semiconductor Memory*, IEEE Press, New York (1972).

Hoeberechts, A. M. E. and G. G. P. Van Gorkom, "Design, Technology, and Behavior of a Silicon Avalanche Cathode," *J. Vac. Sci. Technol.* **4**, 105–107 (1986).

Hoefflinger, B., H. Sibbert, G. Zimmer, E. Kubalek, and E. Menzel, "Model and Performance of Hot-Electron MOS Transistors for High-Speed, Low Power LSI," *Tech. Dig.—Int. Electron Devices Meet.*, 463–467 (1978).

Hoefflinger, B., H. Sibbert, and G. Zimmer, "Model and Performance of Hot-Electron MOS Transistors for VLSI," *IEEE Trans. Electron Devices* **26**, 513–520 (1979).

Hoefflinger, B., "Output Characteristics of Short-Channel Field-Effect Transistors," *IEEE Trans. Electron Devices* **28**, 971–976 (1981).

Hofmann, F. and W. Hansch, "The Charge Pumping Method: Experiment and Complete Simulation," *J. Appl. Phys.* **66**, 3092–3096 (1989).

Hofmann, K. R., "Hot Carrier Injection and Charge Trapping in *n*- and *p*-Channel MOSFETs," *Proc. Int. Conf. Insul. Films Semicond.*, Findhoven, Netherlands, 98–102 (1983).

Hofmann, K. R., W. Weber, C. Werner, and G. Dorda, "Hot Carrier Degradation Mechanism in *n*-MOSFET," *Tech. Dig.—Int. Electron Devices Meet.*, 104–107 (1984).

Hofmann, K. R., C. Werner, W. Weber, and G. Dorda, "Hot-Electron and Hole-Emission Effects in Short *n*-Channel MOSFET's," *IEEE Trans. Electron Devices* **32**, 691–699 (1985).

Hofstein, S. R. and F. P. Heiman, "The Silicon IGFET," *Proc. IEEE* **51**, 1190 (1963).

Hofstein, S. R. and G. Warfield, "Physical Limitation on the Frequency Response of a Semiconductor Surface Inversion Layer," *Solid-State Electron.* **8**, 321 (1965a).

Hofstein, S. R. and G. Warfield, "Carrier Mobility and Current Saturation in the MOS Transistor," *IEEE Trans. Electron Devices* **12**, 129–138 (1965b).

Hohol, T. S. and L. A. Glasser, "Relic: A Reliability Simulator for Integrated Circuits," *Tech. Dig. IEEE ICCAD*, 517–520 (1986).

Holland, S., I. C. Chen, and C. Hu, "Ultra-Thin Silicon-Dioxide Breakdown Characteristics of MOS Devices with n^+ and p^+ Polysilicon Gates," *IEEE Electron Device Lett.* **8**, 572–575 (1987).

Homles, F. E. and C. A. T. Salama, "VMOS—A New MOS Integrated Circuit Technology," *Solid-State Electron.* **17**, 791 (1974).

Honda, M., K. Kondou, H. Mitani, T. Kimura, S. Koshimaru, Y. Nagahashi, and M. Tameda, "25 ns 256 K CMOS SRAM," *Tech. Dig. IEEE Int. Solid-State Circuits Conf.* **29**, 250–251 (1986).

Hong, S. N., G. A. Ruggles, J. J. Wortman, and M. C. Ozturk, "Material and Electrical Properties of Ultra-Shallow p^+–n Junctions Formed by Low-Energy Ion Implantation and Rapid Thermal Annealing," *IEEE Trans. Electron Devices* **38**, 476–486 (1991).

Honlein, W. and G. Landwehr, "Energy Loss of Warm Electrons at the Interface of (100) Silicon MOSFETs," *Surf Sci.* **113**, 260–266 (1982).

Hook, T. B. and T. P. Ma, "Hot-Electron Induced Interface Traps in Metal/SiO_2/Si Capacitors: The Effect of Gate-Induced Strain," *Appl. Phys. Lett.* **48**, 1208–1210 (1986).

Hopfel, R. A., "EVLSI," *IEEE Electron Devices Lett.* **3**, 40–42 (1982).

Hori, T., K. Kurimoto, T. Yabu, and G. Fuse, "A New Submicron MOSFET and LATID (Large-Tilt Angle Implanted Drain) Structure," *Tech. Dig. VLSI Technol. Symp.* **15–16** (1988).

Hori, T., "Demands for Submicron MOSFET's and Nitrided Oxide Gate-Dielectrics," *Conf. Solid State Devices Mater., 21st* Tokyo, 197–200 (1989), invited.

Hori, T., "Drain-Structure Design for Reduced Band-to-Band and Band-to-Defect Tunneling Leakage," *1990 Symp. VLIS Technol., Dig. Tech. Pap.*, 69–70 (1990).

Hori, T., J. Hirase, Y. Odake, and T. Yasui, "Deep-Submicrometer Large-Angle-Tilt Implanted Drain (LATID) Technology," *IEEE Trans. Electron Devices* **39**, 2312–2324 (1992).

Horio, K. and H. Yanai, "Numerical Modeling of Heterojunctions Including the Thermionic Emission Mechanism at the Heterojunction Interface," *IEEE Trans. Electron Devices* **37**, 1093–1098 (1990).

Horiuchi, T., H. Mikoshiba, K. Nakamura, and K. Hamaro, "A Sample Method to Evaluate Device Lifetime Due to Hot-Carrier Effect under Dynamic Stress," *IEEE Electron Devices Lett.* **7**, 337–339 (1986).

Horiuchi, T., T. Homma, Y. Murao, and K. Okumura, "A High Performance Asymmetric LDD MOSFET Using Selective Oxide Deposition Technique," *Tech. Dig. VLSI Tech. Symp.*, 88–89 (1992).

Hosokawa, M. and H. Ikegami, "Characteristics of Hot Electron Ring in a Simple Magnetic Mirror Field," *Jpn. J. Appl. Phys.* **30**, 152–160 (1991).

Hsu, C. C.-H, S. C. S. Pan, and C.-T. Sah, "Trap Generation during Low-Influence Avalanche-Electron Injection in Metal-Oxide-Silicon Capacitors," *J. Appl. Phys.* **58**, 1326–1329 (1985).

Hsu, C. C.-H., D. S. Wen, M. R. Wordeman, Y. Taur, and T. H. Ning, "A Comparative Study of Hot-Carrier Instabilities in *p*- and *n*-type Poly Gate MOSFET's," *Tech. Dig.—Int. Electron Devices Meet.*, 75–78 (1989).

Hsu, C.C.-H., L. K. Wang, M. R. Wordeman, and T. H. Ning, "Hot-Electron-Induced Instability in 0.5 μm *p*-Channel MOSFETs Patterned Using Synchrotron X-Ray Lithography," *IEEE Electron Devices Lett.* **10**, 327–329 (1989).

Hsu, C. C.-H., L. K. Wang, J. Y.-C. Sun, M. R. Wordeman, and T. H. Ning, "Hot-Electron Induced Instability of 0.5 μ CMOS Devices Patterned Using Synchrontron X-Ray Lithography," *Proc. Int. Reliab. Phys. Symp.*, 189–192 (1989).

Hsu, C. C.-H., L. K. Wang, J. Y.-C. Sun, M. R. Wordeman, and T. H. Ning, "Radiation Damage and Its Effect on Hot-Carrier Induced Instability of 0.5 μ CMOS Devices Patterned Using Synchrontron X-Ray Lithography," *Proc. Radiat. —Induced* and/or *Process—Related Electr. Active Defects Semicond.—Insul. Syst.*, Research Triangle Park, NC, 249–261 (1989).

Hsu, C. C.-H., L. K. Wang, J. Y.-C. Sun, M. R. Wordeman, and T. H. Ning, "Radiation Damage and Its Effect on Hot-Carrier Induced Instability of 0.5 μ CMOS Devices Patterned Using Synchrotron X-Ray Lithography," *J. Electron. Mater.* **19**, 721–725 (1990).

Hsu, C. C.-H., L. K. Wang, D. Zicherman, and A. Acovic, "Effect of Hydrogen Annealing on Hot-Carrier Instability of X-Ray Irradiated CMOS Devices," *J. Electron. Mater.* **21**, 769–773 (1992).

Hsu, C. C.-H., B. S. Wu, G. G. Shahidi, B. Davari, W. H. Chang, and A. Acovic, "Understanding of Enhanced Sensitivity to Hot Carrier Degradation in Drain Engineered *n*-FETs," *Ext. Abstr., 1992 Int. Conf. Solid State Devices Mater.*, 512–514 (1992).

Hsu, F.-C., P. K. Ko, S. Tam, C. Hu, and R. S. Muller, "An Analytical Breakdown Model for Short-Channel MOSFETs," *IEEE Trans. Electron Devices* **29**, 1735–1740 (1982).

Hsu, F.-C., R. S. Muller, and C. Hu, "A Simplified Model of Short-Channel MOSFET Characteristics in the Breakdown Mode," *IEEE Trans. Electron Devices* **30**, 571–576 (1983).

Hsu, F.-C., R. S. Muller, C. Hu, and P. K. Ko, "A Simple Punchthrough Model for Short-Channel MOSFETs," *IEEE Trans. Electron Devices* **30**, 1354 (1983).

Hsu, F.-C. and H. R. Grinolds, "Structure-Dependent MOSFET Degradation Due to Hot-Electron Injection," *Tech. Dig.—Int. Electron Devices Meet.*, 742–744 (1983).

Hsu, F.-C. and K. Y. Chiu, "A Comparative Study of Tunneling, Substrate Hot-Electron, and Channel Hot Electron Injection Induced Degradation in Thin Gate MOSFETs," *Tech. Dig.—Int. Electron Devices Meet.*, 96–99 (1984a).

Hsu, F.-C. and K. Y. Chiu, "Hot-Electron Substrate Current Generation during Switching Transients," *Tech. Dig. VLSI Tech. Symp.*, 86–87 (1984b).

Hsu, F.-C. and K.-Y. Chiu, "Temperature Dependence of Hot-Electron-Induced Degradation in MOSFET's," *IEEE Electron Devices Lett.* **5**, 148–150 (1984c).

Hsu, F.-C. and K. Y. Chiu, "Evaluation of LDD MOSFETs Based on Hot-Electron Induced Degradation," *IEEE Electron Devices Lett.* **5**, 162–165 (1984d).

Hsu, F.-C. and S. Tam, "Relationship between MOSFET Degradation and Hot-Electron-Induced Interface-State Generation," *IEEE Electron Devices Lett.* **5**, 50–52 (1984).

Hsu, F.-C. and H. R. Grinolds, "Structure-Enhanced MOSFET Degradation Due to Hot-Electron Injection," *IEEE Electron Devices Lett.* **4**, 71–74 (1984).

Hsu, F.-C., J. Hui, and K. Y. Chiu, "Effect of Final Anneal on Hot-Electron-Induced MOSFET Degradation," *IEEE Electron Devices Lett.* **6**, 369–371 (1985a).

Hsu, F.-C., J. Hui, and K. Y. Chiu, "Hot-Electron Degradation in Submicron VLSI," *Tech. Dig.—Int. Electron Devices Meet.*, 48–51, (1985b).

Hsu, F.-C. and K. Y. Chiu, "Effects of Device Processing on Hot-Electron Induced Device Degradation," *Tech. Dig. VLSI Tech. Symp.*, 108–109 (1985a).

Hsu, F.-C. and K. Y. Chiu, "Hot-Electron Substrate-Current Generation during Switching Transients," *IEEE Trans. Electron Devices* **32**, 394–399 (1985b).

Hsu, W.-J., S. M. Gowda, and B. J. Sheu, "VLSI Circuit Design with Built-in Reliability Using Simulation Techniques," *Proc. Custom Int. Circuits Conf.*, 4 (1990).

Hu, C., "Lucky-Electron Model of Channel Hot-Electron Emission," *Tech. Dig.—Int. Electron Devices Meet.*, 22–25 (1979).

Hu. C., S. Tam, F.-C. Hsu, P. K. Ko, and R. S. Muller, "Correlating the Channel, Substrate, Gate, and Minority-Carrier Currents in MOSFETs," *Tech. Dig. IEEE Int. Solid-State Circuits Conf.* **26**, 88–89 (1983).

Hu, C., "Hot-Electron Effects in MOSFETs," *Tech. Dig.—Int. Electron Devices Meet.*, 176–181 (1983).

Hu, C., S. C. Tam, F.-C. Hsu, P. K. Ko, T.-Y. Chan, and K. W. Terrill, "Hot-Electron-Induced MOSFET Degradation—Model, Monitor, and Improvement," *IEEE Trans. Electron Devices* **32**, 375–385 (1985).

Hu, C., "Hot Electron Effects in VLSI MOSFETs," *Proc. IEEE Int. Symp. VLSI Tech. Syst. Appl.*, 79–84 (1987).

Hu, C., "Reliability Issues of MOS and Bipolar ICs," *IEEE Int. Conf. Comput. Des.: VLSI Comput. Proc.*, Piscataway, NJ, 438–442 (1989).

Hu, C., "Simulating Hot-Carrier Effects on Circuit Performance," *Tutorial Notes, Int. Reliab. Phys. Symp.*, 1c.1–1c.34 (1993).

Hu, J. and J. Moll, "Submicrometer Device Design for Hot-Electron Reliability and Performance," *IEEE Electron Devices Lett.* **6**, 350–352 (1985).

Hu, S. C. and M. P. Brassington, "Direct Measurement of Hot-Carrier Stress Effects on CMOS Circuit Performance," *IEEE Trans. Electron Devices* **36**, 2604–2605 (1989).

Hu, S. C. and M. P. Brassington, "A New Test Structure for Direct Measurement of Hot-Carrier Stress Effects on CMOS Circuit Performance," *IEEE Trans. Electron Devices* **38**, 1958–1959 (1991).

Huang, C.-L., S. A. Grot, G. S. Gildenblat, and V. Bolkhovsky, "Charge Trapping and Dielectric Breakdown in MOS Devices in 77–400 K Temperature Range," *Solid-State Electron.* **32**, 767–775 (1989).

Huang, C.-L., L. Qui, and Z. Ren, "Use of Hot Electrons to Stabilize Plasmas against Low-Frequency Drift Waves," *Chin. Phys.* **10**, 13–20 (1990).

Huang, C.-L., T. Wang, C. N. Chen, M. C. Chang, and J. Fu, "Modeling Hot-Electron Gate Current in Si MOSFETs Using a Coupled Drift-Diffusion and Monte Carlo Method," *IEEE Trans. Electron Devices* **39**, 2562–2568 (1992).

Huang, D. H., E. E. King, J. J. Wang, R. Ormond, and L. J. Palkuti, "Correlation between Channel Hot-Electron Degradation and Radiation-Induced Interface Trapping in *n*-Channel LDD Devices," *IEEE Trans. Nucl. Sci.* **NS-38**, 1336–1341 (1991).

Huang, J. S. T. and G. W. Taylor, " Modeling of an Ion-Implanted Silicon-Gate Depletion-Mode IGFET," *IEEE Trans. Electron Devices* **22**, 995–1001 (1975).

Huang, M. Q., P. T. Lai, Z. J. Ma, H. Wong, and Y. C. Cheng, "Improvement of Punchthrough-Induced Gate-Oxide Breakdown in *n*-Channel Metal-Oxide-Semiconductor Field-Effect Transistors Using Rapid Thermal Nitradation," *Appl. Phys. Lett.* **61**, 453–455 (1992).

Huang, S., M. Khan, Y. Strunk, and T. Batra, "Advanced Dual Metal CMOS Technology for High Density Semicustom Products," *Proc. Custom Int. Circuits Conf.*, Portland, OR, 188–191 (1985).

Huang, T.-Y., "Effects of Channel Shapes on MOSFET Hot-Electron Resistance," *Electron. Lett.* **21**, 211–212 (1985).

Huang, T.-Y. and J. Y. Chen, "Observation of Double-Hump Substrate Current in Funnel-Shape Transistors," *IEEE Electron Devices Lett.* **6**, 510–512 (1985).

Huang, T.-Y., C. Szeta, J. Y. Chen, A. G. Lewis, R. A. Martin, J. Shaw, and M. Koyanagi, "Stress-Induced Double-Hump Substrate Current in MOSFETs," *IEEE Electron Devices Lett.* **7**, 664–666 (1986a).

Huang, T.-Y., I. W. Wu, and J. Y. Chen, "Use of Sacrificial Spacers for Fabricating LDD Transistors in a CMOS Process," *Electron. Lett.* **22**, 430–432 (1986b).

Huang, T.-Y., W. W. Yao, R. A. Martin, A. G. Lewis, M. Koyanagi, and J. Y. Chen, "A Novel Submicron LDD Transistor with Inverse-T Gate Structure," *Tech. Dig.—Int. Electron Devices Meet.*, 742 (1986c).

Huang, T.-Y., W. W. Yao, R. A. Martin, G. Alan, M. Koyanagi, and J. Y. Chen, "New LDD Transistor with Inverse-T Gate Structure," *IEEE Electron Devices Lett.* **8**, 151–153 (1987).

Hubbard, R. F., T. J. Birmingham, and E. W. Hones, "Magnetospheric Electrostatic Emissions and Cold Plasma Densities," *J. Geophys. Res.* **84**, 5828–5838 (1979).

Hui, J., F.-C. Hsu, and J. Moll, "A New Substrate and Gate Current Phenomenon in Short-Channel LDD and Minimum Overlap Devices," *IEEE Electron Devices Lett.* **6**, 135–138 (1985).

Hui, J. and J. Moll, "Submicrometer Device Design for Hot-Electron Reliability and Performance," *IEEE Electron Devices Lett.* **6**, 350–352 (1985).

Hung, K. K., "Electrical Characterization of the Si-SiO_2 Interference for Thin Oxides," Doctoral Thesis, Hong Kong University (1987).

Hung, K. K., P.-K. Ko, C. Hu, and Y. C. Cheng, "Flicker Noise Characteristics of Advanced MOS Technologies," *Tech. Dig.—Int. Electron Devices Meet.*, 34–37 (1988).

Hwang, C. G., D. Y. Cheng, H. R. Yeager, and R. W. Dutton, "Multi-Window Device Analysis of Hot Carrier Transport," *Tech. Dig.—Int. Electron Devices Meet.*, 563–566 (1986).

Hwang, C. G. and R. W. Dutton, "Improved Physical Modeling of Submicron MOSFETs Based on Parameter Extraction Using 2-D Simulation," *IEEE Trans. Comput.-Aided Des.* **8**, 370–379 (1989).

Hwang, C. G. and R. W. Dutton, "Substrate Current Model for Submicrometer MOSFET's Based on Mean Free Path Analysis," *IEEE Trans. Electron Devices* **36**, 1348–1354 (1989).

Hwang, H., W. Ting, D.-L. Kwong, and J. Lee, "Electrical and Reliability Characteristics of Submicrometer *n*MOSFETs with Oxynitride Gate Dielectric Prepared by Thermal Oxidation in N_2O," *IEEE Trans. Electron Devices* **38**, 2712–2713 (1991).

Hwang, H., W. Ting, D.-L. Kwong, and J. Lee, "Improved Reliability Characteristics of Submicrometer *n*-MOSFETs with Oxynitride Gate Dielectric Prepared by Rapid Thermal Oxidation in N_2O," *IEEE Electron Devices Lett.* **12**, 495–497 (1991).

Hwang, H., J. Lee, P. Fazan, and C. Dennison, "Hot-Carrier Reliability Characteristics of Narrow-Width MOSFETs," *Solid-State Electron.* **36**, 665–666 (1993).

Hwang, H., M.-Y. Hao, J. Lee, V. Mathews, P. C. Fazan, and C. Dennison, "Furnace N_2O Oxidation Process for Submicron MOSFET Design Applications," *Solid-State Electron.* **36**, 749–751 (1993).

Hwu, J.-G. and J.-T. Chen, "Improvement of Hot-Electron-Induced Degradation in MOS Capacitors by Repeated Irradiation-then-Anneal Treatments," *IEEE Electron Devices Lett.* **11**, 82–84 (1990).

Igura Y. and E. Takeda, "Hot-Carrier Degradation Mechanism under AC Stress in MOSFET's," *Tech. Dig. VLSI Tech. Symp.*, 47–48 (1987).

Igura, Y., K. Umeda, and E. Takeda, "Initial Stage Degradation Mechanism in Hot-Carrier Effects," *Ext. Abstr., Solid State Devices Mater.*, 31–34 (1987).

Igura, Y., H. Matsuoka, and E. Takeda, "New Device Degradation Due to 'Cold' Carriers Created by Band-to-Band Tunneling," *IEEE Electron Devices Lett.*, 227–229 (1989).

Ihantola, H. K. J. and J. L. Moll, "Design Theory of a Surface FET," *Sold-State Electron.* **7**, 423 (1964).

Iizuka, F. Masuoka, T. Sato, and M. Ishikawa, "Electrically Alterable Avalanche-Injection-Type MOS Read-Only Memory with Stacked-Gate Structures," *IEEE Trans. Electron Devices* **23**, 379 (1976).

Inokawa, H., E. M. Aijimine, and C. Y. Yang, "Degradation and Revovery of MOS Devices Stressed with FN Gate Current," *Jpn. J. Appl. Phys.* **30**, 1931 (1991).

Ioannou, D. E., "Reliability of Short Channel Silicon and SOI VLSI Devices and Circuits," *Semicond. Device Reliab.-Proc. NATO Adv. Res. Workshop*, 507–517 (1990).

Ishida, M., S. Kuono, M. Maeda, and A. Murayama, "A 256 K-Bit CMOS Static RAM," *Sanyo Tech. Rev. (Jpn.)* **20**, 90–94 (1988).

Ishii, H., S. Masuda and Y. Harada, "Metastable Atom Electron Spectroscopy of Clean and Oxidized Si(111) – 7 × 7 Surfaces: Observation of Semiconductor–Insulator Transition," *Surf. Sci.* **239**, 222–226 (1990).

Ito, A., H. A. Swasey, and E. W. George, "Hot Electron Reliability Modeling in VLSI Devices," *Proc. Int. Reliab. Phys. Symp.*, 96–101 (1983).

Ito, S., Y. Homma, E. Sasaki, S. Uchimura, and H. Morishima, "Application of Surface Reformed Thick Spin-on-Glass to MOS Device Planarization," *J. Electrochem. Soc.* **137**, 1212–1218 (1990).

Itoh, K. and H. Sunami, "High Density One Dynamic MOS Memory Cells," *Proc. IEEE* **130**, 127–135 (1983).

Iwai, H., "Hot Carrier–Induced Degradation Modes in Thin-Gate Insulator Dual-Gate MOSFETs," *Proc. 7th Bienn. Euro. Conf.*, 83–92 (1991).

Iwata, S., N. Yamamoto, J. Kobayashi, T. Terada, and T. Mizutani, "A New Tungsten Gate Process for VLSI Applications," *IEEE Trans. Electron Devices* **31**, 1174–1179 (1984).

Iyer, S. S., P. M. Solomon, V. P. Kesan, A. A. Bright, J. I. Freeouf, T. N. Nguyen, and A. C. Warren, "A Gate-Quality Dielectric System for SiGe Metal-Oxide-Semiconductor Device," *IEEE Electron Devices Lett.* **12**, 246–248 (1991).

Izawa, R. and E. Takeda, "The Impact of n^- Drain Length and Gate–Drain/Source Overlap on Submicrometer LDD Devices for VLSI," *IEEE Electron Devices Lett.* **8**, 480–482 (1987).

Izawa, R., T. Kure, and E. Takeda, "Impact of the Gate–Drain Overlapped Device (GOLD) for Deep Submicrometer VLSI," *IEEE Trans. Electron Devices* **35**, 2088–2093 (1988).

Izawa, R., K. Umeda, and E. Takeda, "AC Hot-Carrier Degradation Due to Gate-Pulse-Induced Noise," *Trans. Inst. Electron. Commun. Eng. Jpn.* [*Part*] *C* **J74C-II**, 66–71 (1991).

Izawa, T., K. Watanabe, and S. Kawamura, "Impact of Low-Energy Ion-Implantation on Deep-Submicron Phosphorus-LDD *n*MOSFETs," *Ext. Abstr., 1992 Int. Conf. Solid State Devices Mater.*, 160–161 (1992).

Jackson, D. B., D. A. Bell, B. S. Doyle, B. J. Fishbein, and D. B. Krakauer, "Transistor Hot Carrier Reliability Assurance in CMOS Technologies," *Digital Tech. J.* **4**, 100–113 (1992).

Jacoboni, C. and L. Reggiani, "Bulk Hot-Electron Properties of Cubic Semiconductors," *Adv. Phys.* **28**, 493–553 (1979).

Jacoboni, C., "Hot-Carrier Transient Transport," *J. Lumine.* **30**, 120–143 (1985).

Jai, H. K., H. K. Gi, J. P. Young, and S. M. Hong, "The Fabrication and Characterization of CODE MOSFET," *J. Korean Inst. Telematics Electr.* **27**, 80–85 (1990).

Jain, S., W. Cochran, and M.-L. Chen, "Sloped-Junction LDD (SJLDD) MOSFET Structures for Improved Hot-Carrier Reliability," *IEEE Electron Devices Lett.* **9**, 539–541 (1988).

Jain, V., D. Pramanik, S. R. Nariani, and K. Y. Chang, "Improved Hot Carrier Reliability of Submicron MOS Devices by Modifying the PECVD IMO Film," *Proc. IEEE VLSI Conf., 8th*, 272–278 (1991).

Jain, V., D. Pramanik, S. R. Nariani, and C. Hu, "Internal Passivation for Suppression of Device Instabilities Induced by Backend Processes," *Proc. Int. Reliab. Phys. Symp.*, 11–15 (1992).

Jakubowski, A. and J. Ruzyllo, "Oxide Thin Films for MOS VLSI Circuits," *Elektronika* **25**, 15–19 (1984).

James, F. J. and P. Miller, "Detection of Faint Luminescence from Faulty Action in Integrated Circuits" *Proc. SPIE—Int. Soc. Opt. Eng.* **822**, 150–156 (1987).

Januschewski, F., and H. J. Erzgraeber, "Modeling the Influence of Hot Electrons on the Transfer Characteristic of Short-Channel MOSFETs,' *Phys. Status Solidi A* **99**, 649–656 (1987).

Januschewski, F., H. J. Erzgraeber, and W. Fuessel, "Modeling the Local Damage of Short-Channel MOSFETs Due to Hot Electron Injection Using Results from Photoinjecting Measurements on MOS Capacitors," *Phys. Status Solidi A* **106**, 215–220 (1988).

Jayaraman, R., W. Yang, and C. G. Sodini, "MOS Electrical Characteristics of Low Pressure Re-Oxidized Nitrided-Oxide," *Tech. Dig.—Int. Electron Devices Meet.*, 668–671.

Jeng, M.-C., J. Chung, A. T. Wu, T. Y. Chan, J. Moon, G. May, P.-K. Ko, and C. Hu, "Performance and Hot-Electron Reliability of Deep-Submicron MOSFET's," *Tech. Dig.—Int. Electron Devices Meet.*, 710–713 (1987).

Jeng, M.-C., P.-K. Ko, and C. Hu, "Deep-Submicrometer MOSFET Model for Analog/Digital Circuit Simulations," *Tech. Dig.—Int. Electron Devices Meet.*, 114–117 (1988).

Jeng, M.-C., J. Chung, J. E. Moon, G. May, P.-K. Ko, and C. Hu, "Design Guidelines for Deep-Submicrometer MOSFETs," *Tech. Dig.—Int. Electron Devices Meet.*, 386–389 (1988).

Jeppson, K. O. and C. M. Svensson, "Negative Bias Stress of MOS Devices at High Electric Fields and Degradation of MNOS Devices," *J. Appl. Phys.* **48**, 2004–2014 (1977).

Jiang, C., C. Hu, C. H. Chen, and P. N. Tseng, "Impact of Inter-Metal-Oxide Deposition Condition on NMOS and PMOS Transistor Hot Carrier Effect," *Proc. Int. Reliab. Phys. Symp.*, 122–126 (1992).

Jindal, R. P., "Hot-Electron Effects on Channel Thermal Noise in Fine-Line *n*-MOS Field-Effect Transistors," *IEEE Trans. Electron Devices* **33**, 1395–1397 (1986).

Jindal, R. P., "Noise Phenomena in Submicron Channel Length Silicon *n*-MOS Transistor," *Proc. Int. Conf. 'Noise Phys. Syst.' '1/f Noise'*, Rome, Italy, 199–202 (1986).

Jindal, R. P., "General Noise Considerations for Gigabit-Rate *n*-MOSFET Front-End Design for Optical-Fiber Communication System," *IEEE Trans. Electron Devices* **34**, 305–309 (1987).

Johnson, J. B., "More on the Solid-State Amplifier and Dr. Lilienfeld," *Phys. Today* **17**, 24–26 (1964).

Johnson, W. C. and P. T. Panousis, "The Influence of Debye Length on the *C–V* Measurement of Doping Profiles," *IEEE Trans. Electron Devices* **18**, 965–973 (1971).

Jones, W. T., K. Hess, and G. J. Iafrate, "Hot Electron Diffusion in Fine Line Semiconductor Devices," *Solid-State Electron.* **25**, 1217–1221 (1982).

Joshi, A. B., G. Q. Lo, and D.-L. Kwong, "Process Dependence of Interface State Generation Due to Irradiation and Hot Carrier Stress in Rapid Thermally Nitrided Thin Gate Oxides," *Electron. Lett.* **26**, 1248–1249 (1990).

Joshi, A. B., G. Q. Lo, and D.-L. Kwong, "A Comparison of Radiation and Hot-Electron-Induced Damages in MOS Capacitors with Rapid Thermally Nitrided Thin-Gate Oxides," *Solid-State Electron.* **34**, 1023–1028 (1991).

Joshi, A. B. and D.-L. Kwong, "Endurance of MOSFETs with Rapid Thermally Reoxidized Nitrided Thin Gate Oxides to Hot Carrier-Induced GIDL," *IEEE Trans. Electron Devices* **38**, 2711–2712 (1991).

Joshi, A. B., G. Q. Lo, D.-L. Kwong, and S. Lee, "Improved Performance and Reliability of MOSFETs with Thin Gate Oxides Grown at High Temperature," *Proc. Int. Reliab. Phys. Symp.*, Las Vegas, NV, 316–322 (1991).

Joshi, A. B., D.-L. Kwong, and S. Lee, "Improvement in Performance and Degradation Characteristics of MOSFETs with Thin Gate Oxides Grown at High Temperature," *IEEE Electron Devices Lett.* **12**, 28–30 (1991).

Joshi, A. B., G. Q. Lo, D. K. Shih, and D.-L. Kwong, "Performance and Reliability of Ultrathin Reoxidized Nitrided Oxides Fabricated by Rapid Thermal Processing," *Proc. SPIE—Int. Soc. Opt. Eng.* **1393**, 122–149 (1991).

Joshi, A. B., and D.-L. Kwong, "Study of the Growth Temperature Dependence of Performance and Reliability of Thin MOS Gate Oxides," *IEEE Trans. Electron Devices* **39**, 2099–2107 (1992).

Joshi, A. B. and D.-L. Kwong, "Excellent Immunity of GIDL to Hot-Electron Stress in Reoxidized Nitrided Gate Oxide MOSFETs," *IEEE Electron Devices Lett.* **13**, 47–49 (1992).

Joshi, A. B. and D.-L. Kwong, "Comparison of Neutral Electron Trap Generation by Hot-Carrier Stress in *n*-MOSFETs with Oxide and Oxynitride Gate Dielectrics," *IEEE Electron Devices Lett.* **13**, 360–362 (1992).

Joshi, S. P., R. Lahri, and C. Lage, "Polyemitter Bipolar Hot Carrier Effects in an Advanced BICMOS Technology," *Tech. Dig.—Int. Electron Devices Meet.*, 182–185 (1987).

Joshi, S. P., "Suppression of Poly Emitter Bipolar Hot Carrier Effects in an Advanced BiCMOS Technology," *Proc. 1989 Bipolar Circuits Tech. Meet.*, 144–147 (1989).

Joshi, S. P., "Reliability Imposed Constraints on Current Density of High Performance Poly-Emitter Bipolar Transistors," *Proc. IEEE Bipolar Circuits Tech. Meet.*, 180–183 (1990).

Ju. D. H., R. K. Reich, J. W. Schrankler, M. S. Holt, and G. D. Kirchner, "Transient Substrate Current Generation and Device Degradation in CMOS Circuits at 77 K," *Tech. Dig.—Int. Electron Devices Meet.*, 569–572 (1985).

Jung, K. H., D. L. Kwong, M. S. Swenson, and B. Boeck, "Reliability Comparison of Short-Channel MOSFETs with Cobalt Salicide and Titanium Salicide Structures," *Electron. Lett.* **25**, 96–98 (1989).

Kadary, V. and N. Klein, "Electroluminescence during the Anodic Growth of Tantalum Pentoxide," *J. Electrochem. Soc.* **128**, 749–755 (1981).

Kaga, T. and Y. Sakai, "Effects of Lightly Doped Drain Structure with Optimum Ion Dose on p-Channel MOSFET's," *IEEE Trans. Electron Devices* **35**, 2384–2390 (1988).

Kahng, D. and S. M. Sze, "A Floating Gate and Its Application to Memory Devices," *Bell Syst. Tech. J.* **46**, 1288–1295 (1967).

Kahng, D., W. J. Sundburg, D. M. Boulin, and J. R. Ligenza, "Interfacial Dopants for Dual-Dielectric Charge-Storage Cells," *Bell Syst. Tech. J.* **53**, 1723–1739 (1974).

Kahng, D., "A Historical Perspective on the Development of MOS Transistors and Related Devices," *IEEE Trans. Electron Devices* **23**, 655–657 (1976).

Kakumu, M., M. Kinuigawa, K. Hashimoto, and J. Matsunaga, "Power Supply Voltage for Future CMOS VLSI in Half and Sub Micrometer," *Tech. Dig.—Int. Electron Devices Meet.*, 399–402 (1986).

Kalnitsky, A. and S. Sharma, "Effect of Channel Hot Electron Stress on AC Device Characteristics of MOSFETs," *Solid-State Electron.* **29**, 1053–1057 (1986).

Kamata, K., K. Tanabashi, and K. Kobayashi, "Substrate Current Due to Impact Ionization in MOSFET," *Jpn. J. Appl. Phys.* **15**, 1127–1133 (1976).

Kamienicki, E., "Hot Carriers in Microplasmas and Their Radiation in Germanium and Silicon," *Phys. Status Solidi* **6**, 887 (1964).

Kamocsai, R. L. and W. Porod, "A Monte Carlo Study of the Influence of Traps on High Field Electronic Transport in $ASiO_2$," *Proc. Int. Symp. on Nanostruct. Phys. Fabr.*, College Station, TX, 297–301 (1989).

Kamohara, I., T. Wada, and H. Tango, "Effects of Parasitic Resistance and Hot-Electron-Degraded Transconductance on Lower Submicron *p*- and *n*-MOSFET Characteristics," *Solid State Devices Mater. Conf., 9th*, 35–38 (1987).

Kancleris, Z. and H. Tvardauskas, "Magnetoresistivity of *n*-type InSb and GaAs in Weakly Heating Electric Fields," *Litov. Fiz. Sb.* **25**, 81–87 (1985).

Kaneko, M., I. Narita, and S. Matsumoto, "The Study on Hole Mobility in the Inversion Layer of *p*-Channel MOSFET," *IEEE Electron Devices Lett.* **6**, 575–577 (1985).

Kantsleris, K. and A. Matulis, "A Two-Particle Monte Carlo Method for Thermal Electrons in Covalent Semiconductors," *Litov. Fiz. Sb.* **27**, 120–121 (1987).

Karunasiri, R. P. G. and K. L. Wang, "Quantum Devices Using SiGe/Si Heterostructures," *J. Vac. Sci. Technol. B* **9**, 2064–2071 (1991).

Kasai, N., P. J. Wright, and K. C. Saraswat, "Hot-Carrier-Degradation Characteristics for Fluorine-Incorporated *n*-MOSFET's," *IEEE Trans. Electron Devices* **37**, 1426–1431 (1990).

Kasinski, J. J., L. Gomez-Jahn, L. Min, Q. Bao, and R. J. D. Miller, "Picosecond Dynamics of Electron Transfer at Semiconductor Liquid Junctions," *J. Lumin.* **40–41**, 555–556 (1987).

Kastalsky, A. and S. Luryi, "Novel Real-Space Hot-Electron Transfer Devices," *IEEE Electron Devies Lett.* **4**, 334–336 (1983).

Kastalsky, A., "Novel Real-Space Transfer Devices," *Proc. Int. Conf. High-Speed Electron.: Basic Phys. Phenom. Devices Princ.*, 62–71 (1986).

Katayama, K. and T. Toyabe, "A New Hot Carrier Simulation Method Based on Full 3D Hydrodynamic Equations," *Tech. Dig.—Int. Electron Devices Meet.*, 135–138 (1989).

Kato, I., H. Oka, S. Hijiya, and T. Nakamura, "1.5 μ Gate CMOS Operated at 77 K," *Tech. Dig.—Int. Electron Devices Meet.*, 601–604 (1984).

Kato, I., H. Horie, M. Taguchi, and H. Ishikawa, "Mechanism of Hot Electron Trapping on *p*-MOSFET with p^+ Polysilicon Gate," *Tech. Dig.—Int. Electron Devices Meet.*, 14–17 (1988).

Kato, I., H. Horie, K. Oikawa, and M. Taguchi, "Distribution of Trapped Charges in SiO_2 Film of a p^+-Gate PMOS Structure," *IEEE Trans. Electron Devices* **38**, 1334–1342 (1991).

Kato, K., "Hot Electron Energy Analysis Using a High-Speed Monte Carlo Simulation," *Tech. Dig. VLSI Tech. Symp.*, 79–80 (1987).

Kato, K., "Hot Electron Simulation for MOSFETs Using a High-Speed Monte Carlo Method," *Proc. IEEE Int. Conf. Numer. Anal. Semicond. Devices Int. Circuits*, 249–254 (1987).

Kato, K., "Hot-Carrier Simulation for MOSFETs Using a High-Speed Monte Carlo Method," *IEEE Trans. Electron Devices* **35**, 1344–1350 (1988).

Kato, M., Y. Nishioka, and T. Okabe, "Parasitic MOSFET Degradation Induced by Fowler–Nordheim Injection," *IEEE Electron Devices Lett.* **11**, 590–592 (1990).

Katsube, T., I. Sakata, and T. Ikoma, "Hot Hole Effect on Surface-State Density and Minority-Carrier Generation Rates in Si-MOS Diodes Measured by DLTs," *IEEE Trans. Electron Devices* **27**, 1238–1243 (1980).

Katto, H., K. Okuyama, S. Meguro, R. Nagai, and S. Ikeda, "Hot Carrier Degradation Modes and Optimization of LDD MOSFET's," *Tech. Dig.—Int. Electron Devices Meet.*, 774–777 (1984).

Katto, H., K. Okuyama S. Meguro, and N. Suzuki, "Stress Bias Dependence of Hot Carrier Degradation of MOSFET's," *Tech. Dig. VLSI Tech. Symp.*, 106–107 (1985).

Kawabuchi, K., M. Yoshimi, T. Wada, M. Takahashi, and K. Numata, "Channel-Implanted Dose Dependence of Hot-Carrier Generation and Injection in Submicrometer Buried-Channel *p*-MOSFET's," *IEEE Trans. Electron Devices* **32**, 1685–1687 (1985).

Kawaguchi, Y. and S. Kawaji, "Study of Electron Mobility and Electron–Phonon Interaction in Si MOSFETs by Negative Magnetoresistance Experiments," *Jpn. J. Appl. Phys., Part 2* **21**, 709–711 (1982).

Kawamura, S., T. Makino, and K. Sukegawa, "Hot-Carrier-Induced Photon Emission in Thin SOI/MOSFETs," *IEICE Trans. Electron.* **E75-C**, 1471–1476 (1992).

Kaya, C., D. K. Y. Liu, J. Paterson, and P. Shah, "Buried Source-side Injection (BSSI) for Flash EPROM Programming," *IEEE Electron Devices Lett.* **13**, 465–467 (1992).

Kazerounian, R. and B. Eitan, "Single Poly EPROM for Custom CMOS Logic Applications," *Proc. IEEE Custom Int. Circuits Conf.*, 59–62 (1986).

Kazerounian, R., S. Ali, Y. Ma, and B. Eitan, "5 Volt High Density Poly–Poly Erase Flash EPROM Cell," *Tech. Dig.—Int. Electron Devices Meet.*, 436–439 (1988).

Keeney, S., J. Van Houdt, G. Groeseneken, and A. Mathewson, "Simulation of Enhanced Injection Split Gate Flash EPROM Device Programming," *Microelectron. Eng.* **18**, 253–258 (1992).

Kellner, W., "GaAs Electron Device," *Proc. Eur. Solid State Devices Res. Conf., 19th*, 4–12 (1989).

Kent, W. H., "Charge Distribution in Buried-Channel Charge-Coupled Devices," *Bell Syst. Tech. J.* **52**, 1009–1024 (1973).

Kerr, D. R., S. Logan, P. J. Burkhardt, and W. A. Pliskin, "Stabilization of SiO_2 Passivation Layers with P_2O_5," *IBM J. Res. Dev.* **8**, 376–384 (1964).

Kesner, J., R. S. Post, and D. K. Smith, "Tandem Mirror Hot-Electron Anchors," *Nucl. Fusion* **22**, 577–583 (1982).

Khosru, Q. D. M., N. Yasuda, K. Taniguchi, and C. Hamaguchi, "Interface State Generation and Annihilation in Thin Film *p*-Channel MOSFETs," *Ext. Abstr., 1992 Int. Conf. Solid State Devices Mater.*, 149–151 (1992).

Khurana, N., T. Maloney, and W. Yeh, "ESD on CHMOS Devices—Equivalent Circuits, Physical Models and Failure Mechanism," *Proc. Int. Reliab. Phys. Symp.*, 212–223 (1985).

Khurana, N. and C.-L. Chiang, "Analysis of Product Hot Electron Problems by Gated Emission Microscopy," *Proc. Int. Reliab. Phys. Symp.*, 189–194 (1986).

Kibickas, K., J. Parseliunas, and S. Vasiliauskas, "Transient Hot Electron Phenomena in GaAs n^+–n^+ Structures at 300 K," *Phys. Status Solidi* **102**, K99–K102 (1987).

Kienzler, R., "An Extended Two-Region-MOSFET-Model for Network Analysis," *Arch. Elektrotech. (Berlin)* **69**, 385–398 (1986).

Kilby, J. S., "Invention of the Integrated Circuit," *IEEE Trans. Electron Devices* **23**, 648–654 (1976).

Kim, C. K., "The Physics of Charge-Coupled Devices," In *Charge-Coupled Devices and Systems*, M. J. Howes and D. V. Morgan, eds., Wiley, New York, 1 (1979).

Kim, C. S., K. S. Kim, Y. H. Kim, and J. H. Lee, "Gate Capacitance Measurement on the Small-Geometry MOSFETs with Bias," *J. Korean Inst. Electron. Eng.* **24**, 86–90 (1987).

Kim, C. S., K. S. Kim, Y. H. Kim, B. W. Kim, and J. H. Lee, "Hot-Carrier Induced MOSFET Degradation and Its Lifetime Measurement," *J. Korean Inst. Telemat. Electron.* **25**, 182–187 (1988).

Kim, G. H., C. An, and Y. J. Park, "A Study on the CMOS Device Characteristics Having Intrinsic Polysilicon Gate," *J. Korean Inst. Telemat. Electron.* **28A**, 52–58 (1991).

Kim, K., K. Hess, and F. Capasso, "Real Space Transfer of Two Dimensional Electrons in Double Quantum Well Structures," *Solid-State Electronics* **31**, 351–354 (1987).

Kim, K. and W. H. Choe, "Self-Similar Modeling and Transport Analysis of Laser-Produced Coronas," *IEEE Int. Conf. Plasma Sci.*, 27 (1989).

Kim, S. K., A. Van der Ziel, and S. T. Liu, "Hot Electron Noise Effects in Buried Channel MOSFETs," *Solid-State Electron.* **24**, 425–428 (1981).

Kim, Y., M. S. Kim, and S.-K. Min, "Anomalous Conduction Band Density of States in $Al_xGa_{(1-x)}As$ Alloys," *Solid State Commun.* **68**, 295–299 (1988).

King. T.-J., J. R. Pfiester, and K. C. Saraswat, "A Variable-Work-Function Polycrystalline-$Si_{1-x}Ge_x$ Gate Material for Submicrometer CMOS Technologies," *IEEE Electron Devices Lett.* **12**, 533–535 (1991).

King, T.-J., K. C. Saraswat, and J. R. Pfiester, "PMOS Transistors in LPCVD Polycrystalline Silicon–Germanium Films" *IEEE Electron Devices Lett.* **12**, 584–586 (1991).

Kingston, R. H. and S. F. Neustadler, "Calculation of the Space Charge, Electric Field, and Free Carrier Concentration at the Surface of a Semiconductor," *J. Appl. Phys.* **26**, 718–720 (1955).

Kinugawa, M., M. Kakumu, T. Usami, and J. Matsunaga, "Effects of Silicon Surface Orientation on Submicron CMOS Devices," *Tech. Dig.—Int. Electron Devices Meet.*, 581–584 (1985a).

Kinugawa, M., M. Kakumu, S. Yokogawa, and K. Hashimoto, "Submicron MLDD *n*-MOSFET's for 5 V Operation," *Tech. Dig. VLSI Tech. Symp.*, 116–117 (1985b).

Kinugawa, M., K. Yamada, Y. Katoh, E. Nishimura, M. Kakumu, and J. Matsunaga, "Hot-Carrier-Induced Instability in Totally Down-Scale 1/4 μ CMOS," *Tech. Dig. VLSI Tech. Symp.*, 51–52 (1988).

Kinugawa, M., M. Kakumu, T. Yoshida, T. Nakayama, S. Morita, K. Kubota, F. Matsuoka, H. Oyamatsu, K. Ochii, and K. Maeguchi, "TFT (Thin Film Transistor) Cell Technology for a 4 Mbit and More High Density SRAMs," *1990 Symp VLSI Technol. Dig. Tech. Pap.*, 23–24 (1990).

Kishimoto, A., S. Kuniyoshi, N. Saito, T. Soga, K. Mochiji, and T. Kimura, "Minimization of X-Ray Mask Distortion by Two-Dimensional Finite Element Method Simulation," *Jpn. J. Appl. Phys.* **29**, 2203–2206 (1990).

Kitagawa, Y., N. Miyanaga, Y. Kato, M. Nakatsuka, A. Nishiguchi, T. Yabe, and C. Yamanaka, "Optimum Design of Exploding Pusher Target to Produce Maximum Neutrons," *Jpn. J. Appl. Phys.* **25**, 586–589 (1986).

Klaassen, F. M., "On the Influence of Hot Carrier Effects on the Thermal Noise of Field-Effect-Transistors," *IEEE Trans. Electron Devices* **17**, 858–862 (1980).

Klein, N., "Electrical Breakdown in Solids," *Adv. Electron. Electron Phys.* **26**, 309–424 (1969).

Knoll, M. D., D. Braeunig, and W. R. Fahrner, "Comparative Studies of Tunnel Injection and Irradiation of Metal Oxide Semiconductor Structures," *J. Appl. Phys.* **53**, 6946–6952 (1982).

Knoll, M. D., D. Braeunig, and W. R. Fahrner, "Oxide Charging and Interface State Generation by Tunnel Injection and Co^{60}-Irradiation Experiments on MOS Structures," *Proc. Int. Conf. Insul. Films Semicond.*, Eindhoven, Netherland, 107–111 (1983).

Ko, P.-K., "Approaches to Scaling," In *Advanced MOS Devices Physics*, N. G. Einspruch and G. Gildenblat, eds., Academic Press, San Diego, 1989.

Ko, P.-K., R. S. Muller, and C. Hu, "A Unified Model for Hot-Electron Currents in MOSFETs," *Tech. Dig.—Int. Electron Devices Meet.*, 600–603 (1981).

Ko, P.-K., S. Tam, C. Hu, S. S. Wong, and C. G. Sodini, "Enhancement of Hot Electron Currents in Graded-Gate-Oxide (GGO)-MOSFET," *Tech. Dig.—Int. Electron Devices Meet.*, 88–91 (1984).

Ko, P.-K., "MOS Device Modeling for Circuit Simulation," *Tech. Dig.—Int. Electron Devices Meet.*, 488–491 (1985).

Ko, P.-K., T. Y. Chan, A. T. Wu, and C. Hu, "The Effects of Weak Gate-to-Drain (Source) Overlap on MOSFET Characteristics," *Tech. Dig.—Int. Electron Devices Meet.*, 292–295 (1986).

Kobayashi, T., K. Takahara, T. Kimura, K. Yamamoto, and K. Abe, "High-Field Properties of *n*-INP under High Pressure," *Solid-State Electron.* **21**, 79–82 (1978).

Kobayashi, T., M. Miyake, Y. Okazaki, T. Matsuda, M. Sato, K. Deguchi, S. Ohki, and M. Oda, "8.6 ps/Gate Childed Si E/E *n*-MOS IC," *Tech. Dig.—Int. Electron Devices Meet.*, 881–883 (1988).

Kobeda, E., M. Kellman, and C. M. Osburn, "Rapid Thermal Annealing of Low-Temperature Chemical Vapor Deposited Oxides," *J. Electrochem. Soc.* **138**, 1846–1849 (1991).

Kocevar, P., "Hot Carrier Physics and Higher Order Electron Phonon Dynamics," *Acta Polytech. Scand., Electr. Eng. Ser.*, 102–138 (1983).

Kohyama, S., T. Furuyama, S. Mimura, and H. Iizuka, "Non-Thermal Carrier Generation in MOS Structures," *Jpn. J. Appl. Phys.* **19**, Suppl. 1, 85–92 (1980).

Komori, J., S. Maeda, K. Sugahara, and J. Mitsuhashi, "Evaluation of Hot Carrier Effects in TFT by Emission Microscopy," *30th Proc. Int. Reliab. Phys. Symp.*, 63–67 (1992).

Kooi, E., "Effects of Low Temperature Heat Treatments on the Surface Properties of Oxidized Silicon," *Philips Res. Rep.* **20**, 578–594 (1965).

Kopytkin, B. A., and M. A. Kukin, "Polychromatic High Emission from Submicron MOS Structure," *Mikroelektronika* **1**, 52–54 (1992).

Korman, C. E. and I. D. Mayergoyz, "A Globally Convergent Algorithm for the Solution of the Steady-State Semiconductor Device Equations," *J. Appl. Phys.* **68**, 1324–1334 (1990).

Koryakin, N. D., D. P. Skulachev, I. A. Strukov, and E. F. Yurchuk, "Experimental Study of Noise Generators in the Millimetric Range Using Gunn Diodes," *Meas. Tech.* (*Engl. Transl.*) **29**, 137–141 (1986).

Koshimaru, S., M. Fukuma, T. Tsujide, T. Yamanaka, and Y. Okuto, "An Asymmetric Effect of Short Channel MOSFET's," *Tech. Dig. VLSI Tech. Symp.*, 18–19 (1981).

Kotani, K., T. Shibata, and T. Ohmi, "Hot-Carrier-Immunity Degradation in Metal Oxide Semiconductor Field Effect Transistors Caused by Ion-Bombardment Processes," *Jpn. J. Appl. Phys.* **29**, L2289–L2291 (1990).

Koyanagi, M., H. Kaneko, and S. Shimizu, "Optimum Design of n^+–n^- Double-Diffused Drain MOSFET to Reduce Hot-Carrier Emission," *IEEE Trans. Electron Devices* **32**, 562–570 (1985).

Koyanagi, M., A. G. Lewis, R. A. Martin, T.-Y. Huang, and J. Y. Chen, "Hot Electron Induced Punchthrough in Submicron *p*-MOSFETs," *18th Solid State Devices Mater. Conf.*, 475–478 (1986).

Koyanagi, M., A. Lewis, R. Martin, T. Huang, and J. Chen, "Investigation and Reduction of Hot-Electron Punchthrough (HEIP) Effect in Submicron *p*-MOSFETs," *Tech. Dig.—Int. Electron Devices Meet.*, 722–725 (1986).

Koyanagi, M., A. Lewis, R. Martin, T. Huang, and J. Chen, "Hot-Electron-Induced Punchthrough (HEIP) Effect in Submicron *p*-MOSFETs," *IEEE Trans. Electron Devices* **34**, 839–844 (1987).

Koyanagi, M., A. Lewis, R. Martin, T. Huang, and J. Chen, "Increased Degradation of Half-Micron *p*-MOSFETs Due to Swapped Pulse Stressing," *Tech. Dig.—Int. Electron Devices Meet.*, 844–847 (1987).

Koyanagi, M., T. Huang, A. Lewis, and J. Y. Chen, "Hot-Carrier Reliability in Submicron *p*MOSFETs," *Technol. Syst. Appl. Proc. Tech., IEEE*, 312–316 (1989).

Koyanagi, M., H. Kurino, H. Kiba, H. Mori, T. Hashimoto, Y. Hiruma, T. Fujimori, Y. Yamaguchi, and T. Nishimura, "Hot-Carrier Light Emission in SOI MOSFET Simulated with Coupled Monte Carlo and Energy Transport Analysis," *Tech. Dig.—Int. Electron Devices Meet.*, 45–48 (1991).

Koyanagi, M., H. Kurino, T. Hashimoto, H. Mori, K. Hata, Y. Hiruma, T. Fujimori, I.-W. Wu, and A. G. Lewis, "Relation between Hot-Carrier Light Emission and Kink Effect in Poly-Si Thin Film Transistors," *Tech. Dig.—Int. Electron Devices Meet.*, 571–574 (1991).

Koyanagi, M., H. Kiba, H. Kurino, T. Hashimoto, H. Mori, and K. Yamaguchi, "Coupled Monte Carlo–Energy Transport Simulation with Quasi-Three-Dimensional Temperature Analysis for SOI MOSFET," *IEEE Trans. Electron Devices* **39**, 2640–2641 (1992).

Krautschneider, W. H. "Characterization of MOSFET Gate Oxides by Injection of Controlled Quantities of Electrons," *Proc. Eur. Solid State Devices Res. Conf., 19th*, 691–694 (1989).

Krautschneider, W. H., H. Terletzki, and Q. Wang, "Reliability Problems of Submicron MOS Transistors and Circuits," *Microelectron. Reliab.* **32**, 1499–1508 (1992).

Krick, J. T., P. M. Lenahan, and G. J. Dunn, "Direct Observation of Interfacial Point Defects Generated by Channel Hot Hole Injection in *n*-Channel Metal Oxide Silicon Field Effect Transistors," *Appl. Phys. Lett.* **59**, 3437–3439 (1991).

Krieger, G., P. P. Cuevas, and M. N. Misheloff, "Effect of Impact Ionization Induced Bipolar Action on *n*-Channel Hot-Electron Degradation," *IEEE Electron Devices Lett.* **9**, 26–28 (1988).

Krieger, G., R. Sikora, P. P. Cuevas, and M. N. Misheloff, "Moderately Doped NMOS (M-LDD)—Hot Electron and Current Drive Optimization," *IEEE Trans. Electron Devices* **38**, 121–127 (1991).

Krimigis, S. M., T. P. Armstrong, C. Y. Fan, and E. P. Keath, "Hot Plasma Environment at Jupiter: Voyager 2 Result," *Science* **206**, 977–984 (1979).

Krolevets, K. M. and V. E. Stikanov, "An Investigation of the Longitudinal Charge Distribution of 'Hot' Carriers in MNOS Transistors," *Izv. Vyssh. Vchebn. Zaved. Radioelektron.* **28**, 68–70 (1985).

Ku, Y. H., S. K. Lee, D.-L. Kwong, C.-O. Lee, and J. R. Yeargain, "Effects of Ion-Beam Mixing on the Performance and Reliability of Devices with Self-Aligned Silicide Structure," *IEEE Electron Devices Lett.* **9**, 293–295 (1988).

Kuehne, J., W. Ting, G. Q. Lo, T. Y. Hsieh, and D. L. Kwong, "Radiation and Hot-Electron Effects in MOS Structures with Gate Dielectric Grown by Rapid Thermal Processing in O_2 Diluted with NF_3," *Proc. Electrochem. Soc.* **90**, 364–375 (1990).

Kuei-Shu, C.-L. and J.-G. Hwu, "Effect of Starting Oxide on Electrical Characteristics of Metal-Reoxidized Nitrided Oxide-Semiconductor Devices Prepared by Rapid Thermal Processes," *Jpn. J. Appl. Phys.* **31**, L600–L603 (1992).

Kume, H., E. Takeda, T. Toyabe, and S. Asai, "Dependence of Channel Hot-Electron Injection on MOSFET Structure," *Jpn. J. Appl. Phys.* **21**, 67–71 (1981).

Kuo, J. B., W. C. Lee, and J. H. Sim, "Back Gate Bias Effects on the Pull-Down Transient Behavior in an Ultra-Thin SOL CMOS Inverter Operating at 300 K and 77 K," *Solid-State Electron.* **35**, 1553–1555 (1992).

Kuo, M. M., K. Seki, P. M. Lee, J. Y. Choi, P.-K. Ko, and C. Hu, "Quasi-Static Simulation of Hot-Electron-Induced MOSFET Degradation under AC (Pulse) Stress," *Tech. Dig.—Int. Electron Devices Meet.*, 47–50 (1987).

Kuo, M. M., K. Seki, P. M. Lee, J. Y. Choi, P.-K. Ko, and C. Hu, "Simulation of MOSFET Lifetime under AC Hot-Electron Stress," *IEEE Trans. Electron Devices* **35**, 1004–1011 (1988).

Kurachi, I., T. Yanai, and K. Yoshioka, "Oxide Reliability in Tungsten Polycide Gate Electrode," *IEEE Int. VLSI Multilevel Interconnect. Conf.*, 505 (1989).

Kurino, H., H. Kiba, H. Mori, S. Yokoyama, K. Yamaguchi, and M. Koyonagi, "Coupled Monte Carlo–Energy Relaxation Analysis of Hot-Carrer Light Emission in MOSFET's," *Ext. Abstr. Int. Conf. Solid State Devices Mater.*, 459–461 (1991).

Kusaka, T., Y. Ohji, and K. Mukai, "Breakdown Characteristics of Ultra Thin Silicon Dioxide," *Ext. Abstr. Solid State Devices Mater.*, 463–466 (1986).

Kusunoki, S., M. Inuishi, T. Yamaguchi, K. Tsukamoto, and Y. Akasaka, "Hot-Carrier-Resistant Structure by Re-oxidized Nitrided Oxide Sidewall for Highly Reliable and High Performance LDD MOSFETs," *Tech. Dig.—Int. Electron Devices Meet.*, 649–652 (1991).

Kuznetsov, S. N. and V. A. Gurtov, "Change Storage in MOS Structures as Affected by Avalanche- and Photoinjection," *Appl. Surf. Sci.* **30**, 347–352 (1987).

Kwang, S. Y., J. T. Park, and B. R. Kim, "An Analytical Model for Substrate and Gate Current of Stressed SC-PMOSFET in the Saturation Region," *Ext. Abstr., 1992 Int. Conf. Solid State Devices Mater.*, 164–166 (1992).

Lacaita, A., "Why the Effective Temperature of the Hot Electron Tail Approaches the Lattice Temperature," *Appl. Phys. Lett.* **59**, 1623–1625 (1991).

Lahri, R. and S. P. Joshi, "Engineered Reliability Underpins BiCMOS Process," *Solid State Technol.* **32**, 175–179 (1989).

Lai, F.-S. J., J. Y.-C. Sun, and S. H. Dhong, "Design and Characteristics of a Lightly Doped Drain (LDD) Device Fabricated with Self-Aligned $TiSi_2$," *IEEE Trans. Electron Devices* **33**, 345–353 (1986).

Lai, F.-S. J., L. K. Wang, Y. Taur, J. Y.-C. Sun, K. E. Petrillo, S. K. Chicotka, E. J. Petrillo, M. R. Polcari, T. J. Bucelot, and D. S. Zicherman, "Highly Latchup-Immune 1-μ CMOS Technology Fabricated with 1-MeV Ion Implantation and Self-Aligned $TiSi_2$," *IEEE Trans. Electron Devices* **33**, 1308–1320 (1986).

Lai, S. K., "Two Carrier Nature of Interface State Generation in Hole Trapping and Radiation Damage," *Appl. Phys. Lett.* **39**, 58–60 (1981).

Lam, H. W., A. F. Tasch, T. C. Holloway, K. F. Lee, and J. F. Gibbons, "Ring Oscillators Fabricated in Laser-Annealed Silicon-on-Insulator," *IEEE Electron Devices Lett.* **1**, 99 (1980).

Lance, A. L., "Microwave Power Measurement," *Electron. Test* **7**, 32–34 (1984).

Langer, J. M. and H. Heinrich, "On a Direct Connection of the Transition Metal Impurity Levels to the Band Edge Discontinuities in Semiconductor Hetero-junctions," *Physica B + C (Amsterdam)* **134**, 444–449 (1985).

Lanitsky, A. and S. Sharma, "The Effect of Channel Hot Electron Stress on AC Device Characteristics of MOSFETs," *Solid-State Electron.* **29**, 1053–1057 (1986).

Lanzoni, M., M. Manfredi, L. Selmi, E. Sangiorgi, R. Capellittle, and B. Ricco, "Hot-Electron-Induced Photon Energies in n-Channel MOSFETs Operating at 77 and 300 K," *IEEE Electron Devices Lett.*, 173–176 (1989).

Lanzoni, M., E. Sangiori, C. Fiegna, M. Manfredi, and B. Ricco, "Extended (1.1–2.9 eV) Hot-Carrier-Induced Photon Emission in n-Channel Si MOSFETs," *IEEE Electron Devices Lett.* **12**, 341–343 (1991).

Lapidus, E. M., E. A. Vygovskaya, A. A. Galaev, and P. T. Mustaev, "Influence of Structure Defects on the Characteristics of Warm Electrons and Holes near the Surface of Silicon," *Sov. Phys.—Semicond. (Engl. Transl.)* **16**, 1232–1233 (1982).

Lary, J., S. M. Goodnick, P. Lugli, D. Y. Oberli, and J. Shah, "Intersubband Relaxation of Hot Carriers in Coupled Quantum Wells," *Solid-State Electron.* **32**, 1283-1287 (1989).

Lau, D., G. Gildenblat, C. G. Sodini, and D. E. Nelsen, "Low-Temperature Substrate Current Characterization of n-Channel MOSFETs," *Tech. Dig.—Int. Electron Devices Meet.*, 565–568 (1985).

Law, M. E., E. Solley, M. Liang, and D. E. Burk, "Self-Consistent Model of Minority-Carrier Lifetime, Diffusion Length, and Mobility," *IEEE Electron Devices Lett.* **12**, 401–403 (1991).

Lawler, J. E., E. A. Den Hartog, and W. N. G. Hitchon, "Power Balance of Negative-Glow Electrons," *Phys. Rev. A* **43**, 4427–4437 (1991).

Leblebici, Y., S. M. Kang, C. T. Sah, and T. Nishida, "Modeling and Simulation of Hot Electron Effects for VLSI Reliability," *Tech. Dig.—Int. Conf. Comput.-Aided Des.*, 252–255 (1987).

Leblebici, Y. and S. M. Kang, "Simulation of MOS Circuit Performance Degradation with Emphasis on VLSI Design-for-Reliability," *Proc. IEEE Int. Conf. Comput. Des.: VLSI Compute. Processors*, 492–495 (1989).

Leblebici, Y. and S. M. Kang, "An Integrated Hot-Carrier Degradation Simulator for VLSI Reliability Analysis," *IEEE Conf. Comput.-Aided Des.*, 400–403 (1990).

Leblebici, Y. and S. M. Kang, "A One-Dimensional MOSFET Model for Simulation of Hot-Carrier Induced Device and Circuit Degradation," *Proc. IEEE Symp. Circuits Syst.*, 109–112 (1990).

Leblebici, Y., W. Sun, and S. M. Kang, "Parametric Macro-Modeling of Hot-Carrier-Induced Dynamic Degradation in MOS VLSI Circuits," *IEEE Trans. Electron Devices* **40**, 673–676 (1993).

Leburton, J.-P., H. Gesch, and G. E. Dorda, "Analytical Approach of Hot Electron Transport in Small Size MOSFETs," *Solid-State Electron.* **24**, 763–771 (1981).

Leburton, J.-P. and G. E. Dorda, "The *V–E* Relation and the Field Distribution in Submicron MOSFETs," *J. Phys., Colloq. (Orsay, Fr.)* **42**, 193-200 (1981).

Leburton, J.-P. and G. E. Dorda, "*V–E* Dependence in Small-Sized MOS Transistors," *IEEE Trans. Electron Devices* **29**, 1158–1171 (1982).

Leburton, J.-P. and G. E. Dorda, "Effect of the Electron Temperature on the Gate-Induced Charge in Small Size MOS Transistors," *Solid-State Electron.* **26**, 611–615 (1983).

Leclaire, P., "High Resolution Intrinsic MOS Capacitance-Measurement System," *Proc. Eur. Solid State Devices Res. Conf., 17th*, 131–134 (1988).

Lee, C.-T. and J. A. Burn, "Study of Gate Oxide Leakage and Charge Trapping in ZMR and SIMOX SOI MOSFETs," *IEEE Electron Devices Lett.* **9**, 235–237 (1988).

Lee, J.-I., H.-K. Park, and C.-M. Lee, "Comparison of Hot-Carrier Effect Resistance of LDD, DDD *p*-Drain, and As-Drain Structures," *New Phys. (Korean Phys. Soc.)* **28**, 171–177 (1988).

Lee, K., D. W. Forslund, J. M. Kindel, and E. L. Lindman, "Vacuum Insulation as a Way to Stop Hot Electrons," *Nucl. Fusion* **19**, 1447–1456 (1979).

Lee, K.-H., B. R. Jones, C. Burke, L. V. Tran, J. A. Shimer, and M. L. Chen, "Lightly Doped Drain Structure for Advanced CMOS (Twin-Tub IV)," *Tech. Dig.—Int. Electron Devices Meet.*, 242–245 (1985).

Lee, M. B., J. I. Lee, and K. N. Kang, "The Shift of Threshold Voltage and Subthreshold Current Curve in LDD MOSFET Degraded under Different DC Stress-Biases," *J. Korean Inst. Telematics Electron.* **26**, 46–51 (1989).

Lee, P. M., H. Masuda, and P.-K. Ko, "Transient Substrate Current Delay in *n*-MOSFET," *Tech. Dig.—Int. Electron Devices Meet.*, 502–505 (1987).

Lee, P. M., M. M. Kuo, K. Seki, P.-K. Lo, and C. Hu, "Circuit Aging Simulator (CAS)," *Tech. Dig.—Int. Electron Devices Meet.*, 134–137 (1988).

Lee, P. M., P.-K. Ko, and C. Hu, "Relating CMOS Inverter Lifetime to DC Hot-Carrier Lifetime of *n*-MOSFETS," *IEEE Electron Devices Lett.* **11**, 39–41 (1990).

Lee, S. K., D. K. Shih, D. L. Kwong, N. S. Alvi, N. R. Wu, and H. S. Lee, "Effects of Rapid Thermal Processing on Thermal Oxides of Silicon," *Proc. Mater. Res. Soc. Symp.* **71**, 449–454 (1986).

Lee, S.-W., T.-Y. Chan, and A. T. Wu, "Circuit Performance of CMOS Technologies with Silicon Dioxide and Reoxidized Nitrided Oxide Gate Dielectrics," *IEEE Electron Devices Lett.* **11**, 294–296 (1990).

Lee, S.-W., C. Liang, C.-S. Pan, W. Lin, and J. B. Mark, "A Study on the Physical Mechanism in the Recovery of Gate Capacitance to C_{ax} in Implanted Polysilicon MOS Structure," *IEEE Electron Devices Lett.* **13**, 2–4 (1992).

Lee, W.-H., T. Osakama, K. Asada, and T. Sugano, "Design Methodology and Size Limitations of Submicrometer MOSFETs for DRAM Application," *IEEE Trans. Electron Devices* **35**, 1876–1884 (1988).

Lee, Y.-H., L. D. Yau, E. Hansen, R. Chau, B. Sabi, S. Hossaini, and B. Asakawa, "Hot-Carrier Degradation of Submicrometer *p*-MOSFETs with Thermal/LPCVD Composite Oxide," *IEEE Trans. Electron Devices* **40**, 163–168 (1993).

Lehovec, K. and A. Slobodskoy, "Field-Effect Capacitance Analysis of Surface States on Silicon," *Phys. Status Solidi* **3**, 447 (1963).

Lelis, A. J., T. R. Oldham, H. E. Boesch, and F. B. Mclean, "The Nature of the Trapped Hole Annealing Process," *IEEE Trans. Electron Devices* **36**, 1808–1815 (1989).

Lelis, A. J. and T. J. Oldham, "X-Ray Lithography Effects on MOS Oxides," *IEEE Trans. Nucl. Sci.* **NS-39**, 2204–2210 (1992).

Lepselter, M. P. and S. M. Sze, "SB-IGFET: An Insulated-Gate FET Using Schottky Barrier Contacts as Source and Drain," *Proc. IEEE* **56**, 1088 (1968).

Lepselter, M. P. and S. M. Sze, "DRAM Pricing Trends—The Pi Rule," *IEEE Circuits Devices Mag.* **1**, 53–54 (1985).

Leung, C. C. C. and P. A. Childs, "Prediction of Hot Electron Degradation in MOSFETs: A Comparative Study of Theoretical Energy Distribution," *Proc. 7th Bienn. Eur. Conf.*, 303–306 (1991).

Lewis, E. T., "Design and Performance of 1.25 μ CMOS for Digital Applications," *Proc. IEEE* **73**, 419–432 (1985).

Lewis, T. J., "Role of Electrodes in Conduction and Breakdown Phenomena in Solid Dielectrics," *IEEE Trans. Electr. Insul.* **EI-19**, 210–216 (1984).

Li, X., J.-T. Hsu, P. Aum, D. Chan, J. Rembetski, and C. R. Viswanathan, "Plasma-Damaged Oxide Reliability Study Correlating Both Hot-Carrier Injection and Time-Dependent Dielectric Breakdown," *IEEE Electron Devices Lett.* **14**, 91–93 (1993).

Li, X. M., and M. J. Deen, "Determination of Interface State Density in MOSFETs Using the Spatial Profiling Charge Pumping Technique," *Solid-State Electron.* **35**, 1059–1063 (1992).

Liang, C., H. Gaw, and P. Cheng, "An Analytical Model for Self-Limiting Behavior of Hot-Carrier Degradation in 0.25 μm *n*-MOSFETs," *IEEE Electron Devices Lett.* **13**, 569–571 (1992).

Liang, E. P., "Physics of the Synchrotron Model of Cosmic Gamma Ray Bursts," *Proc. Int. Conf. X-Ray At. Inner-Shell Phys.*, Santa Cruz, CA 597–604 (1984).

Liang, M.-S., C. Chang, W. Yang, C. Hu, and R. W. Brodersen, "Hot Carriers Induced Degradation in Thin Gate Oxide MOSFETs," *Tech. Dig.—Int. Electron Devices Meet.*, 186–189 (1983).

Liang, M.-S., J. Y. Choi, P.-K. Ko, and C. Hu, "Inversion-Layer Capacitance and Mobility of Very Thin Gate-Oxide MOSFETs," *IEEE Trans. Electron Devices*, 409–413 (1986).

Lianshen, X., C. Xueliang, and X. Yuansen, "Characterization of Modeling of LDD MOSFETs," *Chin. J. Semicond.* **8**, 597–603 (1987).

Lidow, A. T. Herman, and H. W. Collins, "Power MOSFET Technology," *Tech. Dig.—Int. Electron Devices Meet.*, 79 (1979).

Lifshitz, N., G. Smolinsky, and J. M. Andrews, "Mobile Changes in a Novel Spin-On Oxide (SOx): Detection of Hydrogen in Dielectrics," *J. Electrochem. Soc.* **136**, 1440–1446 (1989).

Lifshitz, N. and G. Smolinsky, "Hot-Carrier Aging of the MOS Transistor in the Presence of Spin-On Glass as the Interlevel Dielectric," *IEEE Electron Devices Lett.* **12**, 140–142 (1991).

Ligenza, J. R., "Effect of Crystal Orientation on Oxidation Rates of Silicon in High Pressure Steam," *J. Phys. Chem.* **65**, 2011–2014 (1961).

Lin, J.-J., and J.-G. Hwu, "Application of Irradiation-Then-Anneal Treatment on the Improvement of Oxide Properties in Metal-Oxide-Semiconductor Capacitors," *Jpn. J. Appl. Phys., Part 1: Regul. Pap. Short Notes* **31**, 1290–1297 (1992).

Lin, J.-J., K.-C. Lin, and J.-G. Hwu, "Dependence on Hot-Carrier and Radiation Hardnesses of Metal-Oxide-Semiconductor Capacitors on Initial Oxide Resistance Determined by Charge-Then-Decay Method," *Jpn. J. Appl. Phys., Part 2: Lett.* **31**, 1164–1166 (1992).

Lin, P.-S. and C.-H. Chang, "A New Method to Determine the Work-Function Difference and Its Application to Calibrate the Boron-Segregation Coefficient," *IEEE Electron Devices Lett.* **12**, 638–640 (1991).

Lindenfelser, T., D. Fertig, M. Schmidt, and K. Perttula, "12-Volt Analog/Digital BICMOS Process," *Proc. Bipolar Circuits Tech. Meet.*, Minneapolis, MN, 184–187 (1987).

Lindholm, F. A. and C. T. Sah, "Circuit Technique for Semiconductor-Device Analysis with Junction Diode Open Circuit Voltage Decay Example," *Solid State Electron.* **31**, 197–204 (1988).

Lindner, R., "Semiconductor Surface Varactor, "Bell Syst. Tech. J. **41**, 803–831 (1962).

Ling, C. H., Y. T. Yeow, L. K. Ah, and W. H. Yung, "Logarithmic Time Dependence of *p*MOSFET Degradation Observed from Gate Capacitance," *Electron. Lett.* **29**, 418–420 (1993).

Liou, T.-I., C.-S. Teng, and R. B. Merrill, "Hot-Electron-Induced Degradation of Conventional, Minimum Overlap, LDD and DDD *n*-Channel MOSFETs," *IEEE Circuits Devices Mag.*, 9–15 (1988).

Lippens, D., J.-L. Nieruchalski, and E. Constant, "Particle Simulation of Impact Ionization: Application to mm-Wave Impact Devices," *Physica B + C* (*Amsterdam*) **129**, 547–551 (1985).

Liu, B. D. and I. K. Chien, "Comparison of Characteristics of Lightly-Doped Drain MOSFETs," *Solid-State Electron.* **33**, 143–144 (1990).

Liu, B. D., "Comparison of Characteristics of Conventional and LDD Short Channel MOSFETs," *Int. J. Electron.* **71**, 215–225 (1991).

Liu, D. K. Y., C. Kaya, M. Wong, J. Paterson, and P. Shah, "Optimization of a Source-Side-Injection FAMOS Cell for Flash EPROM Applications," *Tech. Dig.—Int. Electron Devices Meet.*, 315–318 (1991).

Liu, S. S., R. J. Smith, R. D. Pashley, J. Shappir, C. H. Fu, and K. R. Kokkonen, "High-Performance MOS Technology for 16 K Static RAM," *Tech. Dig.—Int. Electron Devices Meet.*, 352–354 (1979).

Liu, Z.-H., P. Nee, P.-K. Ko, C. Hu, C. G. Sodini, B. J. Gross, T.-P. Ma, and Y. C. Chang, "Field and Temperature Acceleration of Time-Dependent Dielectric Breakdown for Reoxidized–Nitrided and Fluorinated Oxides," *IEEE Electron Devices Lett.* **13**, 41–43 (1992).

Liu, Z.-H., C. Hu, J.-H. Huang, M.-C. Jeng, P.-K. Ko, and Y. C. Cheng, "Threshold Voltage Model for Deep-Submicrometer MOSFET's," *IEEE Trans. Electron Devices* **40**, 86–93 (1993).

Lo, G. Q., W. C. Ting, D. K. Shih, and D. L. Kwong, "Study of Interface State Generation in Thin Oxynitride Gate Dielectrics under Hot-Electron Stressing," *Electron Lett.* **25**, 1354–1355 (1989).

Lo, G. Q. and D. L. Kwong, "Effects of Reoxidation on the Hot-Carrier Immunity of 0.6 μm MOSFETs with Nitrided Oxides," *Electron Lett.* **26**, 1470–1471 (1990).

Lo, G. Q., W. Ting, D. L. Kwong, J. Kuehne, and C. W. Magee, "MOS Characteristics and Reliability of Thin Gate Dielectrics Grown by Rapid Thermal Processing in O_2 Diluted with NF_3," *Conf. Solid State Devices Mater., 22nd*, Sendai, Japan, 167–170 (1990).

Lo, G. Q., W. C. Ting, D. K. Shih, and D. L. Kwong, "Transconductance Degradation and Interface State Generation in Metal-Oxide-Semiconductor Field-Effect Transistors with Oxynitride Gate Dielectrics under Hot-Carrier Stress," *Appl. Phys. Lett.* **56**, 250–252 (1990).

Lo, G. Q., A. B. Joshi, and D.-L. Kwong, "Hot-Carrier-Stress Effects on Gate-Induced Drain Leakage Current in *n*-Channel MOSFETs," *IEEE Electron Devices Lett.* **12**, 5–7 (1991).

Lo, G. Q. and D.-L. Kwong, "Roles of Electron Trapping and Interface State Generation on Gate-Induced Drain Leakage Current in *p*-MOSFET's," *IEEE Electron Devices Lett.* **12**, 710–712 (1991).

Lo, G. Q., J. Ahn, and D.-L. Kwong, "Improved Hot-Carrier Immunity in CMOS Analog Device with N_2O-Nitrided Gate Oxides," *IEEE Electron Devices Lett.* **13**, 457–459 (1992).

Lo, G. Q., J. Ahn, D.-L. Kwong, and K. K. Young, "Dependence of Hot-Carrier Immunity on Channel Length and Channel Width in MOSFETs with N_2O-grown Gate Oxides," *IEEE Electron Devices Lett.* **13**, 651–653 (1992).

Lombardi, C., P. Olivo, B. Ricco, E. Sangiorgi, and M. Vanzi, "Hot Electrons in MOS Transistors: Lateral Distribution of the Trapped Oxide Charge," *IEEE Electron Devices Lett.* **3**, 215–217 (1982).

Lombardi, C., P. Olivo, B. Ricco, E. Sangiorgi, and M. Vanzi, "Two-Dimensional Effects in Hot-Electron Modified MOSFETs," *IEEE Trans. Electron Devices* **30**, 1416–1419 (1983).

Lorenz, E., J. Gyulai, L. Frey, H. Ryssel, and N. Q. Khanh, "Effect of Oxygen on the Formation of End-of-Range Disorder in Implantation Amorphized Silicon," *J. Mater. Res.* **6**, 1695–1700 (1991).

Lou, L. F. and K. S. Kitazaki, "Photon Emission from Metal-Oxide-Semiconductor Capacitors under Fast-Ramp Conditions," *J. Appl. Phys.* **66**, 5618–5621 (1989).

Lowry, L. E., K. P. MacWilliams, and M. Isaac, "The Influences of Fluorine and Process Variations on Polysilicon Film Stress and MOSFET Hot Carrier Effects," *Soc. Adv. Mater. Process Eng.*, 543–551 (1991).

Lowther, R. E. and J. Johnston, "Fast, Quasi-3D Modeling of Base Resistance for Circuit Simulation," *IEEE Trans. Electron Devices* **38**, 518–526 (1991).

Lu, C.-Y., J. J. Sung, H. C. Kirsch, N.-S. Tsai, R. Liu, A. S. Manocha, and S. J. Hillenius, "High-Performance Salicide Shallow-Junction CMOS Devices for Submicrometer VLSI Application in Twin-Tub VI," *IEEE Trans. Electron Devices* **36**, 2530–2536 (1989).

Lugli, P., "Monte Carlo Studies of Hot Electron," *Phys. Scr.* **T19A**, 190–198 (1987).

Luryi, S., "Coherent vs. Incoherent Resonant Tunneling and Implications for Fast Devices," *Superlattices Microstruct.* **5**, 375–382 (1989).

Lutze, R. S. L., H. P. Vyas, and J. S. T. Huang, "Fully Scaled Submicron Direct Write E-Beam CMOS Technology for VLSI Digital Application," *Tech. Dig.—Int. Electron Devices Meet.*, 590–592 (1984).

Lyo, I.-W. and P. Avouris, "Field-Induced Nanometer- to Atomic-Scale Manipulation of Silicon Surfaces with the STM," *Science* **253**, 173–176 (1991).

Lyumkis, E. D., B. S. Polsky, A. I. Shur, and P. Visocky, "Transient Semiconductor Device Simulation Including Energy Balance Equation," *COMPEL—Int. J. Comput. Math. Elect. Electron. Eng.* **11**, 311–325 (1992).

Ma, T.-P., "Interface Trap Transformation in Radiation or Hot-Electron Damaged MOS Structures," *Semicond. Sci. Technol.* **4**, 1061–1079 (1989).

Ma, T.-P. and P. V. Dressendorfer, eds., *Ionizing Radiation Effects in MOS Devices and Circuits*, Wiley, New York (1989).

Mackens, U. and A. J. Walker, "Effect of SR X-Ray Lithography on MOS Gate Oxide and 0.5 μ *p*-Channel MOSFETs," *Microelectron. Eng.* **11**, 271–274 (1990).

MacWilliams, K. P., L. F. Halle, and T. C. Zietlow, "Improved Hot-Carrier Resistance with Fluorinated Gate Oxides," *IEEE Electron Devices Lett.* **11**, 3–5 (1990).

MacWilliams, K. P., L. E. Lowry, D. J. Swanson, and J. Scarpulla, "Wafer-Mapping of Hot Carrier Lifetime Due to Physical Stress Effects (MOSFET)," *Tech. Dig. VLSI Tech. Symp.*, 100–1001 (1992).

Maes, H. E. and G. Groeseneken, "Effect of Channel Hot Electron Injection on the Spatial Distribution of the Interface Charge in MOS and FAMOS Devices," *11th Eur. Solid State Devices Res. Conf. & 6th SSSDT—Europhys. Conf. Abstr.* **5F**, 49–50 (1981).

Maes, H. E. and G. Groeseneken, "Determination of Spatial Surface State Density Distribution in MOS and SIMOS Transistors after Channel Hot Electron Injection," *Electron. Lett.* **18**, 372–374 (1982).

Maes, H. E., G. Groeseneken, P. Heremans, and R. Bellens, "Understanding of the Hot Carrier Degradation Behavior of MOSFETs by Means of the Charge Pumping Technique," *Appl. Surf. Sci.* **39**, 523–534 (1989).

Maes, H. E., P. Heremans, R. Bellens, and G. Groeseneken, "Hot Carrier Degradation in MOSFETs in the Temperature Range of 77–300 K," *Qual. Reliab. Eng. Int.* **7**, 307–322 (1991).

Mahan, G. D., "Hot Electrons in One Dimension," *J. Appl. Phys.* **58**, 2242–2251 (1985).

Mahnkopf, R., G. Przyrembel, and H. G. Wagemann, "Annealing of Hot-Carrier-Induced MOSFET Degradation," *J. Phys., Colloq. (Orsay, Fr.)* **49**, 771–774 (1988).

Mahnkopf, R., G. Przyrembel, and H. G. Wagemann, "A New Method for the Determination of the Spatial Distribution of Hot Carrier Damage," *J. Phys., Colloq. (Orsay, Fr.)* **49**, 775–778 (1988).

Mahnkopf, R., G. Przyrembel, W. Seifert, and H. G. Wagemann, "Characterization of Hot Carrier Trapping in the Gate Oxide of MOSFETs," *Proc. Eur. Solid State Devices Res. Conf., 19th*, 25–28 (1989).

Mahnkopf, R., G. Przyrembel, and H. G. Wagemann, "Characterization of Hot Carrier Degradation within the Gate Oxide of Short Channel MOSFETs," *Arch. Elektrotech. (Berlin)* **74**, 379–387 (1991).

Majkusiak, B., "Gate Tunnel Current in an MOS Transistor," *IEEE Trans. Electron Devices* **37**, 1087–1092 (1990).

Makino, T., N. Sato, M. Takeda, Y. Furumura, and K. Imaoka, "A Stacked-Source-Drain MOSFET using Selective Epitaxy," *Proc. Int. Symp. Adv. Mater. ULSI*, 113–120 (1988).

Makino, T. and S. Kawamura, "A Study of Hot-Carrier-Effect in SOI-MOSFETs Using Photon Emission," *Ext. Abstr., 1991 Int. Conf. Solid State Devices Mater.*, 20–22 (1991).

Malwah, M. L., J. R. Edwards, and M. Bandali, "Hot Carrier Injection in the Dual Polysilicon Gate Structure and Its Related Reliability Effects on Dynamic RAM Refresh Time," *Proc. Int. Reliab. Phys. Symp.*, 23–26 (1977).

Manchanda, L., "The Effect of $TaSi_2/n_+$ Poly Gate on Hot-Electron Instability of Small Channel MOSFETs," *Proc. Int. Reliab. Phys. Symp.*, 199–204 (1984).

Manchanda, L., "Hot-Electron Trapping and Generic Reliability of p_+ Polysilicon/SiO_2/Si Structures for Fine-Line CMOS Technology," *Proc. Int. Reliab. Phys. Symp.*, 183–188 (1986).

Mano, T., J. Yamada, J. Inoue, and S. Nakajima, "Circuit Techniques for a VLSI Memory," *IEEE J. Solid-State Circuits* **18**, 463–370 (1983).

Marchetaux, J.-C., M. Bourcerie, A. Boudou, and D. Vuillaume, "Application of the Floating-Gate Technique to the Study of the *n*-MOSFET Gate Current Evolution Due to Hot-Carrier Aging," *IEEE Electron Devices Lett.* **11**, 406–408 (1990).

Markiewicz, R. S., "Kinetics of Electron–Hole Droplet Clouds: The Role of Thermalization Phonons," *Phys. Rev. B* **21**, 4674–4691 (1980).

Markiewicz, R. S., "Role of Thermalization Phonons in Producing the Electron–Hole Droplet Cloud," *Solid State Commun.* **33**, 701–705 (1980).

Masuda, H., M. Nakai, and M. Kubo, "Characteristics and Limitation of Scaled-Down MOSFETs Due to Two-Dimensional Field Effect," *IEEE Trans. Electron Devices* **26**, 980–986 (1979).

Masuda, S., M. Aoyama, K. Ohno, and Y. Harada, "Observation of Unusually Enhanced Satellite Band in Penning-Ionization Electron Spectra of Benzene and Toluene," *Phys. Rev. Lett.* **65**, 3257–3260 (1990).

Masuda, S., H. Hayashi, and Y. Harada, "Spatial Distribution of the Wave Functions of a Graphite Surface Studied by Use of Metastable-Atom Electron Spectroscopy," *Phys. Rev. B* **42**, 3582–3585 (1990).

Masuda, S., M. Aoyama, and Y. Harada, "Spatial Distribution of 3*d* Electron in Sandwich Compounds Studied by Penning Ionization Electron Spectroscopy: Ferrocene and Dibenzenechromium," *J. Am. Chem. Soc.* **112**, 6445–6446 (1990).

Masuda, S., H. Ishii, and Y. Harada, "Two-Hole States of the Outermost Surface Layer Studied by Metastable Atom Electron Spectroscopy Si(111)-7 × 7 and Si(100)-2 × 1," *Surf. Sci.* **242**, 400–403 (1991).

Masuhara, T. and R. S. Muller, "Analytical Technique for the Design of DMOS Transistor," *Jpn. J. Appl. Phys.* **16**, 173 (1976).

Masuhara, T., K. Shimohigashi, and E. Takeda, "Present Status and Future of High Performance Submicron Devices," *Trans. Inst. Electron. Inf. Commun. Eng. Jpn.* **72C-11**, 298–311 (1989).

Matsuhashi, H., T. Hayashi, and S. Nishikawa, "New Hot-Carrier Degradation Mode in PMOSFETs with W Gate Electrodes," *IEEE Electron Devices Lett.* **12**, 539–541 (1991).

Matsuhashi, H. and S. Nishikawa, "Effect of Gate Materials on Generation of Interface State by Hot-Carrier Injection," *Ext. Abstr., 1992 Int. Conf. Solid State Devices Mater.*, 515–517 (1992).

Masumoto, H., K. Sawada, S. Asai, M. Hirayama, and K. Nagasawa, "The Effect of Gate Bias on Hot Electron Trapping," *Jpn. J. Appl. Phys.* **19**, L574–L576 (1980).

Matsumoto, H., K. Sawada, S. Asai, M. Hirayama, and K. Nagasawa, "Effect of Long Term Stress on Hot Electron Trapping," *Jpn. J. Appl. Phys.* **20**, Suppl. 1, 255–260 (1981).

Matsumoto, H., K. Miyamoto, Y. Satoh, and S. Ishida, "Comparison of Impact Ionization Current Between *p*-MOS and *n*-MOS," *Jpn. J. Appl. Phys.* **23**, L546–L548 (1984).

Matusmoto, M., Y. Kimura, K. Hirayama, H. Koyama, N. Maki, and H. Matsumoto, "Degradation Mechanism Due to Hot Electron Trapping in High Density CMOS DRAM," *Proc. Int. Symp. Testing Failure Anal.*, Los Angeles, CA, 89–94 (1987).

Matusmoto, Y., T. Higuchi, S. Sawada, S. Shinozaki, and O. Ozawa, "Optimized and Reliable LDD Structure for 1-μ *n*-MOSFET Based on Substrate Current Analysis," *Tech. Dig.—Int. Electron Devices Mater.*, 392–395 (1983).

Matsumoto, Y., T. Higuchi, T. Mizuno, S. Sawada, S. Shinozaki, and O. Ozawa, "Optimized and Reliability LDD Structure for 1-μ *n*-MOSFET Based on Substrate Current Analysis," *IEEE J. Solid-State Circuits* **SC-20**, 349–353 (1985).

Matsunaga, J., "Characterization of Two Step Impact Ionization and Its Influence on *n*-MOS and *p*-MOS VLSI's," *Tech. Dig.—Int. Electron Devices Meet.*, 736 (1980).

Matsunaga, J.-I., S. Kohyama, M. Konaka, and H. Lizuka, "Design Limitations Due to Substrate Currents and Secondary Impact Ionization Electrons in *n*-MOS LSI's," *Jpn. J. Appl. Phys.* **19**, Suppl. 1, 93–97 (1980).

Matsuoka, F., H. Hayashida, K. Hama, Y. Toyoshima, H. Iwai, and K. Maeguchi, "Drain Avalanche Hot Hole Injection Mode on *p*-MOSFETs," *Tech. Dig.—Int. Electron Devices Meet.*, 18–21 (1988).

Matsuoka, F., H. Iwai, H. Hayashida, K. Hama, Y. Toyoshima, and K. Maeguchi, "Analysis of Hot-Carrier-Induced Degradation Mode on *p*-MOSFETs," *IEEE Trans. Electron Devices* **37**, 1487–1495 (1990).

Matsuoka, H., Y. Igura, and E. Takeda, "Device Degradation Due to Band-to-Band Tunneling," *Ext. Abstr., Solid State Devices Mater.*, 589–592 (1988).

Matsuoka, H., T. Ichiguchi, T. Yoshimura, and E. Takeda, "Mobility Modulation in a Quasi-One-Dimensional Si-MOSFET with a Dual-Gate Structure," *IEEE Electron Devices Lett.* **13**, 20–22 (1992).

Matsushita, T., N. Oh-uchi, H. Hayashi, and H. Yamoto, "A Silicon Heterojunction Transistor," *Appl. Phys. Lett.* **35**, 549–550 (1979).

Matsuzaki, N., A. Watanabe, M. Minami, and T. Nagano, "Increased Hot-Carrier Degradation of NMOSFETs under Very Fast Transient Stressing," *IEEE Int. Reliability Physics Symp.*, 129–132 (1991).

Matsuzawa, K., M. Takahashi, M. Yoshimi, and N. Shigyo, "Simulation of Velocity Overshoot and Hot Carrier Effects in Thin-Film SOI-nMOSFETs," *IEICE Trans. Electron.* **E75-C**, 1477–1483 (1992).

May, H. and R. Kienzler, "Limitations of MOS-Memory Cells," *Arch. Elektrotech. (Berlin)* **63**, 327–336 (1981).

May, T. C. and M. H. Woods, "Alpha-Particle-Induced Soft Errors in Dynamic Memories," *IEEE Trans. Electron Devices* **26**, 2–9 (1979).

Mayaram, K., J. Lee, T. Y. Chan, and C. Hu, "An Analytical Perspective of LDD MOSFET's," *Tech. Dig. VLSI Tech. Symp.*, 61–62 (1986).

McAndrew, C. C., E. L. Heasell, and K. Singhal, "Carrier Dynamical VLSI Device Simulation," *Semicond. Sci. Technol.* **3**, 886–894 (1988).

McBrayer, J. D., D. M. Fleetwood, R. A. Pastorek, and R. V. Jones, "Correlation of Hot-Carrier and Radiation Effects in MOS Transistors," *IEEE Trans. Nucl. Sci.* **NS-32**, 3935–3939 (1985).

McBrayer, J. D., R. A. Pastorek, R. V. Jones, and A. Ochoa, "Model Describing Hot-Carrier and Radiation Effects in MOS Transistors," *IEEE Trans. Nucl. Sci.* **NS-34**, 1647–1651 (1987).

McCall, G. H., "Laser-Drive Implosion Experiment," *Plasma Phys.* **25**, 237–285 (1983).

Meguro, S., K. Komori, and Y. Sakai, "Evolution of HI-CMOS Technology," *Hitachi Rev.* **34**, 271–274 (1985).

Meinerzhagen, B., "Consistent Gate and Substrate Current Modeling Based on Energy Transport and the Lucky Electron Concept," *Tech. Dig.—Int. Electron Devices Meet.*, 504–507 (1988).

Melen, R., and D. Buss, eds., *Charge-Coupled Devices: Technology and Applications*, IEEE Press, New York (1977).

Meyer, W. G. and R. B. Fair, "Dynamic Behaviour of the Buildup of Fixed Charge and Interface States during Hot-Carrier Injection in Encapsulated MOSFET's," *IEEE Trans. Electron Devices* **30**, 96–103 (1983).

Meyer-Vernet, N., "Comet Giacobini–Zinner Diagnosis from Radio Measurements," *Adv. Space Res.* **5**, 37–46 (1985).

Mikawa, R. E. and P. M. Lenahan, "Comparison of Ionizing Radiation and Hot Electron Effects in MOS Structures," *IEEE Trans. Nucl. Sci.* **NS-31**, 1573–1575 (1984).

Mikoshiba, H., T. Horiuchi, and K. Hamano, "Comparison of Drain Structures in n-Channel MOSFETs," *IEEE Trans. Electron Devices* **33**, 140–44 (1986).

Miller, R. L., "Stability of EBT in Guiding-Centre Fluid Theory," *Nucl. Fusion* **21**, 1249–1267 (1981).

Mistry, K. R., and B. S. Boyle, "An Empirical Model for the L_{eff} Dependence of Hot-Carrier Lifetimes of n-Channel MOSFETs," *IEEE Electron Devices Lett.* **10**, 500–502 (1989).

Mistry, K. R., and B. S. Doyle, "Hot Carrier Degradation in *n*-MOSFETs Used as Pass Transistors," *IEEE Trans. Electron Devices* **37**, 2415–2416 (1990).

Mistry, K. R., and B. S. Doyle, "The Role of Electron Trap Creation in Enhanced Hot-Carrier Degradation during AC Stress," *IEEE Electron Devices Lett.* **11**, 267–269 (1990).

Mistry, K. R., D. B. Krakauer, and B. S. Doyle, "Impact of Snapback-Induced Hole Injection on Gate Oxide Reliability of *n*-MOSFETs," *IEEE Electron Devices Lett.* **11**, 460–462 (1990).

Mistry, K. R., and B. S. Doyle, "A Model for AC Hot-Carrier Degradation in *n*-Channel MOSFETs," *IEEE Electron Devices Lett.* **12**, 492–494 (1991).

Mistry, K. R., B. S. Doyle, A. Philipossian, and D. B. Jackson, "AC Hot Carrier Lifetimes in Oxide and ROXNOX N-Channel MOSFETs," *Tech. Dig.—Int. Electron Devices Meet.*, 727–730 (1991).

Mistry, K. R., D. B. Krakauer, B. S. Doyle, T. A. Spooner, and D. B. Jackson, "An In-Process Monitor for *n*-Channel MOSFET Hot Carrier Lifetimes," *Proc. Int. Reliab. Phys. Symp.*, 116–121 (1992).

Mistry, K. R., and B. S. Doyle, "AC versus DC Hot-Carrier Degradation in *n*-Channel MOSFET's," *IEEE Trans. Electron Devices* **40**, 96–103 (1993).

Mitsuhashi, J., T. Matsukawa, K. Sugimoto, and S. Kawazu, "Hot Carrier Injection in Submicron MOSFETs Fabricated with Rapid Isothermal Annealing," *Tech. Dig. VLSI Tech. Symp.*, 84–85 (1984).

Mitsuhashi, J., S. Nakao, and T. Matsukawa, "Mechanical Stress and Hydrogen Effects on Hot Carrier Injection," *Tech. Dig.—Int. Electron Devices Meet.*, 386–389 (1986).

Mittl, S. W. and M. J. Hargrove, "Hot Carrier Degradation in *p*-Channel MOSFETs," *Proc. Int. Reliab. Phys. Symp.*, 98–102 (1989).

Miura, Y., "Effect of Orientation on Surface Charge Density at Si–SiO_2 Interface," *Jpn. J. Appl. Phys.* **4**, 958–961 (1965).

Miura-Mattauschi, M. and G. Dorda, "1D Analytical Treatment of Hot-Electron Effects in Short-Channel MOSFET's," *Physica B + C (Amsterdam)* **134**, 77–81 (1985).

Miura-Marrausch M., A. V. Schwerin, W. Weber, C. Werner, and G. Dorda, "Gate Currents in Thin Oxide MOSFETs," *Proc. IEE Solid-State Electron Devices, Part I* **134**, 111–115 (1987).

Miyamoto, K., H. Matsumoto, Y. Ohbayashi, I. Ohkura, and H. Matsumura, "A New Breakdown Phenomenon in Fine Structured VLSI Circuits," *Jpn. J. Appl. Phys.* **20**, L523–L525 (1981).

Mizuno, T., Y. Matsumoto, S. Sawada, S. Shinozaki, and O. Ozawa, "A New Degradation Mechanism of Current Derivability and Reliability of Asymmetrical LDD MOSFETs," *Tech. Dig.—Int. Electron Devices Meet.*, 250–253 (1985).

Mizuno, T., J. Kumagai, Y. Matsumoto, S. Sawada, and S. Shinozaki, "New Degradation Phenomena by Source and Drain Hot-Carriers in Half Micron p-MOSFETs," *Tech. Dig.—Int. Electron Devices Meet.*, 726–729 (1986).

Mizuno, T., S. Sawada, Y. Saitoh, and S. Shinozaki, "Si_3N_4/SiO_2 Spacer Induced High Reliability in LDD MOSFET and Its Simple Degradation Model," *Tech. Dig.—Int. Electron Devices Meet.*, 234–237 (1988).

Mizuno, T., S. Sawada, Y. Saitoh, and T. Tanaka, "Hot-Carrier Injection Suppression Due to the Nitride–Oxide LDD Spacer Structure," *IEEE Trans. Electron Devices* **38**, 584–591 (1991).

Mizutani, Y., S. Taguchi, M. Nakahara, and H. Tango, "Hot Carrier Instability in Submicron $MoSi_2$ Gate MOS/SOS Devices," *Tech. Dig.—Int. Electron Devices Meet.*, 550–553 (1981).

Moazzami, R. and C. Hu, "Projecting Gate Oxide Reliability and Optimizing Reliability Screens," *IEEE Trans. Electron Devices* **37**, 1643–1650 (1990).

Modelli, A., "Valance Band Electron Tunneling in Metal-Oxide–Silicon Structures," *Appl. Surf. Sci.* **30**, 298–303 (1987).

Mogami, T., L. E. G. Johansson, I. Sakai, M. Fukuma, "Hot-Carrier Effects in Surface-Channel PMOSFETs with BF_2- or Boron-Implanted Gates," *Int. Electron Devices Meet., 1991, Tech. Dig.*, 533–536 (1991).

Mohamedi, S. Z., V.-H. Chan, J.-T. Park, F. Nouri, B. W. Scharf, and J. E. Chung, "Hot-Electron-Induced Input Offset Voltage Degradation in CMOS Differential Amplifiers," *30th Ann. Proc. Reliab. Phys.*, 76–80 (1992).

Moll, J. L., "Variable Capacitance with Large Capacity Change," *Wescon Conv. Rec.*, 32 (1959).

Momose, H., M. Saitoh, H. Shibata, T. Maeda, H. Sasaki, K. Satoh, and T. Ohtani, "Performance of CMOS Circuits with LDD-Type *n*-MOSFET's for High Density Static RAM's,' *Tech. Dig.—Int. Electron Devices Meet.*, 304–307 (1984).

Momose, H., H. Shibata, S. Saitoh, J. Miyamoto, K. Kanzaki, and S. Kohyama, "1.0-μm n-Well CMOS/Bipolar Technology," *IEEE J. Solid-State Circuits* **SC-20**, 137–143 (1985).

Momose, H. S., S. Kitagawa, K. Yamabe, and H. Iwai, "Hot Carrier Related Phenomena for *n*- and *p*-MOSFETs with Nitrided Gate Oxide by RTP," *Tech. Dig.—Int. Electron Devices Meet.*, 267–270 (1989).

Momose, H. S., T. Morimoto, Y. Ozawa, M. Tsuchiaki, M. Ono, K. Yamabe, and H. Iwai, "Very Lightly Nitrided Oxide Gate MOSFETs for Deep-Sub-micron CMOS Devices," *Int. Electron Devices Meet., 1991, Tech. Dig.*, 359–362 (1991).

Monkowski, J. R., M. A. Logan, D. W. Freeman, G. A. Brown, and G. A. Ruggles, "High Integrity Thin Silicon Oxides from CVD of Tetraethylorthosilicate," *Proc. Int. Conf. Chem. Vapor Deposition*, Honolulu, HI, 508–517 (1987).

Moon, B.-J., C.-K. Park, K. Lee, and M. Shur, "New Short-Channel n-MOSFET Current–Voltage Model in Strong Inversion and Unified Parameter Extraction Method," *IEEE Trans. Electron Devices* **38**, 592–602 (1991).

Moon, J. E., T. Garfinkel, J. Chung, M. Wong, P.-K. Ko, and C. Hu, "A New LDD Structure: Total Overlap with Polysilicon Spacer (TOPS)," *IEEE Electron Devices Lett.* **11**, 221–223 (1990).

Moore, G., "VLSI: Some Fundamental Challenges," *IEEE Spectrum* **16**, 30 (1979).

Moore, G. E., C. T. Sah, and F. Wanlass, "MOS Field-Effect Devices for Micropower Logic Circuitry," In *Micropower Electronics*, E. Keonjian, ed., pp. 41–55. Pergamon, New York, (1964).

Moravvej-Farshi, M. K. and M. A. Green, "Effects of Interfacial Oxide Layer on Short-Channel Polycrystaline Source and Drain MOSFETs," *IEEE Electron Devices Lett.* **8**, 165–167 (1987).

Morimoto, T. and S. Muramoto, "Two-Dimensional Analysis of Hot-Electron Emission Current in MOSFET," *Electron. Commun. Jpn.* **65**, 116–124 (1982).

Morita, S., T. Noguchi, and N. Mikoshiba, "Peak Shifts of the Photo-Induced Magnetophonon Resonance in *n*-InSb at 4.2 K," *Jpn. J. Phys. Soc.* **45**, 522–527 (1978).

Moslehi, M. M. and K. C. Saraswat, "Studies of Trapping and Conduction in Ultra Thin SiO_2 Gate Insulators," *Tech. Dig.—Int. Electron Devices Meet.*, 157–160 (1984).

Mountain, D. J. and D. M. Burnell, "An Evaluation of Conventional and LDD Devices for Submicron Geometries," *Solid-State Electron.* **33**, 565–570 (1990).

Moynagh, P. B., P. J. Rossner, and R. B. Calligarro, "Rapid Thermal Dielectrics for Gate Applications in VLSI," *Proc. IT Conf.*, Swansea, UK, 540–543 (1988).

Mueller, G. O., R. Mach, G. U. Reinsperger, and G. Schulz, "Hot Electron Cold Cathode for CRTs, etc.," *Conf. Re. 1991 Int. Display Res. Conf.*, 16–19 (1991).

Mueller, R. G., H. Nietsch, B. Roessler, and E. Wolter, "8192-Bit Electrically Alterable ROM Employing a One-Transistor Cell with Floating Gate," *IEEE J. Solid-State Circuits* **12**, 507–514 (1977).

Mühlhoff, H.-M., P. Murkin, M. Orlowski, W. Weber, K. H. Küsters, W. Müller, C. M. Rogers, and H. Wendt, "Submicron *p*-MOSFETs under Static and SWAP Stress," *Tech. Dig. VLSI Tech. Symp.*, 57–58 (1987).

Mühlhoff, H.-M., M. Steimle, and J. Dietl, "The Impact of Hot Carrier Degradation on Scaling of Sub-μm CMOS Processes," *Proc. 7th Bienn. Eur. Conf.*, 93–106 (1991).

Mukherjee, S., T. Chang, R. Pang, M. Knecht, and D. Hu, "Single Transistor EEPROM Cell and Its Implementation in a 512 K CMOS EEPROM," *Tech. Dig.—Int. Electron Devices Meet.*, 616–619 (1985).

Muller, R. S. and T. I. Kamins, *Device Electronics for Integrated Circuits*, 2nd ed. Wiley, New York (1986).

Murai, F., O. Suga, S. Okazaki, and R. Haruta, "The Effect of EB Dose on Hot Carrier Induced Degradation of MOS Transistors," *Proc. SPIE—Int. Soc. Opt. Eng.* **1089**, 367–373 (1989).

Murarka, S. P. "Refractory Silicides for Integrated Circuits," *J. Vac. Sci. Technol.* **17**, 775–792 (1980).

Murray, D. C., A. G. R. Evans, and J. C. Carter, "Shallow Defects Responsible for GR Noise in MOSFETs," *IEEE Trans. Electron Devices* **38**, 407–416 (1991).

Murray, S. J. and N. N. Duncan, "Hot Carrier Effects in *n*-Channel Silicon on Sapphire Transistors," *IEE Colloq. 'Hot Carrier Degradation Short Channel MOS,'* 10.1–10.4 (1987).

Nagai, K., Y. Hayashi, and Y. Tarui, "Carrier Injection into SiO_2 from Surface Driven to Avalanche Breakdown by a Linear RAMP Pulse, and Trapping, Distribution and Thermal Annealing of Injected Holes in SiO_2," *Jpn. J. Appl. Phys.* **14**, 1539–1545 (1975).

Nagai, K., Y. Hayashi, and Y. Tarui, "Avalanche Breakdown of MOS DIODE by Linear RAMP Pulse and Characteristics of Carrier Injected into SiO_2," *Bull. Electrotech. Lab.*, (*Tokyo*) **40**, 263–271 (1976).

Naiman, M. L., F. L. Terry, J. A. Burns, J. I. Raffel, and R. Aucoin, "Properties of thin oxinitride gate dielectrics produced by thermal nitridation of silicon dioxide," *IEEE IEDM Digest Tech. Papers*, 562–564 (1980).

Nakagome, Y., E. Takeda, H. Kume, and S. Asai, "New Observation of Hot-Carrier Injection Phenomena," *Jpn. J. Appl. Phys.* **22**, Suppl. 1, 99–102 (1983).

Nakahara, M., H. Iwasawa, and K. Yasutake, "Anomalous Enhancement of Substrate Terminal Current Beyond Pinch-Off in Silicon *n*-Channel MOS Transistor," *Electron. Commun. Jpn.* **52**, 173–180 (1969).

Nakahara, M., Y. Hitura, T. Noguchi, M. Yoshida, K. Maeguchi, and K. Kanzaki, "Relief of Hot Carrier Constraint on Submicron CMOS Devices by Use of a Buried Channel Structure," *Tech. Dig.—Int. Electron Devices Meet.*, 238–241 (1985).

Nakajima, S., K. Miura, K. Minegishi, and T. Morie, "An Isolation Merged Vertical Capacitor Cell for Large Capacity DRAM," *Tech. Dig.—Int. Electron Devices Meet.*, 240–243 (1984).

Nakamura, A. and C. Weisbuch, "Resonant Raman Scattering vs. Hot Electron Effects in Excitation Spectra of CdTe," *Solid-State Electron.* **21**, 1331–1336 (1978).

Nakamura, H., N. Yasuda, K. Tanaguchi, and C. Hamaguchi, "New Evidence for Double Charged Oxide Trap of Submicron MOSFET's," *Ext. Abstr., Solid State Devices Mater.*, 465–468 (1989).

Nakamura, K., M. Yanagisawa, Y. Nio, K. Okamura, and M. Kikuchi, "Buried Isolation Capactiro (BIC) Cell for Megabit MOS Dynamic RAM," *Tech. Dig.—Int. Electron Devices Meet.*, 236–239 (1984).

Nandakumar, R., P. Bhattacharya, and G. Kousik, "Study of Scanning Electron Microscope Irradiated Damage to Gate Oxides of Metal Oxide Semiconductor Field Effect Transistors," *Jpn. J. Appl. Phys., Part 1: Regul. Pap. Short Notes* **31**, 2651–2655 (1992).

Naruke, K., S. Yamada, E. Obi, S. Taguchi, and M. Wada, "A New Flash-Erase EEPROM Cell with a Sidewall Select-Gate on Its Source Side," *Tech. Dig.—Int. Electron Devices Meet.*, 603–606 (1989).

Nasr, A. I. and R. J. Hollingsworth, "Device and Reliability Optimization in Submicron MOSFET," *Proc. IEEE Int. Comp. Design Conf.*, Port Chester, NY, 555–560 (1984).

Nasr, A. I., G. J. Grula, A. C. Berti, and R. D. Jones, "CMOS-4 Technology for Fast Logic and Dense On-Chip Memory," *Digital Tech. J.* **4**, 39–50 (1992).

Nelson, D. F. and J. A. Cooper, "High-Field Electron Velocities in Silicon Surface Inversion Layers," *Surf. Sci.* **113**, 267–272 (1982).

Neppl, F., J. P. Kotthaus, and J. F. Koch, "Mechanism of Intersubband Resonant Photoresponse," *Phys. Rev. B* **19**, 5240–5250 (1979).

Neugebauer, T., G. Landwehr, and K. Hess, "Negative Differential Resistance in (100) *n*-Channel Silicon Inversion Layers," *Surf. Sci.* **73**, 163–165 (1977).

Neugebauer, T. and G. Landwehr, "Negative Differential Resistance in (100) *n*-Channel Silicon Inversion Layers," *Solid-State Electron.* **21**, 143–146 (1978).

Ng, K. K., G. W. Taylor, and A. K. Sinha, "Instability of MOSFETs Due to Redistribution of Oxide Charges," *IEEE Trans. Electron Devices* **29**, 1323–1330 (1982).

Ng, K. K. and G. W. Taylor, "Effects of Hot-Carrier Trapping in *n*- and *p*-Channel MOSFET's," *IEEE Trans. Electron Devices* **30**, 871–876 (1983).

Ng, K. K. and J. J. Brews, "Measuring the Effective Channel Length of MOSFETs," *Circuits Devices* **6**, 33–38 (1990).

Ng, K. K., C.-S. Pai, W. M. Mansfield, and G. A. Clarke, "Suppression of Hot-Carrier Degradation in Si MOSFETs by Germanium Doping," *IEEE Electron Devices Lett.* **11**, 45–47 (1990).

Nguyen, T. N. and J. D. Plummer, "Physical Mechanisms Responsible for Short Channel Effects in MOS Devices," *Tech. Dig.—Int. Electron Devices Meet.*, 596–599 (1981).

Nguyen-duc, C., S. Cristoloveanu, G. Reimbold, and J. Gautier, "Degradation of Short-Channel MOS Transistors Stressed at Low Temperature," *J. Phys., Colloq. (Orsay, Fr.)* **49**, 661–664 (1988).

Nguyen-duc., C., S. Cristoloveanu, and G. Reimbold, "Effects of Localized Interface Defects Caused by Hot-Carrier Stress in *n*-Channel MOSFET's at Low Temperature," *IEEE Electron Devices Lett.* **9**, 479–481 (1988).

Ni, J. S., "Modeling of Hot Electron Effects on the Device Parameters for Circuit Simulation," *Tech. Dig.—Int. Electron Devices Meet.*, 738–741 (1986).

Nicholas, R. J., "The Magnetophonon Effect," *Prog. Quantum Electron.* **10**, 1–75 (1985).

Nicollian, E. H. and A. Goetzberger, "MOS Conductance Technique for Measuring Surface State Parameters," *Appl. Phys. Lett.* **7**, 216–219 (1965).

Nicollian, E. H. and A. Goetzberger, "The Si–SiO_2 Interface—Electrical Properties as Determined by the MIS Conductance Technique," *Bell Syst. Tech. J.* **46**, 1055–1133 (1967).

Nicollian, E. H., A. Goetzberger, and C. N. Berglund, "Avalanche Injection Currents and Charging Phenomena in Thermal SiO_2," *Appl. Phys. Lett.* **15**, 174–177 (1969).

Nicollian, E. H. and C. N. Berglund, "Avalanche Injection of Electrons into Insulating SiO_2 Using MOS Structures," *J. Appl. Phys.* **41**, 3053–3057 (1970).

Nicollian, E. H., C. M. Berglund, P. F. Schmidt, and J. M. Andrews, "Electrochemical Charging of Thermal SiO_2 Film by Injected Electron Currents," *J. Appl. Phys.* **42**, 5654–5664 (1971).

Nicollian, E. H. and J. R. Brews, *MOS Physics and Technology*, Wiley, New York (1982).

Niehaus, W. C., T. E. Seidel and D. E. Iglesias, "Double Drift IMPATT Diodes near 100GHz," *IEEE Trans. Electron Devices* **20**, 765–771 (1973).

Nihira, H., M. Konaka, H. Iwai, and Y. Nishi, "Anomalous Drain Current in N-MOSFETs and Its Suppression by Deep Ion Implantation," *Tech. Dig.—Int. Electron Devices Meet.*, 487 (1978).

Nikol'skiy, O. A. and V. I. Yudin, "Energy Spectrum of the Electrons of a Plasma in an Electromagnetic Field," *Radio Eng. Electron. Phys. (Engl. Transl.)* **22**, 67–70 (1977).

Ning, T. H. and H. N. Yu, "Optically Induced Injection of Hot Electrons into SiO_2," *J. Appl. Phys.* **45**, 5373–5378 (1974).

Ning, T. H., "High Field Capture of Electrons by Coulomb Attractive Centers in Silicon Dioxides," *J. Appl. Phys.* **47**, 3203–3208 (1976).

Ning, T. H., C. M. Osburn, and H. N. Yu, "Effect of Electron Trapping on IGFET Characteristics," *J. Electron. Mater.* **6**, 65–76 (1977a).

Ning, T. H., C. M. Osburn, and H. N. Yu, "Emission Probability of Hot Electrons from Silicon into Silicon Dioxide," *J. Appl. Phys.* **48**, 286–293 (1977b).

Ning, T. H., "Hot-Carrier Emission Currents in N-Channel IGFETs," *Tech. Dig.—Int. Electron Devices Meet.*, 144–147 (1977).

Ning, T. H., "Hot-Electron Emission from Silicon into Silicon Dioxide," *Solid-State Electron.* **21**, 273–282 (1978).

Ning, T. H., P. W. Cook, R. H. Dennard, C. M. Osburn, S. E. Schuster, and H. N. Yu, "1 μm MOSFET VLSI Technology. Part IV: Hot-Electron Design Constraints," *IEEE Trans. Electron Devices* **26**, 346–353 (1979).

Nishi, Y., "Direction of VLSI CMOS Technology," *J. Hewlett-Packard*, 24–25 (1985).

Nishida, T. and C.-T. Sah, "Physically Based Mobility Model for MOSFET Numerical Simulation," *IEEE Trans. Electron Devices* **34**, 310–320 (1987).

Nishida, Y., T. Nagasawa, and N. Sato, "Resonant Acceleration of Electrons in a Weakly Magnetized Inhomogeneous Plasma," *IEEE Int. Conf. Plasma Sci.*, 98–99 (1986).

Nishimatsu, S., Y. Kawamoto, H. Masuda, R. Hori, and O. Minato, "Grooved Gate MOSFET," *Jpn. J. Appl. Phys.* **16**, Suppl. 1, 17–183 (1976).

Nishimatsu, K. and I. Nishimae, "Reliability of the CMOS Process," *Hitachi Rev.* **33**, 225–228 (1984).

Nishioka, Y., E. F. Da Silva, and T.-P. Ma, "Radiation-Induced Interface Traps in Mo/SiO_2/Si Capacitor," *IEEE Trans. Nucl. Sci.* **NS-34**, 1166–1171 (1987).

Nishioka, Y., E. F. Da Silva, and T.-P. Ma, "Time-Dependent Evolution of Interface Traps in Hot-Electron Damaged Metal/SiO_2/Si Capacitors," *IEEE Electron Devices Lett.* **8**, 566–568 (1987).

Nishioka, Y., E. F. Da Silva, Y. Wang, and T.-P. Yu, "Dramatic Improvement of Hot-Electron-Induced Interface Degradation in MOS Structures Containing F or Cl in SiO_2," *IEEE Electron Devices Lett.* **9**, 38–40 (1988).

Nishioka, Y., E. F. Da Silva, and T.-P. Ma, "Equivalence between Interface Traps in SiO_2/Si Generated by Radiation Damage and Hot-Electron Injection," *Appl. Phys. Lett.* **52**, 720–722 (1988).

Nishioka, Y., E. F. Da Silva, F. Eronides, and T.-P. Ma, "Evidence for (100) Si/SiO_2 Interfacial Defect Transformation after Ionizing Radiation," *IEEE Trans. Nucl. Sci.* **NS-35**, 1227–1233 (1988).

Nishioka, Y., K. Ohyu, Y. Ohji, N. Natuaki, K. Mukai, and T.-P. Ma, "Hot-Electron Hardened Si-Gate MOSFET Utilizing F Implantation," *IEEE Electron Devices Lett.* **10**, 141–143 (1989).

Nishioka, Y., K. Ohyu, and T.-P. Ma, "Channel Length and Width Dependence of Hot-Carrier Hardness in Fluorinated MOSFETs," *IEEE Electron Devices Lett.* **10**, 540–542 (1989).

Nishioka, Y., Y. Ohji, K. Ohyu, K. Mukai, and T.-P. Ma, "Hot Electron Damage-Resistant Si-Gate Submicrometer MOSFETs with a Fluorinated Oxide," *IEEE Trans. Electron Devices* **36**, 2604 (1989).

Nishioka, Y., K. Ohyu, Y. Ohji, E. F. Da Silva, M. Kato, and T.-P. Ma, "Radiation and Hot-Electron Hardened Si-Gate Submicron MOSFETs with a Fluorinated Oxide," *Proc. 2nd Workshop Radia.-Induced Process-Relat. Electr. Active Defects Semicond.-, Insul. Syst.*, pp. 324–327. Research Triangle Park, NC (1989).

Nishioka, Y., Y. Ohji, and T.-P. Ma, "Gate-Oxide Breakdown Accelerated by Large Drain Current in *n*-Channel MOSFET's," *IEEE Electron Devices Lett.* **12**, 134–136 (1991).

Nishioka, Y., T. Itoga, K. Ohyu, and T.-P. Ma, "Improving Hot-Electron Hardness of Narrow Channel MOSFETs by Fluorine Implantation," *Solid-State Electron.* **34**, 1197–2000 (1991).

Nishiuchi, K., H. Oka, T. Nakamura, H. Ishikawa, and M. Shinoda, "A Normally-Off Type Buried Channel MOSFET for VLSI Circuits," *Tech. Dig.—Int. Electron Devices Meet.*, 26 (1978).

Nissan-Cohen, Y., J. Shappir, and D. Frohman-Bentchkowsky, "High-Field and Current-Induced Positive Charge in the Thermal SiO_2 Layers," *J. Appl. Phys.* **57**, 2830–2839 (1985).

Nissan-Cohen, Y., "Analytical Explanation to Double-Hump Substrate Current in Funnel-Shape Transistors," *IEEE Electron Devices Lett.* **7**, 344–346 (1986).

Nissan-Cohen, Y., G. A. Franz, and R. F. Kwasnick, "Measurement and Analysis of Hot-Carrier-Stress Effect on *n*-MOSFET's Using Substrate Current Characterization," *IEEE Electron Devices Lett.* **7**, 451–453 (1986).

Nissan-Cohen, Y., H. H. Woodsbury, T. B. Gorczyca, and C.-Y. Wei, "Effect of Hydrogen on Hot Carrier Immunity, Radiation Hardness and Gate Oxide Reliability in MOS Devices," *Tech. Dig. VLSI Tech. Symp.*, 37–38 (1988).

Nissan-Cohen, Y., "Effect of Hydrogen on Hot Carrier and Radiation Immunity of MOS Devices," *Appl. Surf. Sci.* **39**, 511–522 (1989).

Nitayama, A., N. Takenouchi, T. Hamamoto, and Y. Oowaki, "New Hot-Carrier-Induced Degradation Phenomena in Half-Micrometer MOS Transistors," *IEEE Trans. Electron Devices* **34**, 2384 (1987).

Nogami, K., T. Sakurai, K. Sawada, T. Wada, K. Sato, M. Isobe, M. Kakumu, S. Morita, S. Yokogawa, M. Kinugawa, T. Asami, K. Hashimoto, J.-I. Matsunaga, H. Nozawa, and T. Iizuka, "1-MBit Virtually Static RAM," *IEEE J. Solid-State Circuits* **SC-21**, 662–669 (1986).

Nogami, K., K. Sawada, M. Kinugawa, and T. Sakurai, "VLSI Circuit Reliability under AC Hot-Carrier Stress," *Tech. Dig. VLSI Tech. Symp.*, 13–14 (1987).

Noguchi, T., Y. Asahi, M. Nakahara, K. Maeguchi, and K. Kanzaki, "High Speed CMOS Structure with Optimized Gate Work Function," *Tech. Dig. VLSI Tech. Symp.*, 19–20 (1986).

Noguchi, T., Y. Asahi, N. Ikeda, K. Maeguchi, and K. Kanzaki, "Parasitic Resistance Characterization for Optimum Design of Half Micron MOSFETs," *Tech. Dig.—Int. Electron Devices Meet.*, 730–733 (1986).

Noor Mohammad, S., "Unified Model for Drift Velocities of Electrons and Holes in Semiconductors as a Function of Temperature and Electric Field," *Solid-State Electron.* **35**, 1391–1396 (1992).

Nougier, J. P. and M. Rolland, "Differential Relaxation Times and Diffusivities of Hot Carriers in Isotropic Semiconductors," *J. Appl. Phys.* **48**, 1683–1687 (1977).

Novikov, I. A., "Electrothermal Analogy in Hereditary Media and Its Application," *J. Eng. Phys.* **55**, 1166–1171 (1989).

Noyce, R. N., "Semiconductor Device and Lead Structure," U.S. Patent 2,981,877 (1961).

Noyori, M., Y. Nakata, S. Odanaka, and J. Yasui, "Reduction of V_T Shift Due to Avalanche-Hot-Carrier Injection Using Graded Drain Structures in Submicron *n*-Channel MOSFET," *Proc. Int. Reliab. Phys. Symp.*, 205–209 (1984).

Nunn, D., "Simulation of Nonlinear Wave Particle Interactions in Space Plasmas Using the VHS Technique," *IEEE Int. Conf. Plasma Sci.*, 94–95 (1989).

Odanaka, S., A. Hiroki, K. Ohe, H. Umimoto, and K. Moriyama, "SMART-II: A Three-Dimensional CAD Model for Submicrometer MOSFETs," *Proc. Int. Conf. Number. Anal. Semicond. Devices Int. Circuits, 6th*, 303–310 (1989).

Odanaka, S., A. Hiroki, K. Ohe, K. Moriyama, and H. Umimoto, "SMART-II: A Three-Dimensional CAD Model for Submicrometer MOSFETs," *IEEE Trans. Comput.-Aided Des. Int. Circuits Syst.* **10**, 619–628 (1991).

Odanaka, S. and A. Hiroki, "A Spillover Effect of Avalanche-Generated Electrons in Buried *p*-Channel MOSFETs," *IEEE Electron Devices Lett.* **12**, 224–226 (1991).

O'Dwyer, J. J., *The Theory of Electrical Conduction and Breakdown in Solid Dielectrics*, Clarendon, Oxford (1973).

Ogawa, S. and N. Shiono, "Interface-Trap Generation Induced by Hot-Hole Injection at the Si-SiO_2 Interface," *Appl. Phys. Lett.* **61**, 807–809 (1992).

Ogura, S., P. J. Chang, W. W. Walker, D. L. Critchlow, and J. F. Shepard, "Design and Characteristics of the Lightly-Doped Drain-Source (LDD) Insulated Gate Field Effect Transistor," *IEEE Trans. Electron Devices* **27**, 1359–1367 (1980).

Ogura, S., C. F. Codella, N. Rovedo, J. F. Shepard, and J. Riseman, "A Half Micron MOSFET Using Double Implanted LDD," *Tech. Dig.—Int. Electron Devices Meet.*, 718–721 (1982).

Ogura, S., P. L. Kroesen, C. F. Codella, N. Rovedo, and S. K. Cheung, "Submicron MOSFET Performance at Liquid Nitrogen Temperatures," *IEEE Int. Solid-State Circuits Conf.*, 160–161 (1986).

Ohata, A., A. Toriumi, M. Iwase, and K. Natori, "Observation of Random Telegraph Signals: Anomalous Nature of Defects at the Si/SiO_2 Interface," *J. Appl. Phys.* **68**, 220–204 (1990).

Ohji, Y., Y. Nishioka, K. Yokogawa, K. Mukai, Q. Qiu, E. Arai, and T. Sugano, "Effects of Minute Impurities (H, OH, F) on the SiO_2/Si Interface Investigated by Nuclear Resonant Reaction and Electron Spin Resonant," *IEEE Trans. Electron Devices* **37**, 1635–1642 (1990).

Ohmi, T., K. Kotani, A. Teramoto, and M. Miyashita, "Dependence of Electron Channel Mobility of Si–SiO_2 Interface Microroughness," *IEEE Electron Device Lett.* **12**, 652–654 (1991).

Oh-Uchi, N., H. Hayashi, H. Yamoto, and T. Matsushita, "A New Silicon Heterojunction Transistor Using the Doped SIPOS," *Tech. Dig.—Int. Electron Devices Meet.*, 522–525 (1979).

Oh-Uchi, N., A. Kayanuma, K. Asano, H. Hayashi, and M. Noda, "A New Self-Aligned Transistor Structure for High-Speed and Low-Power Bipolar LSIs" *Tech. Dig.—Int. Electron Devices Meet.*, 55–58 (1983).

Ohyu, K., Y. Nishioka, Y. Ohji, and N. Natsuaki, "Improvement of SiO_2/Si Interface Properties by Fluorine Implantation," *Ext. Abstr., Solid State Devices Mater.*, 607–608 (1988).

Ohyu, K., T. Itoga, Y. Nishioka, and N. Natsuaki, "Improvement of SiO_2/Si Interface Properties Utilizing Fluorine Ion Implantation and Drive-in Diffusion," *Jpn. J. Appl. Phys.* **28**, 1041–1045 (1989).

Okabe, N., H. Takato, Y. Oowaki, T. Hamamoto, and A. Nitayama, "Hot Electron Injection Characteristics in Asymmetrically Structured Submicron MOSFETs," *Jpn. J. Appl. Phys. Lett.* **28**, L21–L23 (1989).

Okada, Y., P. J. Tobin, R. I. Hegde, J. Liao, and P. Rushbrook, "Oxynitride Gate Dielectrics Prepared by Rapid Thermal Processing Using Mixtures of Nitrous Oxide and Oxygen," *Appl. Phys. Lett.* **61**, 3163–3165 (1992).

Okumura, Y., T. Kunikiyo, I. Ogoh, H. Genjo, M. Inuishi, M. Nagatomo, and T. Matsukawa, "Mechanism Analysis of a Highly Reliable Graded Junction Gate/n^- Overlapped Structure in MOS LDD Transistor," *Ext. Abstr., Solid State Devices Mater.*, 477–480 (1989).

Okumura, Y., T. Kunikiyo, I. Ogoh, H. Genjo, M. Inuishi, M. Nagatomo, and T. Matsukawa, "Graded-Junction Gate n^- Overlapped LDD MOSFET Structures for High Hot-Carrier Reliability," *IEEE Trans. Electron Devices* **38**, 2647–2656 (1991).

Olivo, P., J. Suñé, and B. Riccò, "Determination of the Si–SiO_2 Barrier Height from the Fowler–Nordheim Plot," *IEEE Electron Devices Lett.* **12**, 620–622 (1991).

Olivo, P., T. N. Nguyen, and B. Riccò, "Influence of Localized Latent Defects on Electrical Breakdown of Thin Insulators," *IEEE Trans. Electron Devices*, **38**, 527–531 (1991).

Olsen, J. N., "Hot Electron Effect in Line-Radio Temperature Measurements in Laser Plasmas," *IEEE Int. Conf. Plasma Sci.*, 87 (1988).

Omura, Y., S. Nakashima, K. Izumi, and T. Ishii, "0.1-μm-Gate, Ultrathin-Film CMOS Devices Using SIMOX Substrate with 80-nm-Thick Buried Oxide Layer," *Tech. Dig.—Int. Electron Devices Meet.*, 675–678 (1991).

Omura, Y. and K. Izumi, "Hot-Carrier Immunity of a 0.1-μm-Gate Ultrathin-Film MOSFET/SIMOX," *Ext. Abstr., 1992 Int. Conf. Solid State Devices Mater.*, 496–498 (1992).

Omura, Y., S. Nakashima, and K. Izumi, "Investigation on High-Speed Performance of 0.1 μm-Gate, Ultrathin-Film CMOS/SIMOX," *IEICE Trans. Electron.* **E75-C**, 1491–1497 (1992).

Ong, D. G. and R. F. Pierret, "Approximate Formula for Surface Carrier Concentration in Charge-Coupled Devices," *Electron. Lett.* **10**, 6–7 (1974).

Ong, T.-C., S. Tam, P.-K. Ko, and C. Hu, "Width Dependence of Substrate and Gate Currents in MOSFETs," *IEEE Trans. Electron Devices* **32**, 1737–1740 (1985).

Ong, T.-C., P.-K. Ko, and C. Hu, "50-Angstrom Gate-Oxide MOSFETs at 77 K," *IEEE Trans. Electron Devices* **34**, 2129–2135 (1987).

Ong, T.-C., P.-K. Ko, and C. Hu, "Modeling of Substrate Current in *p*-MOSFETs," *IEEE Electron Devices Lett.* **98**, 413–416 (1987).

Ong, T.-C., K. Seki, P.-K. Ko, and C. Hu, "Hot-Carrier-Induced Degradation in *p*-MOSFETs under AC Stress," *IEEE Electron Devices Lett.* **99**, 211–213 (1988).

Ong, T.-C., M. Levi, P.-K. Ko, and C. Hu, "Recovery of Threshold Voltage after Hot-Carrier Stressing," *IEEE Trans. Electron Devices* **35**, 978–984 (1988).

Ong, T.-C., P.-K. Ko, and C. Hu, "Hot-Carrier Effects in Depletion-Mode MOSFETs," *Solid-State Electron.* **32**, 33–36 (1989).

Ong, T.-C., K. Seki, P.-K. Ko, and C. Hu, "*p*-MOSFET Gate Current and Device Degradation," *Proc. Int. Reliab. Phys. Symp.*, 178–182 (1989).

Ong, T.-C., P.-K. Ko, and C. Hu, "Hot-Carrier Current Modeling and Device Degradation in Surface-Channel *p*-MOSFET's," *IEEE Trans. Electron Devices* **37**, 1658–1666 (1990).

Ooka, H., S. Murakami, M. Murayama, K. Yoshida, S. Takao, and O. Kudoh, "High Speed CMOS Technology for ASIC Application," *Tech. Dig.—Int. Electron Devices Meet.*, 240–243 (1986).

Or, B. S. S., L. Forbes, H. Haddad, and W. Richling, "Annealing Effects of Carbon in *n*-Channel LDD MOSFETs," *IEEE Electron Devices Lett.* **12**, 596–598 (1991).

Or, B. S. S., L. Forbes, and H. Haddad, "Thermal Re-emission of Trapped Hot Electrons in NMOS Transistors," *IEEE Trans. Electron Devices* **38**, 2712 (1991).

Orlowski, M., C. Werner, W. Weber, H.-P. Muhlhoff, and P. Murkin, "Unified Model for Electric Fields in LDD-Type MOSFETs," *Tech. Dig. VLSI Tech. Symp.*, 51–52 (1987).

Orlowski, M., C. Mazure, A. Lill, H.-M. Muhlhoff, W. Hansch, A. von Schwerin, and F. Neppl, "Advanced Simulation for Reliability Optimization of Submicron LDD MOSFETs," *Proc. Eur. Solid State Devices Res. Conf., 19th*, 711–714 (1989).

Orlowski, M., C. Werner, and J. P. Klink, "Unified Model for the Electric Fields in LDD MOSFETs. Part I: Field Peaks on the Source Side," *IEEE Trans. Electron Devices* **36**, 375–381 (1989).

Orlowski, M., S. W. Sun, P. Blakey, and R. Subrahmanyan, "The Combined Effects of Band-to-Band Tunneling and Impact Ionization in the Off Regime of an LDD MOSFET," *IEEE Electron Devices Lett.* **12**, 593–595 (1990).

Orlowski, M., C. Mazure, and M. Noell, "New Vertically Layered Elevated Hot Carrier Resistant Source/Drain Structure for Deep Submicron MOSFETs," *Microelectron. Eng.* **15**, 433–436 (1991).

Orlowski, M., C. Mazure, and M. Noell, "Novel Elevated MOSFET Source/Drain Structure," *IEEE Electron Devices Lett.* **12**, 593–595 (1991).

Ortner, W. R. and J. T. Clemens, "Reliability Failure Mode in Dynamic MOS Circ.—The Unopened Metal to Polysilicon Contact Window and the Floating Gate Transistor," *Proc. Int. Reliab. Phys. Symp.*, 16–22 (1977).

Osburn, C. M. and D. W. Ormond, "Dielectric Breakdown in Silicon Dioxide Films on Silicon," *J. Electrochem. Soc.* **119**, 591 (1972).

Ouisse, T., S. Cristoloveanu, and G. Borel, "Hot Carrier–Induced Aging of Short Channel SIMOX Devices," *IEEE SOS/SOI Tech. Conf.*, 38–39 (1990).

Ouisse, T., S. Cristoloveanu, and G. Borel, "Hot-Carrier-Induced Degradation of the Back Interface in Short-Channel Silicon-on-Insulator MOSFETs," *IEEE Electron Devices Lett.* **12**, 290–292 (1991).

Ouisse, T., S. Cristoloveanu, and G. Borel, "Electron Trapping in Irradiated SIMOX Buried Oxides," *IEEE Electron Devices Lett.* **12**, 312–314 (1991).

Ouisse, T., A. J. Auberton-Hervé, B. Giffard, and G. Reimbold, "Hot Carrier–Induced Degradation Mechanisms in Short-Channel SIMOX *p*-MOSFETs," *Microelectron. Eng.* **19**, 473–476 (1992).

Ozawa, Y., M. Iwase, and A. Toriumi, "Experimental Evidence for Hole-Induced Interface State Generation under High Field Tunneling Current Stressing," *Ext. Abstr., Jpn. Soc. Appl. Phys.*, 161–164 (1988).

Pagaduan, F. E., C. Y. Yang, N. R. Wu, S. Chiao, and C. Wang, "Effects of Thermal Annealing on Generated Electron Traps in Thin Oxide," *Proc. J. Electrochem. Soc. Symp. Reduced Temp. Process. VLSI* **86-5**, 95–107 (1986).

Pagaduan, F. E., A. Hamada, C. Y. Yang, and E. Takeda, "Hot-Carrier Detrapping Mechanism in MOS Devices," *Jpn. J. Appl. Phys.* **28**, L2047–L2049 (1989a).

Pagaduan, F. E., A. Hamada, C. Y. Yang, and E. Takeda, "Hot-Carrier Detrapping in Post-Stress Behavior of MOS Devices," *Ext. Abstr., Solid State Devices Mater.*, 469–472 (1989b).

Pagaduan, F. E., C. Y. Yang, T. Toyabe, Y. Nishioka, A. Hamada, Y. Igura, and E. Takeda, "Simulation of Substrate Hot-Electron Injection," *IEEE Trans. Electron Devices* **37**, 994–998 (1990).

Palkuti, L. J., R. D. Ormond, C. Hu, and J. Chung, "Correlation between Channel Hot-Electron Degradation and Radiation-Induced Interface Trapping in MOS Devices," *IEEE Trans. Nucl. Sci.* **NS-36**, 2140–2146 (1989).

Pao, H. C. and C. T. Sah, "Effects of Diffusion Current on Characteristics of MOS Transistors," *IEEE Trans. Electron Devices* **12**, 139 (1965).

Papadopoulos, K., "Review of Anomalous Resistivity for the Ionosphere," *Rev. Geophys. Space Phys.* **15**, 113–127 (1977).

Park, C.-K., C.-Y. Lee, K. Lee, B.-J. Moon, Y. H. Byun, and M. Shur, "A Unified Current–Voltage Model for Long-Channel *n*MOSFETs," *IEEE Trans. Electron Devices* **38**, 399–406 (1991).

Park, H.-J., K. Lee, and C.-K. Kim, "A New CMOS NAND Logic Circuit for Reducing Hot-Carrier Problems," *IEEE J. Solid-State Circuits* **24**, 1041–1047 (1989).

Park, H. S., A. Van Der Ziel, R. J. J. Zijlstra, and S. T. Liu, "Discrimination between Two Noise Models in Metal-Oxide-Semiconductor Field-Effect Transistors," *J. Appl. Phys.* **52**, 296–299 (1981).

Parrillo, L. C., S. J. Cosentino, R. W. Mauntel, B. A. Bergami, P. J. Tobin, F. K. Baker, S. Poon, Y. Yoshii, F. Pintchovski, S. W. Sun, F. T. Liou, J. Alvis, M. Kearney, and M. Swenson, "Versatile, High-Performance, Double-Level-Poly Double-Level Metal, 1.2-Micron CMOS Technology," *Tech. Dig.—Int. Electron Devices Meet.*, 244–247 (1986).

Parrillo, L. C., "CMOS Active and Field Device Fabrication," *Semicond. Int.* **11**, 64–70 (1988).

Parrillo, L. C., J. R. Pfiester, J.-H. Lin, E. O. Travis, R. D. Sivan, and C. D. Gunderson, "Disposable Polysilicon LDD Spacer Technology," *IEEE Trans. Electron Devices* **38**, 39–46 (1991).

Parrillo, L. C., J. R. Pfiester, M. P. Woo, B. Roman, W. Ray, J. Ko, and C. Gunderson, "The Effect of Biased Spacers on LDD MOSFET Behavior," *IEEE Electron Devices Lett.*, 542–545 (1991).

Parris, M. C., K. P. Roenker, T. T. Yuliasto, and J. H. Nevin, "Hot Carrier Effects on Submicron MOSFETs and Their Implications for VLSI Circuit Design," *Proc. IEEE Natl. Aerosp. Electron. Conf.*, 2–9 (1984).

Parshin, D. A. and A. R. Shabaev, "Inverted Distributions and Carrier Bunching in Momentum Space in a Magnetic Field Crossed with an Alternating Electric Field," *Sov. Phys.—JETP (Engl. Transl.)* **59**, 388–393 (1984).

Pearson, G. L. and W. H. Brattain, "History of Semiconductor Research," *Proc. IRE* **43**, 1794–1806 (1955).

Peinke, J., J. Parisi, B. Rohricht, K. M. Mayer, U. Rau, and R. P. Huebener, "Spatio-Temporal Instabilities in the Electric Breakdown of *p*-Germanium," *Solid-State Electron.* **31**, 817–820 (1988).

Pell, E. M. and G. M. Roe, "Reverse Current and Carrier Lifetime as a Function of Temperature in Germanium Junction Diodes," *J. Appl. Phys.* **26**, 658–665 (1955).

Pelle, B., J. Kispeter, and J. Peiszner, "Barrier-Height and Hot Electron Attenuation Length Measurements in Au–Si, Ag–Si, and Al–Si Diodes between 280–350 K," *Acta Phys. Chem.* **30**, 39–52 (1984).

Perkin, W. A., "Beam-Injected and Lorentz-Trapped Hot Electron Plasma," *Phys. Fluids* **12**, 713–719 (1969).

Perkins, H. A. and J. D. Schmidt, "An Integrated Semiconductor Memory System," *IEEE Proc. Fall J. Comput. Conf.*, 1053–1064 (1965).

Petrova, R. S., R. S. Kamburova, and P. K. Vitanov, "Hot Carriers' Effects in Short Channel Devices," *Microelectron. Reliab.* **26**, 155–162 (1986).

Petrova, R. S., R. S. Kamburova, P. K. Vitanov, and E. N. Stefanov, "Hot Electron Effects in Narrow Width MOS Devices," *J. Microelectron.* **18**, 25–30 (1987).

Pfann, W. G., "Principles of Zone-Refining," *Trans. AIME* **194**, 747–799 (1952).

Pfann, W. G., "Technique of Zone Melting and Crystal Growth," *Solid State Phys.* **1**, 423–451 (1957).

Pfann, W. G. and C. G. B. Garrett, "Semiconductor Varactor Using Space-Charge Layers," *Proc. IRE* **47**, 2011 (1959).

Pfann, W. G., "The Semiconductor Revolution," *J. Electrochem. Soc.* **121**, 9–15 (1974).

Pfiester, J. R. and F. K. Baker, "Asymmetrical High Field Effects in Submicron MOSFETs," *Tech. Dig.—Int. Electron Devices Meet.*, 51–54 (1987).

Pfiester, J. R., F. K. Baker, R. D. Sivan, N. Crain, J.-H. Lin, M. Liaw, C. Seelbach, C. Gunderson, and D. Denning, "A Self-Aligned LDD/Channel Implanted ITLDD Process with Selectively-Deposited Poly Gates for CMOS VLSI," *Tech. Dig.—Int. Electron Devices Meet.*, 769–772 (1989).

Pfiester, J. R., F. K. Baker, R. D. Sivan, N. Crain, J.-H. Lin, M. Liaw, C. Seelbach, and C. Gunderson, "Selectively Deposited Poly-Gate ITLDD Process with Self-Aligned LDD/Channel Implantation," *IEEE Electron Devices Lett.* **11**, 253–255 (1990).

Pfiester, J. R., N. Crain, J.-H. Lin, C. D. Gunderson, and V. Kaushik, "A Poly-Framed LDD Sub-Half-Micrometer CMOS Technology," *IEEE Electron Devices Lett.* **11**, 529–531 (1990).

Pfiester, J. R., R. D. Sivan, C. D. Gunderson, N. E. Crain, J.-H. Lin, H. M. Liaw, C. A. Seelbach, and F. K. Baker, "An ITLDD CMOS Process with Self-Aligned Reverse-Sequence LDD/Channel Implantation," *IEEE Trans. Electron Devices* **38**, 2460–2464 (1991).

Pilkuhn, M. H., ed., *International Conference on the Physics of Semiconductors, 12th Proceeding, 1974*, Teubner, Stuttgart (1974).

Pimbley, J. M., G. Gildenblat, and F.-C. Hsu, "Comments on 'Structure-Enhanced MOSFET Degradation Due to Hot-Electron Injection' (and Reply)," *IEEE Electron Devices Lett.* **4**, 256–260 (1984).

Pimbley, J. M. and G. Gildenblat, "Effect of Hot-Electron Stress on Low Frequency MOSFET Noise," *IEEE Electron Devices Lett.* **5**, 345–347 (1984).

Pimbley, J. M., "Measurement Method for the Increase of Digital Switching Time Due to Hot-Electron Stress," *IEEE Electron Devices Lett.* **6**, 366–368 (1985).

Placencia, I., F. Martin, J. Sune, and X. Aymerich, "On the Dissipation of Energy by Hot Electrons in SiO_2," *J. Phys. (Appl. Phys.)* **23**, 1576–1581 (1990).

Plossu, C., C. Choquet, V. Lubowiecki, and B. Balland, "Spatial Distributions of Surface States in MOS Transistors," *Solid State Commun.* **65**, 1231–1235 (1988).

Pocha, M. D., A. G. Gonzalez, and R. W. Dutton, "Threshold Voltage Controllability in Double-Diffused MOS Transistors," *IEEE Trans. Electron Devices* **21**, 778 (1974).

Poetzl, H. W., "Contributions to Device and Process Simulation," *Siemens Forsch.- Entwicklungsber.* **17**, 308–313 (1988).

Poindexter, E. H., "Electron Paramagnetic Resonance Studies of Interface Defects in Oxidized Silicon," *Z. Phys. Chem.* (*Wiesbaden*) [N. S.] **151**, 165–176 (1987).

Poirier, R. and J. Olivier, "Hot Electron Emission from Silicon into Silicon Dioxide by Surface Avalanche," *Appl. Phys. Lett.* **15**, 364–365 (1969).

Poorter, T. and D. R. Wolters, "Intrinsic Oxide Breakdown at Near Zero Electric Fields," *Proc. Int. Conf. Insul. Films Semicond.*, Eindhoven, Netherlands, 266–269 (1983).

Poorter, T. and P. Zoestbergen, "Hot Carrier Effects in MOS Transistors," *Tech. Dig.—Int. Electron Devices Meet.*, 100–103 (1984).

Post, R. F., "Scaling Laws for Mirrortron Ion Accelerators," *IEEE Int. Conf. Plasma Sci.*, 206 (1990).

Pramanik, D., V. Jain, and K. Y. Chang, "A High Reliability Triple Metal Process for High Performance Application Specific Circuits," *Proc. Int. IEEE VLSI Multilevel Interconnect. Conf., 8th*, 27–33 (1991).

Przyrembel, G., R. Mahnkopf, and H. G. Wagemann, "Hot Carrier Sensitivity of MOSFETs Exposed to Synchrotron-Light," *J. Phys., Colloq.* (*Orsay, Fr.*) **49**, 767–770 (1988).

Pshaenich, A., "MOS Thysistor Improves Power-Switching Circuits," *Electron. Des.*, 165–170 (1983).

Pugh, W., D. L. Critchlow, R. A. Henie, and L. A. Russell, "Solid State Memory Development at IBM," *IBM J. Res. Dev.* **25**, 585–602 (1981).

Purbo, O. W. and C. R. Selvakumar, "High-Gain SOI Polysilicon Emitter Transistors," *IEEE Electron Devices Lett.* **12**, 635–637 (1991).

Qiu-Yi, Y., A. Zrenner, and F. Koch, "Interface Degradation in Si-Metal-Oxide-Semiconductor Structures by Homogeneous, Microwave Heating of Channel Carriers," *Appl. Phys. Lett.* **52**, 561–563 (1988).

Quader, K. N., C. Li, R. Tu, E. Rosenbaum, P.-K. Ko, and C. Hu, "A New Approach for Simulation of Circuit Degradation Due to Hot-Electron Damage in NMOSFETs," *Tech. Dig.—Int. Electron Device Meet.*, 337–340 (1991).

Quader, K. N., P.-K. Ko, C. Hu, P. Fang, and J. T. Yue, "Simulations of CMOS Circuit Degradation Due to Hot-Carrier Effects," *Proc. Int. Reliab. Phys. Symp.*, 16–23, (1992).

Quazi, D. M. K., N. Yasuda, A. Maruyama, K. Taniguchi, and C. Hamaguchi, 'Spatial Distribution of Trapped Holes in the Oxide of MOSFETs after Uniform Hot-Hole Injection," *Ext. Abstr. 1991 Int. Conf. Solid State Devices Mater.*, 14–16 (1991).

Quon, B. H. and R. A. Dandl, "Preferential Electron-Cyclotron Heating of Hot Electrons and Formation of Overdense Plasmas," *Phys. Fluid B* **1**, 2010–2017 (1989).

Radojcic, R., "Some Aspects of Hot-Electron Aging in MOSFETs," *IEEE Trans. Electron Devices* **31**, 1381–1386 (1984).

Radojcic, R., "Hot-Electron Aging in *p*-Channel MOSFETs for VLSI CMOS," *IEEE Trans. Electron Devices* **31**, 1896–1898 (1984).

Raguotis, R., and A. Reklaitis, "Dynamical Response of Electrons in GaAs in a Higher Electric Field," *Phys. Status Solidi A* **62**, 399–405 (1980).

Rahmat, K., J. White, and D. A. Antoniadis, "Computation of Drain and Substrate Currents in Ultra-Short-Channel *n*MOSFETs Using the Hydrodynamic Model," *Tech. Dig.—Int. Electron Devices Meet.*, 115–118, (1991).

Rakkhit, R., M. C. Peckerar, and C. T. Yao, "Investigation of the Time Dependence of Current Degradation in MOS Devices," *Proc. Int. Reliab. Phys. Symp.*, 103–109 (1989).

Rakkhit, R., S. Haddad, C. Chang, and J. Yue, "Drain-Avalanche Induced Hole Injection and Generation of Interface Traps in Thin Oxide MOS Devices," *Proc. Int. Reliab. Phys. Symp.*, 150–153 (1990).

Rakkhit, R. and J. T Yue, "A Comparison of Inverter-Type Circuit Lifetime and Quasi-Static Analysis of NMOSFET Lifetime," *Proc. Int. Reliab. Phys. Symp.*, 112–117 (1991).

Ramesh, K., A. N. Chandorkar, and J. Vasi, "Study of Electron Traps in Silicon Dioxide Due to Mobile Sodium Ions at the $Si-SiO_2$ Interface," *J. Inst. Electron. Telecommun. Eng.* **33**, 38–40 (1987).

Ratnam, P. and A. Naem, "Drain Engineering of Hot-Carrier-Resistant MOSFETs Using Concave Silicon Surfaces for Deep Submicron VLSI Technology," *Solid-State Electron.* **33**, 1163–1168 (1990).

Reczek, W., F. Bonner, and B. Murphy, "Reliability of Latch-up Characterization Procedures," *Proc. Int. Conf. Microelectron Test Struct. 1990*, 51–54 (1990).

Rees, H. D., "Hot Electron Devices for Millimetre and Submillimetre Applications," *J. Phys., Colloq. (Orsay, Fr.)* **42**, 157–170 (1981).

Reggiani, L., R. Brunetti, and C. Jacoboni, "Monte Carlo Calculation of Hot-Electron Noise in Si at 77 K," *Proc. Int. Conf. Noise Phys. Syst.* Gaithersburg, MD, 414–416 (1981).

Reich, R. K., J. W. Schrankler, D.-H. Ju, M. S. Holt, and G. D. Kirchner, "Radiation-Dependent Hot-Carrier Effects," *IEEE Electron Devices Lett.* **7**, 235–237 (1986).

Reimbold, G., F. Paviet-Salomon, H. Haddara, G. Guegan, and S. Cristoloveanu, "Hot Electron Reliability of Deep Submicron MOS Transistors," *J. Phys., Colloq. (Orsay, Fr.)* **49**, 665–668 (1988).

Reimbold, G., P. Saint Bonnet, B. Giffard, and A.-J. Auberton-Hervé, "Influence of LDD on Aging of SOI NMOS Transistors," *IEEE SOS/SOI Tech. Conf.*, 36–37 (1990).

Reimbold, G., and A.-J. Auberton-Hervé, "Aging Analysis of nMOS of a 1.3-μm Partially Depleted SIMOX SOI Technology Comparison with a 1.3-μm Bulk Technology," *IEEE Tran. Electron Devices* **40**, 364–370 (1993).

Reizman, F. and W. Van Gelder, "Optical Thickness Measurement of $SiO_2-Si_3N_4$ Films on Silicon," *Solid-State Electron.* **10**, 625 (1967).

Rho, K.-M., K. Lee, M. Shur, and T. A. Fjeldly, "Unified Quasi-Static MOSFET Capacitance Model," *IEEE Trans. Electron Devices* **40**, 131–135 (1993).

Rhoderick, E. H., *Metal–Semiconductor Contacts*, Oxford Univ. Press, New York, (1978).

Rhoderick, E. H., "Solid State Devices," *Proc. 13th Eur. Solid State Devices Res. Conf. 8th SSSDT—Inst. Phys. Conf. Series*, No. 69 (1984).

Ricco, B., E. Sangiorgi, and D. Cantarelli, "Low Voltage Hot-Electron Effects in Short Channel MOSFETs," *Tech. Dig.—Int. Electron Devices Meet.*, 92–95 (1984).

Riccon, R., E. Sangiorgi, F. Venturi, and P. Lugli, "Monte-Carlo Modeling of Hot Electron Gate Current in MOSFETs," *Tech. Dig.—Int. Electron Devices Meet.*, 559–562 (1986).

Richman, P., *MOSFETs and Integrated Circuits*, Wiley, New York (1973).

Rideout, V. L., F. G. Gaensslen, and A. LeBlanc, "Device Design Consideration for Ion Implanted N-Channel MOSFETs," *IBM J. Res. Dev.* **19**, 50 (1975).

Ridley, B. K., *Quantum Processes in Semiconductor*, Clarendon, Oxford, (1982).

Ridley, B. K., "On the Properties of Hot Electrons in Semiconductor Quantum Wells," *J. Phys. C* **17**, 5357–5365 (1984).

Rinerson, D., M. Ahrens, J. Lien, B. Venkatesh, T. Lin, P. Song, S. Longcor, L. Shen, D. Rogers, and M. Briner, "512 K EPROMs," *Proc. IEEE Int. Solid-State Circuits Conf.*, 136–137, 327 (1984).

Risch, L. and L. Weber, "Dynamic Hot-Carrier Degradation of Fast-Switching CMOS Inverters with Different Duty Cycles," *J. Phys., Colloq. (Orsay, Fr.)* **49**, 657–660 (1988).

Roberts, J. W., S. G. Chamberlain, and J. R. F. McMacken, "Energy–Momentum Transport–Based Simulator Adapted to the General Purpose Semiconductor Device Simulation Development Tool CHORD," *Can. J. Phys.* **69**, 217–223 (1991).

Roberts, J. W. and S. G. Chamberlain, "An Experimental Procedure for Measuring Silicon Lattice Heating Due to Hot Carriers in MOSFETs," *Solid-State Electron.* **36**, 351–360 (1993).

Roblin, P., A. Samman, and S. Bibyk, "Simulation of Hot Electron Trapping and Aging of *n*-MOSFETs," *IEEE Trans. Electron Devices* **35**, 2229–2237 (1988).

Rodder, M. and D. A. Antoniadis, "Hot-Carrier Effects in Hydrogen-Passivated *p*-Channel Polycrystalline-Si MOSFETs," *IEEE Trans. Electron Devices* **34**, 1079–10083 (1987).

Rodder, M., "Leakage-Current-Induced Hot-Carrier Degradation of *p*-Channel MOSFETs," *IEEE Electron Devices Lett.* **9**, 573–535 (1988).

Rodder, M., "ON/OFF Current Ratio in P-Channel Poly-Si MOSFETs: Dependence on Hot-Carrier Stress Conditions," *IEEE Electron Devices Lett.* **11**, 346–348 (1990).

Roenker, K. P., M. C. Parris, A. Gupta, and K. Raol, "Role of Interface State Generation in Hot Electron Induced MOSFET Parameter Shifts," *Ext. Abstr. Electrochem. Soc.* **85-2**, 314–315 (1985).

Roenker, K. P., M. C. Parris, and A. Gupta, "Understanding Hot Electron Induced Degradation in Insulated Gate Field Effect Transistors," *Proc. 6th Bienn. Univ./Gov./Ind. Microelectron. Symp.*, 194–197 (1985).

Roessler, B. and R. G. Mueller, "Erasable and Electrically Reprogrammable Read-Only Memory Using the *n*-Channel SIMOS One-Transistor Cell," *Siemens Forsch.- Entwicklungsber.* **4**, 345–351 (1975).

Rofan, R. and C. Hu, "Stress-Induced Oxide Leakage," *IEEE Electron Devices Lett.* **12**, 632–634 (1991).

Rosenbaum, E., P. M. Lee, R. Moazzami, P.-K. Ko, and C. Hu, "Circuit Reliability Simulator—Oxide Breakdown Module," *Tech. Dig.—Int. Electron Devices Meet.*, 331–334 (1989).

Rosenbaum, E., R. Rofan, and C. Hu, "Effect of Hot-Carrier Injection on *n*- and *p*MOSFET Gate Oxide Integrity," *IEEE Electron Devices Lett.* **12**, 599–601 (1991).

Rosenberg, S. J., D. L. Dwight, and B. L. Euzent, "H-MOS Reliability," *IEEE Trans. Electron Devices* **26**, 48–51 (1979).

Rossler, B., "Electrically Erasable and Reprogrammable Read-Only Memory Using the *n*-Channel SIMOS One-Transistor Cell," *IEEE Trans. Electron Devices* **24**, 606–610 (1977).

Roth, I., "Excitation of High Frequency, Electrostatic Waves in the Magnetospheres of the Outer Planets," *IEEE Int. Conf. Plasma Sci.*, 97 (1989).

Roy, A. and M. H. White, "A New Approach to Study Electron and Hole Charge Separation at the Semiconductor–Insulator Interface," *IEEE Trans. Electron Devices* **37**, 1504–1513 (1990).

Runyon, S., "Hot National Builds in Reliability for its BICMOS," *Electronics* **61**, 61–62 (1988).

Rusu, A. and C. Bulucea, "Deep-Depletion Breakdown Voltage of SiO_2/Si MOS Capacitors," *IEEE Trans. Electron Devices* **26**, 201 (1979).

Sabnis, A. G. and J. T. Clemens, "Characterization of the Electron Mobility in the Inverted ⟨100⟩ Si Surface," *Tech. Dig.—Int. Electron Devices Meet.*, 18 (1979).

Sabnis, A. G. and J. T. Nelson, "Physical Model for Degradation of DRAMs during Accelerated Stress Aging," *Proc. Int. Reliab. Phys. Symp.*, 90–95 (1983).

Sabnis, A. G., "Impact of Advances in Technology on the Properties of Si/SiO_2 Interface," *Proc. Int. Reliab. Phys. Symp.*, 156–160 (1984).

Sabnis, A. G. and J. T. Nelson, "Characterization of Si/SiO_2 Interface Degradation Due to Hot-Carrier Injection," *Tech. Dig.—Int. Electron Devices Meet.*, 52–55 (1985).

Sabnis, A. G., "Process Dependent Build-up of Interface States in Irradiated *n*-Channel MOSFETs," *IEEE Trans. Nucl. Sci.* **NS-32** (6), 3905 (1985).

Sabnis, A. G., "Comparison between Hot-Carrier Drift and Radiation Damage in MOS Devices," *Proc. Eur. Solid State Devices Res. Conf., 17th*, 247–251 (1987).

Sah, C.-T., R. N. Noyce, and W. Shockley, "Carrier Generation and Recombination in P–N Junction and P–N Junction Characteristics," *Proc. IRE* **45**, 1228–1243 (1957).

Sah, C.-T., H. Sello, and D. A. Tremere, "Diffusion of Phosphorus in Silicon Dioxide Film," *J. Phys. Chem. Solids* **11**, 288–298 (1959).

Sah, C.-T., "A New Semiconductor Tetrode, the Surface-Potential Controlled Transistor," *Proc. IRE* **49**, 1632–1634 (1961).

Sah, C.-T., "Effect of Surface Recombination and Channel on P–N Junction and Transistor Characteristics," *IRE Trans. Electron Devices* **9**, 94–108 (1962).

Sah, C.-T., "Transistors," In *1963 McGraw-Hill Yearbook of Science and Technology*, McGraw-Hill, New York, 560–562 (1963).

Sah, C.-T., "Characteristics of the MOS Transistors," *IEEE Trans. Electron Devices* **11**, 324–345 (1964).

Sah, C.-T., "The Equivalent Circuit Model in Solid-State Electronics. Part I: The Single Energy Level Defect Centers," *Proc. IEEE* **55**, 654–771 (1967).

Sah, C.-T., "The Equivalent Circuit Model in Solid-State Electronics. Part II: The Multiple Energy Level Impurity Centers," *Proc. IEEE* **55**, 672–684 (1967).

Sah, C.-T., "The Equivalent Circuit Model in Solid-State Electronics. Part III (Conduction and Displacement Currents)," *Solid-State Electron.* **13**, 1547–1575 (1970).

Sah, C.-T., "Equivalent Circuit Models in Semiconductor Transport for Thermal, Optical, Auger-Impact and Tunneling Recombination–Generation–Trapping Processes," *Phys. Status Solidi* **7**, 541–559 (1971).

Sah, C.-T., J. Y.-C. Sun, and J. J.-T. Tzou, "Effects of Avalanche Injection Currents on the Endurance of Si MOS Devices," *Tech. Dig.—Int. Electron Devices Meet.*, 753–756 (1982).

Sah, C.-T., J. Y.-C. Sun, and J. J.-T. Tzou, "Generation Annealing Kinetics of Interface States on Oxidized Silicon Activated by 10.2-eV Photohole Injection," *J. Appl. Phys.* **53**, 8886–8893 (1982).

Sah, C.-T., J. Y.-C. Sun, and J. J.-T. Tzou, "Study of the Atomic Models of Three Donor-like Traps on Oxidized Silicon with Aluminum Gate from Their Processing Dependencies," *J. Appl. Phys.* **54**, 5864–5679 (1983).

Sah, C.-T., "Hydrogenation and Dehydrogenation of Shallow Acceptors and Donors in Si: Fundamental Phenomena and Survey of the Literature," In *Properties of Silicon*, ISPEC, IEE, London, Sect. 17-16, 584–604 (1988a).

Sah, C.-T., "Hydrogenation and Dehydrogenation of Shallow Acceptors and Donors in Si: Kinetic Rate Data," In *Properties of Silicon*, ISPEC, IEE, London, Sect. 17-17, 604–613 (1988b).

Sah, C.-T., "Interface Traps on Si Surface," In *Properties of Silicon*, ISPEC, IEE, London, Sect. 17.1, 499–507 (1988c).

Sah, C.-T., M. S. C. Luo, C. C. H. Hsu, T. Nishida, and A. J. Chen, "Interface Traps on Oxidized Si From Two-Terminal, Dark Capacitance–Voltage Measurements on MOS Capacitors," In *Properties of Silicon*, ISPEC, IEE, London, Sect. 17-4, 521–531 (1988).

Sah, C.-T., "Interface Traps on Oxidized Si from X-Ray Photoemission Spectroscopy, MOS Diode Admittance, MOS Transistor, and Photogeneration Measurements," In *Properties of Silicon*, ISPEC, IEE, London, Sect. 17-3, 512–520 (1988).

Sah, C.-T., "Evolution of the MOS Transistor—From Conception to VLSI," Proc. of the IEEE **76**, 1280–1326 (1988d).

Sah, C.-T. and C. C. H. Hsu, "Oxide Traps on Oxidized Si," In *Properties of Silicon*, ISPEC, IEE, London, Sect. 17-5, 532–547 (1988).

Sah, C.-T., "Models and Experiments on Degradation of Oxidized Silicon," *Solid-State Electron.* **33**, 147–167 (1990).

Sai-Halasz, G. A., M. R. Wordeman, D. P. Kern, E. Granin, S. Rishton, H. Y. Ng, D. S. Zicherman, D. Moy, T. H. P. Chang, and R. H. Dennard, "Experimental Technology and Characterization of Self-Aligned 0.1 μm-Gate-Length Low-Temperature Operation NMOS Devices," *Tech. Dig.—Int. Electron Devices Meet.*, 397–400 (1987).

Saito, K., T. Morase, S. Sato, and U. Harada, "A New Short-Channel MOSFET with Lightly-Doped Drain," *Denshi Tsushin Rengo Taikai*, 220 (1987).

Saito, K. and A. Yoshii, "Unified Analysis on Hot Carrier Generation in *p*-Channel and *n*-Channel MOSFETs," *Jpn. J. Appl. Phys., Part 2 Lett.* **27**, 2398–2400 (1988).

Saitoh, M., M. Kinugawa, and H. Hashimoto, "Hot-Carrier-Induced Degradation in MOSFETs Studied by Recovery Temperature Spectroscopy (RTS)," *IEEE Trans. Electron Devices* **34**, 2384 (1987).

Sak, N. and M. G. Ancona, "Spatial Uniformity of Interface Trap Distribution in MOSFETs," *IEEE Trans. Electron Devices* **37**, 1057–1063 (1990).

Sakai, Y., N. Hashimoto, O. Minato, and T. Masuhara, "Advanced CMOS Technology for VLSI," In *Semiconductor Technologies* (J. Nishizawa ed.) North-Holland (Publ.) Amsterdam, 97–112 (1982).

Sakashita, K., T. Arakawa, H. Takagi, K. Sugizaki, S. Asai, and I. Ohkura, "A 10 K Gate CMOS Gate Array with Gate Isolation Configuration," *IEEE Custom Int. Circuits Conf.*, 14–18 (1983).

Sakata, I. and T. Ikoma, "Anomalous Gate Current Oscillation Due to Hot Carrier Effect in MOS Diodes," *Jpn. J. Appl. Phys.* **17**, 583–584 (1978).

Saks, N. S., P. L. Heremans, L. Van Den Hove, H. E. Maes, R. F. De Keersmaecker, and G. J. Gilbert, "Observation of Hot-Hole Injection in n-MOS Transistors Using a Modified Floating-Gate Technique," *IEEE Trans. Electron Devices* **33**, 1529–1534 (1986).

Saks, N. S., D. M. McCarthy, and M. G. Ancona, "Comparison of Hot Carrier Degradation at 295 and 77 K," *Proc. Low Temp. Electron. High Temp. Supercond. Symp.*, Honolulu, HI, 136–141 (1988).

Saks, N. S. and M. G. Ancona, "Spatial Uniformity of Interface Trap Distribution in MOSFETs," *IEEE Trans. Electron Devices* **37**, 1057–1063 (1990).

Saks, N. S., R. B. Klein, S. Yoon, and D. L. Griscom, "Formation of Interface Traps in Metal-Oxide-Semiconductor Devices During Isochronal Annealing after Irradiation at 78 K," *J. Appl. Phys.* **70**, 7434–7442 (1991).

Sakurai, T. and T. Sugano, "Theory of Continuously Distributed Trap States at Si–SiO_2 Interfaces," *J. Appl. Phys.* **52**, 2889–2896 (1981).

Sakurai, T., M. Kakumu, and T. Iizuka, "Hot-Carrier Suppressed VLSI with Submicron Geometry," *Tech. Dig. IEEE Int. Solid-State Circuits Conf.*, 272–273 (1985).

Sakurai, T., K. Nogami, M. Kakumu, and T. Iizuka, "Hot-Carrier Generation in Submicrometer VLSI Environment," *IEEE J. Solid-State Circuits* **SC-21**, 187–192 (1986).

Salama, C. A. T., "A New Short Channel MOSFET Structure (UMOST)," *Solid-State Electron.* **20**, 1003 (1977).

Saleh, N., A. El-Hennawy, and S. El-Hennawy, "Simulation of a Non-Avalanche Injection Based CMOS EEPROM Memory Cell Compatible with Scaling-Down Trends," *Dig.—IEEE Int. Electr. Electron. Conf. Expos.*, 590–593 (1985).

Samuilov, V. A., E. A. Bondarionok, D. Shulman, and D. L. Pulfrey, "Memory Switching Effects in *a*-Si/*c*-Si Heterojunction Bipolar Structures," *IEEE Electron Devices Lett.* **13**, 396–398 (1992).

San, K. T. and T.-P. Ma, "Determination of Trapped Oxide Charge in Flash EPROMs and MOSFETs with Thin Oxides," *IEEE Electron Devices Lett.* **13**, 439–441 (1992).

Sanchez, J. J., K. K. Hsueh, and T. A. DeMassa, "Drain-Engineered Hot-Electron-Resistant Devices Structures: A Review," *IEEE Trans. Electron Devices* **36**, 1125–1132 (1989).

Sanchez, J. J., K. K. Hsueh, and T. A. DeMassa, "Hot-Electron Resistant Device Processing and Design: A Review," *IEEE Trans. Semicond. Manuf.* **SM-2**, 1–8 (1989).

Sanchez, J. J. and T. A. DeMassa, "Review of Carrier Injection in the Silicon/Silicon-Dioxide System," *IEEE Proc. G (Circuits, Devices Syst.)* **138**, 377–389 (1991).

Sangiorgi, E., B. Ricco, and P. Olivo, "Hot Electrons and Holes in MOSFETs Biased below the Si–SiO_2 Interfacial Barrier," *IEEE Electron Devices Lett.* **6**, 513–515 (1985).

Sangiorgi, E., E. A. Hofstatter, R. K. Smith, P. F. Bechtold, and W. Fitchner, "Scaling Issues Related to High Field Phenomena in Submicrometer MOSFETs," *IEEE Electron Devices Lett.* **7**, 115–118 (1986).

Sangiorgi, E., M. R. Pinto, F. Venturi, and W. Fichtner, "Hot-Carrier Analysis of Submicrometer MOSFET's," *IEEE Electron Devices Lett.* **9**, 13–16 (1988).

Sarace, J. C., R. E. Kerwin, D. L. Klein, and R. Edwards, "Metal–Nitride–Oxide–Silicon FET with Self-Aligned Gates," *Solid-State Electron.* **11**, 653–660 (1968).

Saraswat, K. C. and J. D. Meindl, "HV Silicon-Gate MOS Integrated Circuit for Driving Piezoelectric Tactile Display," *Tech. Dig. IEEE Int. Solid State Circuits Conf.*, 164–165 (1974).

Sasai, H., M. Saitoh, and K. Hashimoto, "Hot-Carrier Induced Drain Leakage Current in *n*-channel MOSFET," *Tech. Dig.—Int. Electron Devices Meet.*, 726–729 (1987).

Satake, H., T. Hamasaki, T. Maeda, and M. Norishima, "Collector Profile Design for High-Performance Dynamic Operation of Bipolar Transistors at Liquid Nitrogen Temperature," *Ext. Abstr., Jpn. Soc. Appl. Phys.*, 821–824 (1990).

Satake, H. and T. Hamasaki, "Low-Temperature (77 K) BJT Model with Temperature Dependences on the Injected Condition and Base Resistance," *IEEE Trans. Electron Devices* **37**, 1688–1696 (1990).

Sato, M., K. Yoshikawa, S. Mori, and K. Kanzaki, "Gate Current Injection in Submicron EPROM Cells," *Tech. Dig. VLSI Tech. Symp.*, 82–83 (1984).

Sato, M., K. Yoshikawa, S. Atsumi, S. Mori, K. Makita, A. Omichi, and H. Nozawa, "Characterization and Performance Analysis of Masked LDD Transistor for CMOS VLSIs," *17th Solid State Devices Mater. Conf.*, 25–28 (1985).

Satoh, S.-I., Y. Ohbayashi, K. Mizuguchi, M. Yoneda, and H. Abe, "Self-Aligned Graded-Drain Structure for Short Channel MOS Transistor," *Tech. Dig. VLSI Tech. Symp.*, 38–39 (1982).

Satoh, S.-I. and H. Abe, "Self-Aligned Graded-Drain Structure for VLSI," *Jpn. Annu. Rev. Electron. Comput. Telecommun.* **13**, 121–135 (1984).

Satoh, S.-L., T. Eimori, and H. Matsumoto, "Hot Electron Improvement in MOS RAMs Based on Epitaxial Substrate," *Jpn. J. Appl. Phys.*, **24**, L184–L186 (1985).

Satoh, Y., K. Miyamoto, and H. Matsumoto, "Effect of Temperature of Hot Electron Trapping MOSFETs," *Jpn. J. Appl. Phys.*, **22**, L221–L222 (1983).

Sawada, K., T. Sakurai, K. Nogami, K. Sato, T. Shirotori, M. Kakuma, S. Morita, M. Kinugawa, T. Asami, K. Narita, J. Matsunaga, A. Higuchi, M. Isobe, and T. Iizuka, "A 30-μA Data-Retention Pseudostatic RAM with Virtually Static RAM Mode," *IEEE J. Solid-State Circuits* **SC-23**, 12–19 (1988).

Sawada, S., Y. Matsumoto, S. Shinozaki, and O. Ozawa, "Effects of Field Boron Dose on Substrate Current in Narrow Channel LDD MOSFETs," *Tech. Dig.—Int. Electron Devices Meet.*, 778–781 (1984).

Scheibe, A. and W. Krauss, "Two-Transistor SIMOS EAROM Cell," *IEEE J. Solid-State Circuits* **SC-15**, 353–357 (1980).

Schmidt, J. D., "Integrated MOS Transistor RAM," *Solid State Des.*, 21–25 (1971).

Schmitt-Landsiedel, S. and G. Dorda, "Interface States in MOSFETs Due to Hot-Electron Injection Determined by the Charge Pumping Technique," *Electron. Lett.* **17**, 761–763 (1981).

Schmitt-Landsiedel, S. and G. Dorda, "Spatial Distribution of Hot Electrons as a Physical Limit to MOS Transistor Performance," *Electron. Lett.* **18**, 1041–1043 (1982).

Schmitt-Landsiedel, S. and G. Dorda, "The Influence of Hot Channel Electrons on the Surface Potential in MOSFETs," *Proc. Int. Conf. At. Inner-Shell Phys.*, Palo Alto, CA, 167–171 (1984).

Schmitt-Landsiedel, S. and G. Dorda, "Novel Hot-Electron Effects in the Channel of MOSFETs Observed by Capacitance Measurements," *IEEE Trans. Electron Devices* **32**, 1294–1301 (1985).

Schnable, G. L., K. M. Schlesier, G. A. Swartz, and C. P. Wu, "Impact of Anomalous Short-Channel MOS Transistors on VLSI Circuit Reliability," *Microelectron. Reliab.* **33**, 565–582 (1993).

Schroder, D. K. and H. C. Nathanson, "On the Separation of Bulk and Surface Components of Lifetime Using the Pulsed MOS Capacitor," *Solid-State Electron.* **13**, 577–582 (1970).

Schröter, M., "Simulation and Modeling of the Low-Frequency Base Resistance of Bipolar Transistors and Its Dependence on Current and Geometry," *IEEE Trans. Electron Devices*, **38**, 538–544 (1991).

Schuetz, A. and C. Werner, "State-of-the-Art of MOS Modeling," *Tech. Dig.—Int. Electron Devices Meet.*, 766–769 (1984).

Schwalke, U., W. Hansch, and A. Lill, "Comparative Study of HC-Degradation of NMOS and PMOS Devices with n^+ and p^+ Gate: Experiments and Simulation," *Proc. Conf. Solid State Devices Mater., 22nd*, 307–310 (1990).

Schwarz, S. A. and S. E. Russek, "Semi-Empirical Equations for Electron Velocity in Silicon: Part II. MOS Inversion Layer," *IEEE Trans. Electron Devices* **30**, 1634–1639 (1983).

Schwerin, A., V. W. Hansch, and W. Weber, "*n*- and *p*-Channel MOSFET Degradation: Experiment and Simulation," *IEE Colloq. 'Hot Carrier Degradation Short Channel MOS' (Dig.)*, 13.1–13.3 (1987).

Schwerin, A. V. and M. M. Heyns, "Oxide Field Dependence of Bulk and Interface Trap Generation in SiO_2 Due to Electron Injection," *Proc. 7th Bienn. Eur. Conf.*, 263–266 (1991).

Schwerin, A. V. and M. M. Heyns, "Homogeneous Hole Injection into Gate Oxide Layers of MOSFETs: Injection Efficiency, Hole Trapping and $SiSiO_2$ Interface State Generation," *Proc. 7th Bienn. Eur. Conf.*, 283–286 (1991).

Seilmeier, A., H.-J. Hubner, M. Woerner, G. Abstreiter, G. Weimann, and W. Schlapp, "Direct Observation of Intersubband Relaxation in Narrow Multiple Quantum Well Structures," *Solid-State Electron.* **31**, 767–770 (1988).

Seilmeier, A., H.-J. Hubner, M. Woerner, G. Abstreiter, G. Weimann, and W. Schlapp, "Normal and Hot Electron Magneto-Phonon Resonance in a GaAs-Heterostructure," *Solid-State Electron.* **31**, 771–775 (1988).

Sekido, M., K. Taniguchi, and C. Hamaguchi, "Direct Observation of Gaussian-Type Energy Distribution for Hot Electrons in Silicon," *Jpn. J. Appl. Phys.* **30**, 1149–1153 (1991).

Selberherr, S., A. Schutz, and H. W. Potzl, "MINIMOS—A Two-Dimensional MOS Transistor Analyzer," *IEEE Trans. Electron Devices* **27**, 1540–1550 (1980).

Selberherr, S., "Low Temperature MOS Device Modeling," *Proc. Symp Low Temp. Electron. High Temp. Supercond.*, Honolulu, HI, 70–86 (1988).

Selberherr, S., "MOS Device Modeling at Liquid-Nitrogen Temperature," *Tech. Dig.—Int. Electron Devices Meet.*, 496–499 (1988).

Selberherr, S., "MOS Device Modeling at 77 K," *IEEE Trans. Electron Devices* **36**, 1464–1474 (1989).

Selberherr, S., "Device Modeling and Physics," *Phys. Scrip.* **T35**, 293–298 (1991).

Selmi, L., E. Sangiorgi, and B. Ricco, "Parameter Extraction from *I–V* Characteristics of Single MOSFETs," *IEEE Trans. Electron Devices* **36**, 1094–1101 (1989).

Selmi, L., M. Lanzoni, S. Bigliardi, and E. Sangiorgi, "Photon Emission from Sub-Micron *p*-Channel MOSFETs Biased at High Fields," *Microelectron. Eng.* **19**, 747–750 (1992).

Selvakumar, C. R. and B. Hecht, "SiGe-Channel *n*-MOSFET by Germanium Implantation," *IEEE Electron Devices Lett.* **12**, 444–446 (1991).

Sequin, C. H., M. F. Tompsett, and G. E. Smith, "Experimental Verification of the Charge Coupled Diode Concept," *Bell Syst. Tech. J.* **49**, 593–600 (1970).

Serack, J. A., J. M. Robertson, and A. J. Walton, "The Application of a Novel Experimental Technique to Investigate Hot Carriers in MOSFETs," *IEE Colloq. 'Hot Carrier Degradation Short Channel MOS'* (*Dig.*), 4.1–4.4 (1987).

Shabde, S., A. Bhattacharyya, R. S. Kao, and R. S. Muller, "Analysis of MOSFET Degradation Due to Hot-Electron Stress in Terms of Interface-State and Fixed Charge Generation," *Solid-State Electron.* **31**, 1603–1610 (1988).

Shah, J., "Photoexcited Hot Carriers: From Cw to 6 Fs in 20 Years," *Solid-State Electron.* **32**, 1051–1056 (1989).

Shahidi, G. G., D. A. Antoniadis, and H. I. Smith, "Reduction of Hot-Electron-Generated Substrate Current in Sub-100-nm Channel Length Si MOSFETs," *IEEE Trans. Electron Devices* **35**, 2430 (1988).

Shahriary, I., J. R. Schwank, and F. G. Allen, "Energy Loss and Escape Depth of Hot Electrons from Shallow *p–n* Junctions in Silicon," *J. Appl. Phys.* **50**, 1428–1438 (1979).

Shannon, J. M., "Hot Electron Diodes and Transistors," *Conf. Ser.—Inst. Phys.* **69**, 45–62 (1984).

Sharma, D., S. Goodwin-Johansson, D.-S. Wen, C. K. Kim, and C. M. Osburn, "A 1 μm CMOS Technology with Low Temperature Processing," *Proc. Int. ULSI Sci. Tech. Symp.*, Philadelphia, 49–63 (1987).

Shaw, J.-J. and K. Wu, "Determination of Spatial Distribution of Interface States on Submicron, Lightly Doped Drain Transistors by Charge Pumping Measurement," *Tech. Dig.—Int. Electron Devices Meet.*, 83–86 (1989).

Sher, A., S. Krishnamurthy, and A.-B. Chen, "Transport in Submicron Devices," *Microelectron. Eng.* **9**, 377–380 (1989).

Sheu, B. J., C. Hu, P.-K. Ko, and F. C. Hsu, "Source-and-Drain Series Resistance of LDD MOSFETs." *IEEE Electron Devices Lett.* **5**, 356–367 (1984).

Sheu, B. J., W.-J. Hsu, and V. C. Tyree, "Modeling Requirements for Computer-Aided VLSI Circuit Reliability Assessment," *Proc. 8th Bienn. Univ./Gov./Ind. Microelectron. Symp.*, Westborough, MA, 199–204 (1989).

Shewchun, J. and L. Y. Wei, "Mechanism for Reverse-Bias Breakdown Radiation in *p–n* Junctions," *Solid-State Electron.* **8**, 485 (1965).

Shi, Z. M., J. P. Mieville, and M. Dutoit, "Effect of Electron Heating on RTS in Deep Submicron *n*-MOSFETs," *Microelectron. Eng.* **19**, 751–754 (1992).

Shibata, T., R. Nakayama, K. Kurosawa, S. Onga, M. Konaka, and H. Iizuka, "A Simplified Box (Buried-Oxide) Isolation Technology for Megabit Dynamic Memories," *Tech. Dig.—Int. Electron Devices Meet.*, 27–30 (1983).

Shibata, T. T. Moriya, K. Kurosawa, T. Mitsuno, K. Okumura, Y. Horiike, K. Yamada, and M. Muromachi, "*n*-MOS Process Integration for a 1 M Word × 1 Bit DRAM," *Tech. Dig.—Int. Electron Devices Meet.*, 75–78 (1984).

Shigyo, N., S. Onga, M. Yoshimi, and K. Taniguchi, "Three-Dimensional Simulation of Hot Carrier Effects in Submicron MOSFETs," *Trans. Inst. Electron. Commun. Eng. Jpn.*, 248–250 (1986).

Shigyo, N. and R. Dang, Chapter 9: "Three-Dimensional Device Simulation Using a Mixed Process/Device Simulator," In *Process and Device Modeling*, W. L. Eng., ed., Elsevier/North-Holland, Amsterdam, 301–327 (1986).

Shigyo, N., S. Fukuda, T. Wada, K. Hieda, T. Hamamoto, H. Watanabe, K. Sunouchi, and H. Tango, "Three-Dimensional Analysis of Subthreshold Swing and Transconductance and Fully Recessed Oxide (Trench) Isolated 1/4-μm-Width MOSFET's," *IEEE Trans. Electron Devices* **35**, 945–950 (1988).

Shigyo, N., T. Wada, and S. Yasuda, "Discretization Problem for Multidimensional Current Flow," *IEEE Trans. Comput.-Aided Des.* **8**, 1046–1050 (1989).

Shigyo, N., S. Fukuda, and K. Kato, "The Influence of Boundary Locations on Wiring Capacitance Simulation," *IEEE Trans. Electron Devices* **36**, 1171–1174 (1989).

Shigyo, N., H. Tanimoto, M. Norishima, and S. Yasuda, "Minority Carrier Mobility Model for Device Simulation," *Solid-State Electron.* **33**, 727–731 (1990).

Shih, D. K., G. Q. Lo, W. Ting, and D. L. Kwong, "Performance and Reliability of Short-Channel MOSFETs with Superior Oxynitride Gate Dielectrics Fabricated Using Multiple Rapid Thermal Processing," *Int. Symp. VLSI Technol. Syst. Appl. Proc. Tech. Pap.*, 197–201 (1989).

Shih, D. K., D. L. Kwong, and S. Lee, "Short-Channel MOSFETs with Oxynitride Gate Dielectrics Fabricated Using Multiple Rapid Thermal Processing," *Electron. Lett.* **25**, 190–191 (1989).

Shih, Y.-H., Y. Leblebici, and S. M. Kang, "New Simulation Methods for MOS VLSI Timing and Reliability," *Int. Conf. Comput.-Aided Des. Dig. Tech. Pap.*, 162–165 (1991).

Shimaya, M., N. Shiono, O. Nakajima, C. Hashimoto, and Y. Sakakibara, "Electron Beam Induced Damage in Poly-Si Gate MOS Structures and Its Effect on Long-Term Stability," *J. Electrochem. Soc.* **130**, 945–950 (1983).

Shimaya, M. and N. Shiono, "The Effect of Post-Oxidation Annealing on Hot Carrier Trapping Characteristics in SiO_2," *Ext. Abstr., Solid State Devices Mater.*, 471–474 (1986).

Shimizu, A., N. Ohki, H. Ishida, T. Yamanaka, N. Hashimoto, T. Hashimoto, and E. Takeda, "High Drivability and High Reliability MOSFETs with Non-doped Poly-Si Spacer LDD Structure (SLDD)," *Tech. Dig. VLSI Tech. Symp.*, 90–91 (1992).

Shimizu, S., S. Kusunoki, M. Inuishi, K. Tsukamoto, and Y. Akasaka, "Hot Carrier Effect of AC Stress on *p*-MOSFETs," *Ext. Abstr., Int. Conf. Solid State Dev. Mater.*, 231–233 (1991).

Shimokawa, K., T. Usami, S. Tokitou, N. Hirashita, M. Yoshimaru, and M. Ino, "Suppression of the MOS Transistor Hot Carrier Degradation Caused by Water Desorbed from Intermetal Dielectric," *Tech. Dig. VLSI Tech. Symp.*, 96–97 (1992).

Shimoyama, N., K. Machida, K. Murase, and T. Tsuchiya, "Enhanced Hot-Carrier Degradation Due to Water in TEOS/O_3-oxide and Water Blocking Effect of ECR-SiO_2," *Tech. Dig. VLSI Tech. Symp.*, 94–95 (1992).

Shimoyama, N. and T. Tsuchiya, "AC Hot-Carrier-Degradation Mechanism in LDDMOSFETs," *Ext. Abstra., Int. Conf. Solid State Dev. Mater.*, 518–520 (1992).

Shin, H., A. F. Tasch, C. M. Maziar, and S. K. Banerjee, "A New Approach to Verify and Derive a Transverse-Field-Dependent Mobility Model for Electrons in MOS Inversion Layers," *IEEE Trans. Electron Devices* **36**, 1117–1124 (1989).

Shin, H., A. F. Tasch, Jr., T. J. Bordelon, and C. M. Maziar, "MOSFET Drain Englineering Analysis or Deep-Submicrometer Dimensions: A New Structural Approach," *IEEE Trans. Electron Devices* **39**, 1922–1927 (1992).

Shinozaki, S., "Device Structure Design Technology," *J. Inst. Electr. Eng. Jpn.* **112**, 8–12 (1992).

Shiono, N. and C. Hashimoto, "Hot-Electron Limited Operating Voltages for 0.8 μm MOSFETs," *IEEE Trans. Electron Devices* **29**, 1630–1632 (1982).

Shiono, N., O. Nakajima, and C. Hashimoto, "Interface Trapped Charge Generation and Electron Trapping Kinetics in MOSFETs Due to Substrate Hot-Electron Injection into Gate SiO_2," *J. Electrochem. Soc.* **129**, 1760–1764 (1982).

Shirahata, M., K. Taniguchi, and C. Hamaguchi, "Self-Consistent Monte Carlo Simulation for Two-Dimensional Electron Transport in MOS Inversion Layer," *Jpn. J. Appl. Phys.* **26**, 1447–1452 (1987).

Shiraki, Y., "Photoconductivity of Silicon Inversion Layers," *J. Phys. C* **10**, 4539–4544 (1977).

Shirota, R. and T. Yamaguchi, "A New Analytical Model for Low Voltage Hot Electron Taking Auger Recombination as Well as Phonon Scattering Process into Account," *Tech. Dig.—Int. Electron Device Meet.*, 123–126 (1991).

Shive, J. N., "The Double Surface Transistor," *Phys. Rev.* **75**, 689–690 (1949).

Shockley, W., "On the Surface States Associated with a Periodic Potential," *Phys. Rev.* **56**, 317–323 (1939).

Shockley, W. and G. L. Pearson, "Modulation of Conductance of Thin Films of Semiconductors by Surface Charges," *Phys. Res.* **74**, 232–233 (1948).

Shockley, W., "Theory of P–N Junctions in Semiconductors and P–N Junction Transistors," *Bell. Syst. Tech. J.* **28**, 436–489 (1949).

Shockley, W., "Hot Electrons in Germanium and Ohm's Law," *Bell Syst. Tech. J.* **30**, 990–1034 (1951).

Shockley, W. and W. T. Read, "Statistics of Recombination of Holes and Electrons," *Phys. Rev.* **87**, 835–842 (1952).

Shockley, W., "A Unipolar FET," *Proc. IRE* **40**, 1365–1376 (1952).

Shockley, W., "Electrons, Holes, and Traps," *Proc. IRE* **46**, 973–990 (1958).

Shockley, W., "Problems Related to P–N Junctions in Silicon," *Solid-State Electron.* **2**, 35–67 (1967).

Shockley, W., "The Path to the Conception of the Junction Transistor," *IEEE Trans. Electron Devices* **23**, 597–620 (1976).

Shoemaker, P. A. and R. Shimabukuro, "Modifiable Weight Circuit for Use in Adaptive Neuromorphic Networks," *Neural Networks*, **1**, 409 (1988).

Shone, F., K. Wu, J. Shaw, E. Hokelek, S. Mittal, and A. Haranahalli, "Gate Oxide Charging and Its Elimination for Metal Antenna Capacitor and Transistor in VLSI CMOS Double Layer Metal Technology," *Tech. Dig. VLSI Tech. Symp.*, 73–74 (1989).

Shono, K., K. Ishida, and N. Yamada, "Reliability on Short-Channel MOSLSIs," *Fujitsu Sci. Tech. J.* **24**, 446–455 (1988).

Shukuri, S., Y. Onose, N. Ohki, T. Nishida, M. Hirao, and E. Takeda, "A 0.3 μm BiCMOS Technology with Highly Self-Aligned Structures for Deep Submicron ULSIs," *Ext. Abstr. Int. Conf. Solid State Devices Mater.*, 162–164 (1991).

Sibbert, H., B. Hofflinger, and G. Zimmer, "An Analytic Model for the Power Handling Capacity and Scaling of the Smallest MOSTS for VLSI Digital Circuits," *NTG-Fachber.* **68**, 128–134 (1979).

Siegel, P. H., A. R. Kerr, and R. J. Mattauch, "A Comparison of the Measured and Theoretical Performance of a 140–220 GHz Schottky Diode Mixer," *Proc. IEEE Int. Microwave Symp.—Expanding Microwave Horiz.*, 549–551 (1984).

Silard, A. P. and M. J. Dut, "A New Expression for Breakdown Voltage of Practical Linearly Graded p–n Junction," *IEEE Trans. Electron Devices* **38**, 422–424 (1991).

Simoen, E., M.-H. Gao, J.-P. Colinge, and C. Claeys, "Study of Hot-Carrier Stress Effects on the DC Characteristics of SOI NMOST's Operating at 4.2 K," *1991 IEEE Int. SOI Conf.*, Vail Valley, CO 2–3 (1991).

Simoen, E., B. Dierickx, and C. Claeys, "Random Telegraph Signal Noise: A Probe for Hot-Carrier Degradation Effects in Submicrometer MOSFETs," *Microelectron. Eng.* **19**, 605–608 (1992).

Simoen, E. and C. Claeys, "Hot-Carrier Degradation of *n*MOSTs Stressed at 4.2 K," *Solid-State Electron.* **36**, 527–532 (1993).

Simoen, E., M.-H. Gao, J.-P. Colinge, and C. Claeys, "Metastable Charge-Trapping Effect in SOI *n*MOSTs at 4.2 K," *Semicond. Sci. Technol.* **8**, 423–428 (1993).

Sing, R., G. Kumar, and T. S. Jayadev, "On the Scaling of MOSFETs for Submicron VLSI," *Ext. Abstr. Electrochem. Soc.* **85-1**, 366 (1985).

Sing, R., "High Speed Electronics using High Temperature Superconducting Thin Films," *Rev. Solid State Sci.* **2**, 473–494 (1988).

Sing, Y. W. and B. Sudlow, "Modeling and VLSI Design Constraints of Substrate Current," *Tech. Dig.—Int. Electron Devices Meet.*, 732–735 (1980).

Singer, P., "Transistor: 40 Years Later," *Semicond. Int.* **11**, 74–77 (1988).

Singh, R., "Growth of Thin Thermal Silicon Dioxide Films and Low Defect Density," *Microelectron. J.* **23**, 273–281 (1992).

Singh, R. J. and R. S. Srivastava, "Interface State Generation by Altering Voltage Stress under Visible Irradiation at the Silicon–Silicon Dioxide Interface," *J. Appl. Phys.* **54**, 1162–1164 (1983).

Singh, R. S., C. S. Korman, D. J. Kaputa, and E. P. Surowiec, "Total-Dose and Charge-Trapping Effects in Gate Oxides for CMOS LSI Devices," *IEEE Trans. Nucl. Sci.* **NS-31**, 1518–1523 (1984).

Singh, S. N. and P. K. Singh, "Modeling of Minority-Carrier Surface Recombination Velocity at Low–High Junction of an n^+–p–p^+ Silicon Diode," *IEEE Trans. Electron Devices* **38**, 337–343 (1991).

Sinha, A. K., "Thin-Film Tungsten for Silicon I. C. Applications," *Jpn. J. Appl. Phys., Suppl.* **2**, Part I, 487 (1974).

Sinha, A. K., W. S. Lindenberger, W. D. Powell, and E. I. Pavilonis, "Avalanche-Induced Hot Electron Injection and Trapping in Gate Oxides on *p*-Si," *J. Electrochem. Soc.* **127**, 2046–2049 (1980).

Sinha, A. K., K. Ashok, W. S. Lindenberger, D. B. Fraser, S. P. Murarka, and E. N. Fuls, "MOS Compatibility of High-Conductivity $TaSi_2/n^+$ Poly-Si Gates," *IEEE Trans. Electron Devices* **27**, 1425–1430 (1980).

Sinha, A. K., "Refractory Metal Silicides for VLSI Applications," *J. Vac. Sci. Technol.* **19**, 778–785 (1981).

Skardon, J., "MOS and CMOS as Drivers of CV Technology," *Solid State Technol.* **33**, 93–95 (1990).

Smith, D. K., K. Brau, J. Casey, J. W. Coleman, M. Gaudreau, M. Gerver, S. Golovato, W. Guss, S. Horne, J. Irby, J. Kesner, B. G. Lane, M. Mauel, R. Myer, R. S. Post, E. Sevillano, J. D. Sullivan, and R. Torti, "Recent Results from the Tara Tandem Mirror Experiment," *IEEE Int. Conf. Plasma Sci.*, 85 (1986).

Smith, H. I. and D. A. Antoniadis, "Mesoscopic Devices: Will They Supersede Transistors in ULSI?" *Ext. Abstr. Int. Conf. Solid State Devices Mater.*, 485–486 (1992).

Smith, R. A., *Semiconductors*, 2nd ed., Cambridge Univ. Press, Cambridge, UK, 271 (1978).

Snow, E. H., B. E. Deal, A. S. Grove, and C. T. Sah, "Ion Transport Phenomena in Insulating Films Using the MOS Structure," *J. Appl. Phys.* **36**, 1664–1673 (1965).

Snow, E. H. and B. E. Deal, "Polarization Phenomena and Other Properties of Phosphosilicate Glass Films on Silicon," *J. Electrochem. Soc.* **113**, 2631 (1966).

Snow, E. H., A. S. Grove, and D. J. Fitzgerald, "Effect of Ionization Radiation on Oxidized Silicon Surfaces and Planar Devices," *Proc. IEEE* **55**, 1168 (1967).

Sodini, C., P.-K. Ko, and J. L. Moll, "The Effect of High Fields on MOS Device and Circuit Performance," *IEEE Trans. Electron Devices* **31**, 1386–1393 (1984).

Song, D., J. Lee, H. S. Min, and Y. J. Park, "Theory of Thermal Noise in MOS Transistors," *Noise Phys. Syst. 1/f Fluctuations*, 269–272 (1992).

Song, M., J. S. Cable, J. C. S. Woo, and K. P. MacWilliams, "Optimization of LDD Devices for Cryogenic Operation," *IEEE Electron Devices Lett.* **12**, 375–378 (1991).

Sparks, D. R., J. Fruth, J. Christenson, and D. Adams, "Practical Aspects of Lateral and Substrate PNP Fabrication with a BiCMOS Process," *Microelectron. Eng.* **18**, 225–235 (1992).

Speckbacker, P., A. Asenov, M. Bollu, F. Koch, and W. Weber, "Hot-Carrier-Induced Deep-Level Defects from Gate-Diode Measurements on MOSFETs," *IEEE Electron Devices Lett.* **11**, 95–97 (1990).

Spooner, H. and N. R. Couch, "Advances in Hot Electron Injector Gunn Diodes," *GEC J. Res.* **7**, 34–45 (1989).

Srinivasan, V. and J. J. Barnes, "Small Width Effects on MOSFET Hot-Electron Reliability," *Tech. Dig.—Int. Electron Devices Meet.*, 740–745 (1980).

Srivastava, A. and R. Nema, "Development of Master CMOS Test Vehicle Chip for Space Radiation Effect Studies using CAMD X-Ray Synchrotron Radiation," *Nucl. Instrum. Methods Phys. Res., Sect. A* **A319**, 350–357 (1992).

Stegherr, M., "Flicker Noise in Hot Electron Degraded Short Channel MOSFETs," *Solid-State Electron.* **27**, 1055–1056 (1984).

Steimle, M. and H.-M. Muhlhoff, "Limitations of Digital CMOS-Processes for Analog Applications Due to Channel Length Modulation and Hot Carrier Degradation," *Microelectron Eng.* **15**, 429–432 (1991).

Stinson, M. G. and C. M. Osburn, "Effects of Ion Implantation on Deep-Submicrometer, Drain-Engineered MOSFET Technologies," *IEEE Trans. Electron Devices* **38**, 487–497 (1991).

Stirling, W. L. and I. Alexeff, "Turbulently Heated Electron Plasma," *Plasma Phys.* **12**, 489–502 (1970).

Stoev, I. and F. Bauer, "Field Stress Degradation in Depletion-Mode *n*-Channel MOSFETs," *Bulg. J. Phys.* **12**, 506–510 (1985).

Stoev, I., P. Balk, and F. Bauer, "Hot-Hole-Induced Degradation in Depletion-Mode *n*-Channel MOSFETs," *Electron. Lett.* **21**, 30–31 (1985).

Stoev, I. G., "Characteristics of Short-Channel MOSFET with Buried Channel," *C. R. Acad. Bulg. Sci.* **38**, 855–857 (1985).

Strangeway, R. J., "On the Applicability of Relativistic Dispersion to Auroral Zone Electron Distributions," *J. Geophys. Res.* **91**, 3152–3166 (1986).

Strzalkowski, I., M. Marczewski, M. Kowalski, and J. Wislowski, "Low Field dc Investigation of Hot Carrier Trapping in Silicon Dioxide films," *Appl. Phys. A* **51**, 19–22 (1990).

Strzalkowski, I., M. Marczewski, M. Kowalski, C. Jastrzebski, and A. Bakowski, "On Trap Generation in SiO_2 Films of Si MOSFETs by Hot Electrons," *Acta Phys. Pol. A* **82**, 685–688 (1992).

Sturm, J. C., "Performance Advantages of Submicron Silicon-on-Insulator Devices for VLSI," *Proc. Silicon-on-Insul. Buried Met. Semicond. Symp.*, Boston, MA, 295–307 (1988).

Su, H. Q., C. C. Wei, and T.-P. Ma, "Mobility Degradation in Very Thin Oxide *p*-Channel MOSFETs," *IEEE Trans. Electron Devices* **32**, 559–561 (1985).

Su, L. T., J. A. Yasaitis, and D. A. Antoniadis, "A High-Performance Scalable Submicron MOSFET for Mixed Analog/Digital Applications," *Tech. Dig.—Int. Electron Devices Meet.*, 367–370 (1991).

Subrahmaniam, R., J. Y. Chen, and A. H. Johnston, "MOSFET Degradation Due to Hot-Carrier Effect at High Frequencies," *IEEE Electron Devices Lett.* **11**, 21–23 (1990).

Suehle, J. S., T. J. Russell, and K. F. Galloway, "Interface Trap Effects on the Hot-Carrier Induced Degradation of MOSFETs during Dynamic Stress," *IEEE Trans. Nucl. Sci.* **NS-34**, 1359–1365 (1987).

Suehle, J. S., "The Effects of Localized Hot-Carrier-Induced Charge in VLSI Switching Circuits," *Proc. 17th Microelectron. Conf.*, Nis, Yugoslavia **2**, 805–812 (1989).

Sugano, T., "Low Temperature Electronics Research in Japan," *Proc. Symp Low Temp. Electron. High Temp. Supercond.*, Honolulu, HI, 18–29 (1987).

Sukegawa, K. and S. Kawamura, "Effects of Hot Electron Trapping in Ultra-Thin-Film SOI/SIMOX pMOSFETs," *IEICE Trans. Electron.* **E75-C**, 1484–1490 (1992).

Sun, C. C., J. M. Xu, A. Hagley, R. Surridge, and A. S. Thorpe, "Electron Mobility Measurement in Short-Channel FET's Using the Cutoff Frequency Method," *IEEE Electron Devices Lett.* **11**, 382–384 (1990).

Sun, E., J. Moll, J. Berger, and B. Alders, "Breakdown Mechanism in Short-Channel MOS Transistors," *Tech. Dig.—Int. Electron Devices Meet.*, 478 (1978).

Sun, J. Y.-C., Y. Taur, R. H. Dennard, S. P. Klepner, and L. K. Wang, "0.5 μm-Channel CMOS Technology Optimized for Liquid-Nitrogen-Temperature Operation," *Tech. Dig.—Int. Electron Devices Meet.*, 1425–1430 (1980).

Sun, J. Y.-C., J. R. Maldonado, M. D. Rodriguez, J. Laskar, and D. S. Zicherman, "Effects of X-Ray Irradiation of the Channel Hot-Carrier Reliability of Thin-Oxide *n*-Channel MOSFETs," *18th Solid State Devices Mater. Conf.*, 479–482 (1986).

Sun, J. Y.-C., Y. Taur, R. H. Dennard, S. P. Klepner, and L. K. Wang, "0.5-μm-Channel CMOS Technology Optimized for Liquid-Nitrogen-Temperature Operation," *Tech. Dig.—Int. Electron Devices Meet.*, 236–239 (1986).

Sun, R. C., J. T. Clemens, and J. T. Nelson, "Effects of Silicon Nitride Encapsulation on MOS Device Stability," *Proc. Int. Reliab. Phys. Symp.*, 244–251 (1980).

Sun, S. C. and J. D. Plummer, "Electron Mobility in Inversion and Accumulation Layers on Thermally Oxidized Silicon Surfaces," *IEEE Trans. Electron Devices* **27**, 1497–1508 (1980).

Sun, S. W., M. Swenson, J. R. Yeargain, C.-O. Lee, C. Swift, J. R. Pfiester, W. Bibeau, and W. Atwell, "Dual-Poly (n^+/p^+) Gate, Ti-Salicide, Double-Metal Technology for Submicron CMOS ASIC and Logic Applications," *Proc. IEEE Custom Int. Circuits Conf.*, 18.7–18.7.4 (1989).

Sun, S. W., K.-Y. Fu, C. T. Swift, and J. R. Yeargain, "Oxide Charge Trapping and HCl Susceptibility of a Submicron CMOS Dual-Poly (n^+/p^+) Gate Technology," *Proc. Int. Reliab. Phys. Symp.*, 183–188 (1989).

Sun, S. W., M. Orlowski, and K.-Y. Fu, "Parameter Correlation and Modeling of the Power-Law Relationship in MOSFET Hot-Carrier Degradation," *IEEE Electron Devices Lett.* **11**, 297–299 (1990).

Sunami, H., M. Koyanagi, and N. Hashimoto, "Intermediate Oxide Formation in Double-Polysilicon Gate MOS Structure," *J. Electrochem. Soc.* **27**, 2499 (1980).

Svensson, C. M., "The Defect Structure of Si/SiO_2 Interface—A Model Based on Trivalent Silicon and Its Hydrogen Compounds," In *The Physics of SiO_2 and its Interfaces*, S. T. Pantelides, ed., Pergamon, Oxford, 329–332 (1978).

Svoboda, V., "1.5 μm and 1.2 μm CMOS Process for ASIC Production," *Proc. Microelectron. Conf.*, 45–49 (1991).

Sze, S. M. and G. Gibbons, "Effects of Junction Curvature on Breakdown Voltage in Semiconductors," *Solid-State Electron.* **9**, 831 (1966).

Sze, S. M., "Current Transport and Maximum Dielectric Strength of Silicon Nitride Films," *J. Appl. Phys.* **38**, 2951 (1967).

Sze, S. M., *Physics of Semiconductor Devices*, 2nd ed., Wiley, New York (1981).

Tada, Y., C. Nagata, and M. Iwahashi, "Band-to-Band Tunneling Current as a Probe for the Hot-Carrier Effects (MOSFETs)," *Proc. Conf. Solid State Devices Mater., 22nd*, 315–318 (1990).

Takacs, D. and M. Steger, "Characterization of the Hot Carrier Related MOS Parameters Using Negative Feedback Circuits in VLSI CMOS," *Proc. Eur. Solid State Devices Res. Conf., 19th*, 723–726 (1989).

Takacs, D., "Performance of Extrinsic and Intrinsic MOSFETs in Deep Submicron VLSI," *Proc. Conf. Solid State Devices Mater., 22nd*, 263–266 (1990).

Takagi, K. and A. Van Der Ziel, "Excess High Frequency Noise and Flicker Noise in MOSFETs," *Solid-State Electron.* **22**, 289–292 (1979).

Takagi, S., M. Iwase and A. Toriumi, "On the Universality of Inversion-Layer Mobility in N- and P-channel MOSFET's," *Tech. Dig.—Int. Electron Devices Meet.*, 398–401 (1988).

Takagi, S., M. Iwase and A. Toriumi, "Effects of Surface Orientation on the Universality of Inversion-Layer Mobility in Si MOSFETs," *Ext. Abstr., Solid State Devices Mater.*, 275–278 (1990).

Takagi, T., I. Yamada, J. Ishikawa, T. H. Stix, K. Weisemann, G. Fuchs, R. L. Darling, D. L. McShan, R. H. Davis, G. F. Tonon, M. Friedman, A. Van Der Woude, I. Alexeff, and W. D. Jones, "International Conference on Multiply-Charged Heavy Ion Sources and Accelerating Systems Oct. 25–28 1971," *IEEE Trans. Nucl. Sci.* **NS-19**, 1–320 (1972).

Takahashi, T. B. B. Triplett, K. Yokogawa, and T. Sugano, "Electron Spin Resonance Observation of the Creation, Annihilation, and Charge State of the 74-Gauss Doublet in Device Oxides Damaged by Soft X Rays," *Appl. Phys. Lett.* **51**, 1334–1336 (1987).

Takeda, E., H. Kume, T. Toyabe, and S. Asai, "Submicron MOSFET Structure for Minimizing Channel Hot-Electron Injection," *Tech. Dig. VLSI Tech. Symp.*, 22–23 (1981).

Takeda, E., H. Kume, Y. Nakagome, and S. Asai, "As–P(n^+–n^-) Double Diffused Drain MOSFET for VLSIs," *Tech. Dig. VLSI Tech. Symp.*, 40–41 (1982a).

Takeda, E., N. Suzuki, and T. Hagiwara, "Device Performance Degradation Due to Hot-Carrier Injection at Energies Below the Si–SiO_2 Energy Barrier," *Tech. Dig.—Int. Electron Devices Meet.*, 396–399 (1982b).

Takeda, E., H. Kume, T. Toyabe, and S. Asai, "Submicron MOSFET Structure for Minimizing Hot-Carrier Generation," *IEEE Trans. Electron Devices* **29**, 611–618 (1982c).

Takeda, E., H. Kume, Y. Nakagome, T. Makino, A. Shimizu, and S. Asai, "An As–P(n^+–n^-) Double Diffused Drain MOSFET for VLSI's," *IEEE Trans. Electron Devices* **30**, 652–657 (1983a).

Takeda, E., Y. Nakagome, H. Kume, N. Suzuki, and S. Asai, "Comparison of Characteristics of n-Channel and p-Channel MOSFET's VLSI's," *IEEE Trans. Electron Devices* **30**, 675–680 (1983b).

Takeda, E. and N. Suzuki, "An Empirical Model for Device Degradation Due to Hot-Carrier Injection," *IEEE Electron Devices Lett.* **4**, 111–113 (1983).

Takeda, E., T. Makino, and T. Hagiwara, "The Impact of Drain Impurity Profile and Junction Depth on Submicron MOSFETs," *Ext. Abstr., Solid State Devices Mater.*, 261–264 (1983c).

Takeda, E., K. Kume, and S. Asai, "New Grooved-Gate MOSFET with Drain Separated from Channel Implanted Region (DSC)," *IEEE Trans. Electron Devices* **30**, 681–686 (1983d).

Takeda, E., Y. Nakagome, H. Kume, and S. Asai, "New Hot-Carrier Injection and Device Degradation in Submicron MOSFETs," *Proc. Inst. Electr. Eng.* **130**, Part I, 144–150 (1983e).

Takeda, E., A. Shimizu, and T. Hagiwara, "Role of Hot-Hole Injection in Hot-Carrier Effects and the Small Degraded Channel Region in MOSFETs," *IEEE Electron Device Lett.* **4**, 329–331 (1983f).

Takeda, E., H. Kume, Y. Nakagome, N. Suzuki, S. Asai, and T. Hagiwara, "Hot-Carrier Effects in Submicron VLSIs," *Tech. Dig. VLSI Tech. Symp.* 104–105 (1983g).

Takeda, E., N. Suzuki, Y. Igura, and T. Hagiwara, "'Distortion' of Scaling Laws in 0.5 μm Effective Channel Length MOSFETs," *Jpn. Trans. Inst. Electron. Commun. Eng. Jpn.* **66C**, 1042–1049 (1983h).

Takeda, E., T. Hagiwara, and N. Suzuki, "Device Performance Degradation Due to Hot Carriers Having Energies below the Si–SiO_2 Energy Barrier," *J. Appl. Phys.* **55**, 3180–3182 (1984).

Takeda, E., "Hot-Carrier Effects in Submicrometer MOS VLSIs," *Proc. Inst. Electr. Eng.* **131**, Part I, 153–162 (1984).

Takeda, E., Y. Ohji, and H. Kume, "High Field Effects in MOSFETs," *Tech. Dig.—Int. Electron Devices Meet.*, 60–63 (1985a).

Takeda, E., G. A. C. Jones, and H. Ahmed, "Constraints on the Application of 0.5-μm MOSFETs to ULSI Systems," *IEEE Trans. Electron Devices* **32**, 322–327 (1985b).

Takeda, E., K. Takeuchi, A. Hiraiwa, T. Toyabe, H. Sunami, and K. Itoh, "Three Dimensional Leakage Current in Corrugated Capacitor Cells," *Ext. Abstr., Solid State Devices Mater.*, 37–40 (1985c).

Takeda, E., "Hot-Carrier and Wear-Out Phenomena in Submicron VLSIs," *Tech. Dig. VLSI Tech. Symp.*, 2–5 (1985).

Takeda, E., K. Takeuchi, E. Yamasaki, T. Toyabe, K. Ohshima, and K. Itoh, "The Scaling Law of Alpha-Particle Induced Soft Errors for VLSIs," *Proc. Int. Electron Devices Meet.*, 542–545 (1986a).

Takeda, E., K. Takeuchi, E. Yamasaki, T. Toyabe, and K. Itoh, "Effective Tunneling Length in Alpha-Ray Induced Soft Error," *Conf. Solid State Devices Mater.*, 311–314 (1986b).

Takeda, E., A. Shimizu, H. Kume, and K. Itoh, "Role of Source n^- Region in LDD MOSFETs," *IEEE Trans. Electron Devices* **33**, 869–870 (1986c).

Takeda, E., H. Matsuoka, Y. Igura, and S. Asai, "Band to Band Tunneling MOS Device (B^2T-MOSFET—A Kind of 'Si Quantum Device'," *Tech. Dig.—Int. Electron Devices Meet.*, 402–405 (1988).

Takeda, E., "Cross Section of Hot-Carrier Phenomena in MOS ULSIs," *Proc. Int. Conf. Insul. Films Semicond., 6th*, Garching, West Germany, 535–551 (1989).

Takeda, E., "Perspectives on VLSI MOS Devices and Advances towards Si Quantum Devices," *Ext. Abstr., Solid State Devices Mater.*, 521–524 (1989).

Takeda, E., R. Izawa, K. Umeda, and R. Nagai, "AC Hot-Carrier Effects in Scaled MOS Devices," *IEEE 1991 Int. Reliab. Phys. Symp.*, 118–122 (1991).

Takeishi, Y., "Some Recent Advances in Basic VLSI Technology," *Symp. Recent Top. Semicond. Phys. Commem. 60th Birthday Prof. Y. Uemura*, 207–234 (1983).

Takemae, Y., T. Ema, M. Nakano, F. Baba, T. Yabu, K. Miyasaka, and K. Shirai, "1 MB DRAM with 3-dimensional Stacked Capacitor Cells," *Tech. Dig. IEEE Int. Solid-State Circuits Conf.*, 250–251 (1985).

Takeshima, M., "Analysis of Temperature Sensitive Operation in 1.6 μm $In_{0.53}Ga_{0.47}As$ Laser," *J. Appl. Phys.* **56**, 691–695 (1984).

Tam, S., P.-K. Ko. C. Hu, and R. S. Muller, "Correlation between Substrate and Gate Currents in MOSFETs," *IEEE Trans. Electron Devices* **29**, 1740–1744 (1982a).

Tam, S., F.-C. Hsu, P.-K. Ko, C. Hu, and R. S. Muller, "Hot-Electron Induced Excess Carriers in MOSFETs," *IEEE Electron Devices Lett.* **3**, 376–378 (1982b).

Tam, S., F.-C. Hsu, C. Hu, R. S. Muller, and P.-K. Ko, "Hot-Electron Currents in Very Short-Channel MOSFETs," *IEEE Electron Devices Lett.* **4**, 249–251 (1983).

Tam, S., P.-K. Ko, and C. Hu, "Lucky-Electron Model of Electron Injection in MOSFETs," *IEEE Trans. Electron Devices* **31**, 1116–1125 (1984).

Tam, S. and C. Hu, "Hot Electron Induced Photon and Photo-Carrier Generation in Silicon MOSFETs," *IEEE Trans. Electron Devices*, 1264–1273 (1984).

Tam, S., S. Sachdev, M. Chi, G. Verma, J. Ziller, G. Tsau, S. Lai, and V. Dham, "High Density CMOS 1-T Electrical Erasable Non-Volatile (Flash) Memory Technology," *Tech. Dig. VLSI Tech. Symp.*, 31–32 (1988).

Tamaki, Y., T. Shiba, N. Honma, S. Mizuo, and A. Hayasaka, "New U-Groove Isolation Technology for High Speed Bipolar Memory," *Tech. Dig. VLSI Tech. Symp.*, 24–25 (1983).

Tanaka, S. and M. Ishikawa, "One-Dimensional Writing Model of N-Channel Floating Gate Ionization–Injection MOS (FIMOS)," *IEEE Trans. Electron Devices* **28**, 1190–1197 (1981).

Tanaka, S. and S. Watanabe, "Model for the Relation between Substrate and Gate Currents in *n*-Channel MOSFETs," *IEEE Trans. Electron Devices* **30**, 668–675 (1983).

Tanaka, S., Saito, S. Atsumi, and K. Yoshikawa, "Self-Consistent Pseudo-Two-Dimensional Model for Hot-Electron Current in MOST'S," *IEEE Trans. Electron Devices* **33**, 743–753 (1986).

Tanaka, S., "A Lucky Drift Model, Including a Soft Threshold Energy, for the Relation between Gate and Substrate Currents in MOSFETs," *Solid-State Electron.* **32**, 935–946 (1989).

Tandon, P., A. Barakji, R. K. Pancholy, M. Khan, and T. Batra, "CMOS IV: A 1.3 Micron CMOS Technology," *Ext. Abstr., Electrochem. Soc.* **85-2**, 444 (1985).

Tang, J. Y., and K. Hess, "Real Space Transfer Noise in Buried-Channel Devices," *IEEE Trans. Electron Devices* **28**, 285–289 (1981).

Tang, J. Y. and K. Hess, "Theory of Hot Electron Emission from Silicon into Silicon Dioxide," *J. Appl. Phys.* **54**, 5145–5151 (1983).

Tang, T.-W., "Physics and Modeling of Hot-Carriers in Silicon Submicrometer Devices," *Proc. Int. Workshop Phys. Semicond. Dev.*, Madras, India, 253–262 (1985).

Tang, Y., D. M. Kim, Y.-H. Lee, and B. Sabi, "Unified Characterization of Two-Region Gate Bias Stress in Submicrometer *p*-Channel MOSFET's," *IEEE Electron Devices Lett.* **11**, 203–205 (1990).

Taniguchi, K., Y. Shibata, and C. Hamaguchi, "Universal Model for Impurity Diffusion and Oxidation of Silicon," *Ext. Abstr., Solid State Devices Mater.*, 161–164 (1989).

Tanimoto, S., T. Mihara, K. Asada, and T. Sugano, "A Possible Mechanism of Electron Injection for the Threshold Voltage Shift of Metal-Oxide-Semiconductor Field-Effect Transistors at Low Voltage," *J. Appl. Phys.* **65**, 4061–4065 (1989).

Tannenbaum, M. and D. E. Thomas, "Diffused Emitter and Base Silicon Transistors," *Bell Syst. Tech. J.* **35**, 1–15 (1956).

Tao, T. F., J. R. Ellis, L. Kost, and A. Doshier, "Feasibility Study of PbTe and PbSnTe Infrared Charge Coupled Imager," *Proc. Appl. Conf., Charge Coupled Devices, Nav. Electron. Lab. Cent.*, San Diego, 259 (1973).

Tarui, Y., "Diffusion Self-Aligned Enhance-Depletion MOS-IC," *J. Jpn. Soc. Appl. Phys.* **40**, 193 (1971).

Tasch, A. F., "Metal Oxide Semiconductor (MOS) Technology Scaling Issues and Their Relation to Submicron Lithography," *Proc. SPIE—Int. Soc. Opt. Eng.* **333**, 68–75 (1982).

Tasch, A. F., H. Shin, T. J. Bordelon, and C. M. Maziar, "Limitations of LDD Types of Structures in Deep-Submicrometer MOS Technology," *IEEE Electron Devices Lett.* **11**, 517–519 (1990).

Tasch, A. F., H. Shin, and C. M. Maziar, "A New Structural Approach for Reducing Hot Carrier Generation in Deep Submicron MOSFETs," *1990 Symp. VLSI Technol., Dig. Tech. Pap.*, 43–44 (1990).

Tasch, A. F., H. Shin, and C. M. Maziar, "New Submicron MOSFET Structural Concept for Suppression of Hot Carriers," *Electron. Lett.* **26**, 39–41 (1990).

Tasch, A. F. and H. Shin, "Scaling the MOS Transistor to Its Limit in ULSI," *Proc.—Electrochem. Soc.* **90**, 3–11 (1990).

Tasch, A. F., Jr., "Best Structures for Deep Submicrometer (0.1–0.3 μm) MOS Devices," *IEEE Trans. Electron Devices* **38**, 2688 (1991).

Taylor, G. W., "The Effects of Two-Dimensional Charge Sharing on the above Threshold Characteristics of Short-Channel IGFET's," *Solid-State Electron.* **22**, 701 (1979).

Terashima, K., C. Hamaguchi, and K. Taniguchi, "Monte Carlo Simulation of Two-Dimensional Hot Electrons in *n*-Type Si Inversion Layers," *Superlattices Microstruct.* **1**, 15–19 (1985).

Terman, L. M., "An Investigation of Surface States at a Silicon/Silicon Oxide Interface Employing Metal-Oxide-Silicon Diodes," *Solid-State Electron.* **5**, 285–299 (1962).

Terrill, K. W., C. Hu, and P.-K. Ko, "An Analytical Model for the Channel Electric Field in MOSFET's with Graded-Drain Structures," *IEEE Electron Devices Lett.* **5**, 440–442 (1984).

Theuerer, H. C., "Removal of Boron from Silicon by Hydrogen Water Vapor Treatment," *Trans. AIME* **206**, 1316–1319 (1956).

Theuerer, H. C., J. J. Kleimack, H. H. Loar, and H. Christenson, "Epitaxial Diffused Transistors," *Proc. IRE* **48**, 1642–1643 (1960).

Thewes, R., M. Broz, G. Tempel, W. Weber, and K. Goser, "Channel-Length-Independent Hot-Carrier Degradation in Analog *p*-MOS Operation," *IEEE Electron Devices Lett.* **13**, 590–592 (1992).

Thomas, I. P., "Fabrication and Initial Characterization of a Range of *n*-Channel Lightly Doped Drain (LDD) Devices," *IEE Colloq. Dig.* **15**, 8.1–8.4 (1987).

Thornber, K. K., "Relation of Drift Velocity to Low-Field Mobility and High Field Saturation Velocity," *J. Appl. Phys.* **51**, 2127 (1980).

Thurner, M. and S. Selberherr, "Extension of MINIMOS to a 3-Dimensional Simulation Program," *Proc. IEEE Int. Conf. Numer. Anal. Semicond. Devices Int. Circuits*, 327–332 (1987).

Thurner, M. and S. Selberherr, "Comparison of Long- and Short-Channel MOSFETs Carried out by 3D-MINIMOS," *Proc. Eur. Solid State Devices Res. Conf., 17th*, 73–76 (1988).

Tihanyi, J. and H. Schlotterer, "Influence of the Floating Substrate Potential on the Characteristics of ESFI MOS Transistors," *Solid-State Electron.* **18**, 309 (1975).

Tihanyi, J. and D. Widmann, "DIMOS—A Novel IC Technology with Submicron Effective Channel MOSFET's," *Tech. Dig.—Int. Electron Devices Meet.*, 399 (1977).

Tihanyi, J., "Integrated Power Devices," *Tech. Dig.—Int. Electron Devices Meet.*, 6 (1982).

Ting, C. S., M. Liu, and D. Y. Xing, "Balance Equation Approach to Hot Electron Transport in Many-Valley Semiconductors: Comparing with the Monte Carlo Results for *n*-Type Si," *Solid-State Electron.* **31**, 551–554 (1988).

Ting, C. S., M. Liu, and D. Y. Xing, "Ballistic Electron Transport in GaAs/AlGaAs Tunneling Junctions with Optical Phonon Emission," *Solid-State Electron.* **31**, 555–558 (1988).

Ting, W., G. Q. Lo, J. Ahn, T. Y. Chu, and D. L. Kwong, "MOS Characteristics of Ultrathin SiO_2 Prepared by Oxidizing Si in N_2O," *IEEE Electron Devices Lett.* **12**, 416–418 (1991).

Titcomb, S. L., K. T. Paskiet, and A. E. Grass, "Hot-Carrier Degradation of LDD *n*-Channel MOSFETs at 77 K," *Proc. IEEE Workshop Low Temp. Semicond. Electron.*, 114–117 (1989).

Tolstikhin, V. I., "Effect of Nonuniformity of Carrier Heating in an Inversion Layer of an MOS Transistor in Carrier Emission into the Dielectric," *Sov. Microelectron.* (*Engl. Transl.*) **14**, 147–153 (1985).

Tominaga, Y. and Y. Kosa, "Features of High-Performance CMOS and Its Applications," *Hitachi Rev.* **33**, 219–224 (1984).

Tomizawa, K. and D. Pavlidis, "Transport Equation Approach for Heterojunction Bipolar Transistors," *IEEE Trans. Electron Devices* **37**, 519–529 (1990).

Tompsett, M. F., "The Quantitative Effect of Interface States on the Performance of Charge-Coupled Devices," *IEEE Trans. Electron Devices* **20**, 45–55 (1973).

Tompsett, M. F., "Video-Signal Generation," In *Electron Imaging*, T. P. McLean and P. Schagen, eds., Academic Press, New York, 55 (1979).

Tonti, W. R., W. P. Noble, W. W. Abadeer, S. W. Mittl, and W. E. Haensch, "Doping Profile Design for Substrate Hot Carrier Reliability in Deep Submicron Field Effect Transistors," *29th Ann. Proc. Reliab. Phys.*, Las Vegas, NV, 306–309 (1991).

Torii, K., T. Kaga, and E. Takeda, "Three Dimensional Effects on Submicrometer Diagonal MOSFETs," *Ext. Abstr., Solid State Devices Mater.*, 101–104 (1989).

Toriumi, A., M. Yoshimi, M. Iwase, and K. Taniguchi, "Experimental Determination of Hot-Carrier Energy Distribution and Minority Carrier Generation Mechanism Due to Hot Carrier Effects," *Tech. Dig.—Int. Electron Devices Meet.*, 56–59 (1985a).

Toriumi, A., M. Yoshimi, and K. Taniguchi, "Study of Gate Current and Reliability in Ultra-Thin Gate Oxide MOSFETs," *Tech. Dig. VLSI Tech. Symp.*, 110–111 (1985b).

Toriumi, A. and M. Iwase, "Substrate Hole-Current Generation Due to Electron Tunneling in Metal-Oxide-Semiconductor System," *Ext. Abstr., 19th Conf. Solid State Devices Mater.*, 351–354 (1987).

Toriumi, A., M. Yoshimi, M. Iwase, Y. Akiyama, and K. Taniguchi, "Study of Photon Emission from *n*-Channel MOSFETs," *IEEE Trans. Electron Devices* **34**, 1501–1508 (1987).

Toriumi, A., "Experimental Study of Hot Carriers in Small Size Si-MOSFETs," *Solid-State Electron.* **32**, 1519–1525 (1989).

Torrey, H. C. and C. A. Whitmer, *Crystal Rectifiers*, McGraw-Hill, New York (1948).

Tosi, M., L. Baldi, P. Caprara, and C. Bergonzoni, "Hot Carriers and Gate Current in CMOS Devices," *IEE Colloq. Dig.* **15**, 12.1–12.3 (1987).

Tove, P. A., K. E. Bohlin, H. Norde, U. Magnusson, J. Tiren, A. Soderbarg, M. Rosling, F. Masszi, and J. Nylander, "Silicon IC Technology Using Complementary MESFETs," *Proc. Eur. Solid State Devices Res. Conf., 17th*, 607–609 (1987).

Toyabe, T. and H. Kodera, "A Theory for Inter-Valley Transfer Effect in Two-Valley Semiconductor," *Jpn. J. Appl. Phys.* **13**, 1404–1413 (1974).

Toyabe, T., K. Yamaguchi, S. Asai, and M. S. Mock, "A Numerical Model of Avalanche Breakdown in MOSFETs," *IEEE Trans. Electron Devices* **25**, 825–832 (1978).

Toyabe, T. and S. Asai, "Analytical Models of Threshold Voltage and Breakdown Voltage of Short Channel MOSFETs Derived from Two-Dimensional Analysis," *IEEE Trans. Electron Devices* **26**, 453–461 (1979).

Toyabe, T., H. Masuda, Y. Aoki, H. Sukuri, and T. Hagiwara, "Three-Dimensional Device Simulator CADDETH with Highly Convergent Matrix Solution Algorithms," *IEEE Trans. (Comput.-Aided Res.)* **4**, 482 (1985).

Toyabe, T., H. Masuda, S. Yanamoto, H. Masuda, "Determination of Doping Profile in Sub-Debye-Length," *Int. Electron Devices Meet., 1989, Tech. Dig.* 699–701 (1989).

Toyoshima, Y., H. Nihira, M. Wada, and K. Kanzaki, "Mechanism of Hot-Electron Induced Degradation in LDD *n*-MOSFET," *Tech. Dig.—Int. Electron Devices Meet.*, 786–789 (1984).

Toyoshima, Y., N. Nishira, and K. Kanzaki, "Profiled Lightly Doped Drain (PLDD) Structure for High Reliable *n*-MOSFETs," *Tech. Dig. VLSI Tech. Symp.*, 118–119 (1985).

Toyoshima, Y., F. Matsuoka, H. Hayashida, H. Iwai, and K. Kanzaki, "Study on Gate Oxide Thickness Dependence of Hot Carrier Induced Degradation for *n*-MOSFET," *Tech. Dig. VLSI Tech. Symp.*, 39–40 (1988).

Toyoshima, Y., H. Iwai, F. Matsuoka, H. Hayashida, K. Maeguchi, and K. Kanzaki, "Analysis on Gate-Oxide Thickness Dependence of Hot-Carrier-Induced Degradation in Thin-Gate Oxide *n*-MOSFETs," *IEEE Trans. Electron Devices* **37**, 1496–1503 (1990).

Tran, L. V., R. A. Ashton, B. R. Jones, C. W. Lawrence, and D. A. McGillis, "Device Characterization of a 1.0 μm CMOS Technology for Logic and Custom VLSI Applications," *Proc. IEEE Custom Int. Circuits Conf.*, Rochester, NY, 46–50 (1986).

Triplett, B. B., T. Takahashi, and T. Sugano, "Electron Spin Resonance Observation of Defects in Device Oxides Damaged by Soft X Rays," *Appl. Phys. Lett.* **50**, 1663–1665 (1987).

Trocino, M. R., K.-Y. Fu, and K.-W. Teng, "Stress-Bias Dependence of Hot-Carrier-Induced Degradation in MOSFETs," *Solid-State Electron.* **31**, 873–875 (1988).

Trofimenkoff, F. N., J. W. Haslett, and R. E. Smallwood, "Hot Electron Thermal Noise Models for FETs," *Int. J. Electron.* **44**, 257–272 (1978).

Troutman, R. R., "Subthreshold Design Considerations for IGFET," *IEEE J. Solid-State Circuits*, **SC-9**, 55 (1974).

Troutman, R. R., "Low-Level Avalanche Multiplication in IGFET's," *IEEE Trans. Electron Devices* **23**, 419–425 (1976).

Troutman, R. R., "Silicon Surface Emission of Hot Electrons," *Solid-State Electron.* **21**, 283–289 (1978).

Troutman, R. R., "VLSI Limitation from Drain-Induced Barrier Lowering," *IEEE Trans. Electron Devices* **26**, 461 (1979).

Tsang, P. J., S. Ogura, W. W. Walker, J. F. Shepard, and D. L. Critchlow, "Fabrication of High-Performance LDD-FET's with Oxide Sidewall-Spacer Technology," *IEEE Trans. Electron Devices* **29**, 590–596 (1982).

Tseng, H.-H., P. J. Tobin, J. D. Hayden, K.-M. Chang, and W. Miller, "A Comparison of CVD Stacked Gate Oxide and Thermal Gate Oxide for 0.5-μm Transistors Subjected to Process-Induced Damage," *IEEE Trans. Electron Devices* **40**, 613–618 (1993).

Tsubouchi, N., "VLSI Oxidation Technology," In *Hardware and Software Concepts in VLSI*, (G. Rabbat ed.), Van Nostrand Reinhold (1983), pp. 330–351.

Tsuchiya, T. and S. Nakajima, "Novel High Alpha-Particle-Immunity and High Density d-RAM Cell," *Jpn. J. Appl. Phys.* **21**, 79–84 (1981).

Tsuchiya, T. and S. Nakajima, "Miniaturization Degree of Dynamic MOS RAM Cells with Readout Signal Gain," *IEEE Trans. Electron Devices* **29**, 1713–1717 (1982).

Tsuchiya, T. and M. Itsumi, "New Dynamic RAM Cell for VLSI Memories," *IEEE Electron Devices Lett.* **3**, 7–10 (1982).

Tsuchiya, T., T. Kobayashi, and S. Nakajima, "Hot-Carrier Degradation Mechanism in Si *n*-MOSFETs," *Ext. Abstr., Solid State Devices Mater.*, 21–24 (1985).

Tsuchiya, T. and S. Nakajima, "Emission Mechanism and Bias-Dependent Emission Efficiency of Photons Induced by Drain Avalanche in Si MOSFETs," *IEEE Trans. Electron. Devices* **32**, 405–412 (1985).

Tsuchiya, T. and J. Frey, "Relationship between Hot-Electrons/Holes and Degradation of *p*- and *n*-Channel MOSFETs," *IEEE Electron Device Lett.* **6**, 8–11 (1985).

Tsuchiya, T., T. Kobayashi, and S. Nakajima, "Hot-Carrier-Injected Oxide Region and Hot-Electron Trapping as the Main Cause in Si *n*-MOSFET Degradation," *IEEE Trans. Electron Devices* **34**, 386–391 (1987).

Tsuchiya, T., "Mechanism of Hot-Electron-Induced MOSFET's Degradation," *Tech. Dig. VLSI Tech. Symp.*, 53–54 (1987a).

Tsuchiya, T., "Trapped-Electron and Generated Interface-Trap Effects in Hot-Electron-Induced MOSFET Degradation," *IEEE Trans. Electron Devices* **34**, 2291–2296 (1987b).

Tsuchiya, T., "A New Enhanced Degradation Phenomenon in MOSFETs under AC Stress: The Effect of Band-to-Band Tunneling," *Tech. Dig. VLSI Tech. Symp.*, 79–80 (1989).

Tsuchiya, T., Y. Okazaki, M. Miyake, and T. Kobayashi, "Hot-Carrier Degradation Mode and Prediction Method of DC Lifetime in Deep-Submicron PMOSFET," *Proc. Int. Conf. Solid State Devices Mater., 22nd*, 291–194 (1990).

Tsuchiya, T., M. Harada, K. Deguchi, and T. Matsuda, "The Influence of Synchrotron X-Ray Damage on Hot-Carrier-Induced Degradation in Subquarter-Micron NMOSFETS," *Ext. Abstr., Int. Conf. Solid State Devices Mater.*, 17–19 (1991).

Tsui, P. G. Y., L. Howington, P. M. Lee, T. Tiwald, B. Mowry, F. K. Baker, J. D. Hayden, B. B. Feaster, and B. Garbs, "An Integrated System for Circuit Level Hot-Carrier Evaluation," *Proc. IEEE 1990 Custom Int. Circuits Conf. 12th*, 4 (1990).

Tsui, P. G. Y., P. M. Lee, F. K. Baker, J. D. Hayden, L. Howington, L. Tiwald, and B. Mowry, "A Circuit Level Hot-Carrier Evaluation System," *IEEE J. Solid-State Circuits* **26**, 410–414 (1991).

Tsunekawa, S., H. Kume, and Y. Homma, "Influence of Two-Level Planar Interconnection Processes Using Bias-Sputtered SiO_2 for MOSFETs," *J. Electrochem. Soc.* **136**, 2632–2637 (1989).

Tsutsu, N., Y. Uraoka, Y. Nakata, S. Akiyama, and H. Esaki, "New Detection Method of Hot-Carrier Degradation Using Photon Spectrum Analysis of Weak Luminescence on CMOS VLSI," *Proc. Int. Conf. Microelectron. Test Struct.*, 143–148 (1990).

Tsutsu, N., Y. Uraoka, T. Morii, and K. Tsuji, "Life Time Evaluation of MOSFET in ULSIs Using Photon Emission Method," *Proc. Int. Conf. Microelectron. Test Struct.*, 94–99 (1992).

Twardowski, A. and C. Hermann, "Polarized Hot-Electron Photoluminescence in Highly Doped GaAs," *Phys. Rev. B* **32**, 8253–8257 (1985).

Tzou, J. J., C. C. Yao, R. Cheung, and H. Chan, "Hot-Electron-Induced MOSFET Degradation at Low Temperatures," *IEEE Electron Devices Lett.* **6**, 450–452 (1985).

Tzou, J. J., C. C. Yao, R. Cheung, and H. Chan, "Process Dependence of Hot Electron Injection in MOSFETs," *Ext. Abstr. Electrochem. Soc.* **85-1**, 290 (1985).

Tzou, J. J., C. C. Yao, and H. W. K. Chan, "Hot-Carrier-Induced Degradation in *p*-Channel LDD MOSFETs," *IEEE Electron Devices Lett.* **7**, 5–7 (1986).

Uchida, H., S. Inomata, and T. Ajioka, "Effects of Interface Traps and Bulk Traps in SiO_2 on Hot-Carrier-Induced Degradation," *Proc. IEEE Int. Conf. Microelectron. Test Struct.* **2**, 103–108 (1989).

Uchida, Y., N. Endo, S. Saito, and Y. Nishi, "Avalanche-Tunnel Injection in MNOS Transistor," *IEEE Trans. Electron Devices* **24**, 688–693 (1977).

Uchimura, A. and Y. Uemura, "Hot Electron Effects in Landau Levels of MOS Inversion Layers," *J. Phys. Soc. Jpn.* **47**, 1417–1425 (1979).

Uchiyama, A., H. Fukuda, T. Hayashi, T. Iwabuchi, and S. Ohno, "High Performance p^+-Gate *p*MOSFETs with N_2O-Nitrided SiO_2 Gate Films," *Electron. Lett.* **26**, 1932–1933 (1990).

Ushirokawa, A., E. Suzuki, and M. Warashina, "Avalanche Injection Effects in MIS Structures and Realization of *n*-Channel Enhancement Type MOSFETs," *Jpn. J. Appl. Phys.* **12**, 398–407 (1973).

Vadasz, L. L., and A. S. Grove, "Temperature Dependence of MOS Transistor Characteristics below Saturation," *IEEE Trans. Electron Devices* **13**, 863 (1966).

Vadasz, L. L., A. S. Grove, T. A. Rowe, and G. E. Moore, "Silicon Gate Technology," *IEEE Spectrum* **6**, 28–35 (1969).

Vaidya, S., D. K. Atwood, E. H. Kung, and K. K. Ng, "Comparison of 20 kV and 50 kV e-Beam Lithography for MOS Circuit Fabrication," *Microelectron. Eng.* **9**, 213–216 (1989).

Van den Bosch, G., G. Groeseneken, P. Heremans, M. Heyns, and H. E. Maes, "Hole Trapping and Hot-Hole-Induced Interface Trap Generation in MOSFETs at Different Temperatures," *Microelectron. Eng.* **19**, 477–480 (1992).

Van Der Pol, J. A. and J. J. Koomen, "Relation between the Hot Carrier Lifetime of Transistors and CMOS SRAM Products," *Proc. Int. Reliab. Phys. Symp.*, 178–185 (1990).

Van Der Ziel, A., "The State of Solid Device Noise Research," *Physica B + C (Amsterdam)* **83**, 41–51 (1976).

Van Der Ziel, A., "Limiting Noise in Solid State Devices," *Proc. Int. Conf. Noise Phys. Syst.*, Bad Nauheim, Germany, 2–12 (1978).

Van Der Ziel, A. and E. N. Wu, "Thermal Noise in High Electron Mobility Transistors," *Solid-State Electron.* **26**, 383–384 (1983).

Van Houdt, J., P. Hermans, J. S. Witters, G. Groeseneken, and H. E. Maes, "Study of the Enhanced Hot-Electron Injection in Split-Gate Transistor Structures," *20th Eur. Solid State Devices Res. Conf.*, 261–264 (1990).

Van Zanten, A. T., W. C. H. Gubbels, and F. J. List, "Static Memories: An Overview and Trends," *NTG-Fachber.* **102**, 187–196 (1988).

Varker, C. J., D. Pettengill, W.-T. Shiau, and B. Reuss, "Hot Carrier Induced Hfe Degradation in BiCMOS Transistors," *Proc. Int. Reliab. Phys. Symp.*, 58–62 (1992).

Veeraraghavan, S. and J. G. Fossum, "Short-Channel Effects in SOI MOSFETs," *IEEE Trans. Electron Devices* **36**, 522–528 (1989).

Venturi, F., E. Sangiorgi, R. Brunetti, C. Jacoboni, and B. Ricco, "Monte Carlo Simulation of Electron Heating in Scaled Deep Submicron MOSFETs," *Tech. Dig.—Int. Electron Devices Meet.*, 485–488 (1989).

Venturi, F., E. Sangiorgi, and B. Ricco, "The Impact of Voltage Scaling on Electron Heating and Device Performance of Submicrometer MOSFETs," *IEEE Trans. Electron Devices* **38**, 1895–1904 (1991).

Verdonckt-Vanderbroek, S., E. Crabbé, B. Meyerson, D. Harame, P. Restle, J. Stork, A. Megdanis, C. Stanis, A. Bright, G. Kroesen, and A. Warren, "High-Mobility Modulation-Doped Graded SiGe-Channel *p*-MOSFETs," *IEEE Electron Devices Lett.* **EDL-12**, 447–449 (1991).

Verwery, J. F., "Nonavalanche Injection of Hot Carriers into SiO_2," *J. Appl. Phys.* **44**, 2681–2687 (1973).

Verwey, J. F., R. P. Kramer, and B. J. de Maagt, "Mean Free Path of Hot Electrons at the Surface of Boron-Doped Silicon," *J. Appl. Phys.* **46**, 2612–2619 (1975).

Verwey, J. F., "Nonvolatile Semiconductor Memory Devices," *Adv. Electron. Electron Phys.* **41**, 249–309 (1976).

Villeneuve, D. M., G. D. Enright, and M. C. Richardson, "Features of Lateral Energy Transport in CO_2-Laser-Irradiated Microdisk Targets," *Phys. Rev. A* **27**, 2656–2662 (1983).

Vishnubhotla, L., T.-P. Ma, H.-H. Tseng, and P. J. Tobin, "Interface Trap Generation and Electron Trapping in Fluorinated SiO_2," *Appl. Phys. Lett.* **59**, 3595–3597 (1991).

Vishnubhotla, L., P. M. Tso, H. T. Hsing, and P. J. Tobin, "Effects of Avalanche Hole Injection in Fluorinated SiO_2 MOS Capacitors," *IEEE Electron Devices Lett.* **EDL-14**, 196–198 (1993).

Voldman, S. H., "Test Structures for Analysis and Parameter Extraction of Secondary Photon-Induced Leakage Currents in CMOS DRAM Technology," *Proc. Int. Conf. Microelectron. Test Struct.*, 39–43 (1992).

von Bruns, S. L. and R. L. Anderson, "Hot-Electron-Induced Interface State Generation in *n*-Channel MOSFET's at 77 K," *IEEE Trans. Electron Devices* **34**, 75–82 (1987).

von Klitzing, K., "Impurity Spectroscopy on Tellurium by Means of Magnetoresistance Measurements under Nonohmic Conditions," *Solid-State Electron.* **21**, 223–228 (1978).

Voss, P. H., L. Pflennings, C. G. Phelan, C. M. O'Connell, J. Thomas, H. Ontrop, S. A. Bell, and R. H. W. Salters, "14-ns 256 K Multiplied by 1 CMOS SRAM with Multiple Test Modes," *IEEE J. Solid-State Circuits* **SC-24**, 874–880 (1989).

Voves, J. and J. Vesely, "MOSFET Gate Current Modeling Using Monte Carlo Method," *J. Phys., Colloq. (Orsay, Fr.)*, **49**, 791–794 (1988).

Vuillaume, D., J. C. Marchetaux, and A. Boudou, "Evidence of Acceptor-Like Oxide Defects Created by Hot-Carrier Injection in *n*-MOSFETs: A Charge-Pumping Study," *IEEE Electron Devices Lett.* **12**, 60–62 (1991).

Vuillaume, D., and B. S. Doyle, "Properties of Hot Carrier Induced Traps in MOSFETs Characterized by the Floating-Gate Technique," *Solid-State Electron.* **35**, 1099–1107 (1992).

Vuillaume, D., J.-C. Marchetaux, P.-E. Lippens, A. Bravaix, and A. Boudou, "A Coupled Study by Floating-Gate and Charge-Pumping Techniques of Hot Carrier–Induced Defects in Submicrometer LDD *n*-MOSFETs," *IEEE Trans. Electron Devices* **40**, 773–781 (1993).

Wada, M., T. Shibata, M. Konaka, H. Iizuka, and R. L. M. Dang, "Two-Dimensional Computer Simulation of Hot Carrier Effects in MOSFETs," *Tech. Dig.—Int. Electron Devices Meet.*, 223–226 (1981).

Wada, M., T. Shibata, H. Iizuka, and R. L. M. Dang, "Modeling and Simulation of Hot Carrier Effects in MOSFETs," *Tech. Dig. VLSI Tech. Symp.*, 102–103 (1982).

Wada, M., K. Hieda, and S. Watanabe, "A Folded Capacitor Cell (FCC) for Future Megabit DRAMs," *Tech. Dig.—Int. Electron Devices Meet.*, 244–247 (1984).

Wada, M., H. Iizuka, T. Shibata, and R. Dang, "Modelling of Hot-Carrier Effects in Small-Geometry MOSFETs," *Electron. Commun. Jpn.* **67**, 96–103 (1984).

Walden, R. H., R. H. Krambeck, R. J. Strain, J. McKenna, N. L. Schryer, and G. E. Smith, "The Buried Channel Charge Coupled Device," *Bell Syst. Tech. J.* **51**, 1635 (1972).

Wallmark, J. T. and H. Johnson, *Field Effect Transistors, Physics, Technology, and Applications*, Prentice-Hall, Englewood Cliffs, NJ (1966).

Wang, C. H. and A. Neugroschell, "Minority-Carrier Transport Parameters in Degenerate *n*-Type Silicon," *IEEE Electron Devices Lett.* **EDL-12**, 576–578 (1990).

Wang, C. M., J. J. Tzou, and C. Y. Yang, "Hot-Carrier-Induced Latchup and Trapping/Detrapping Phenomena," *Proc. Int. Reliab. Phys. Symp.*, 110–113 (1989).

Wang, C. T., "Software Program 'SUPERMOS' for Characterizing Hot Carrier Effects of Short-Channel MOSFET's," *Proc. 7th Bienn. Univ./Gov./Ind. Microelectron. Symp.*, 13–18 (1987).

Wang, C. T., "Improved Hot-Electron-Emission Model for Simulating the Gate-Current Characteristic of MOSFETs," *Solid-State Electron.* **31**, 229–231 (1988).

Wang, H., M. Davis, and R. Lahri, "Transient Substrate Current Effects on *n*-Channel MOSFET Device Lifetime," *Tech. Dig.—Int. Electron Devices Meet.*, 216–219 (1988).

Wang, L. K., "X-ray Lithography Induced Radiation Damage in CMOS and Bipolar Devices," *J. Electron. Mater.* **21**, 753–756 (1992).

Wang, Q., W. H. Krautschneider, W. Weber, and D. Schmitt-Landsiedel, "Influence of MOSFET *I–V* Characteristics on Switching Delay Time of CMOS Inverters after Hot-Carriers Stress," *IEEE Electron Devices Lett.* **12**, 238–240 (1991).

Wang, Q., W. H. Krautschneider, M. Brox, and W. Weber, "Time Dependence of Hot-Carrier Degradation in LDD *n*-MOSFETs," *Microelectron. Eng.* **15**, 441–444 (1991).

Wang, R., J. Dunkley, T. A. DeMassa, and L. F. Jelsma, "Threshold Voltage Variations with Temperature in MOS Transistors," *IEEE Trans. Electron Devices* **18**, 386 (1971).

Wang, X.-W., A. Balasinski, T.-P. Ma, and Y. Nishioka, "Pre-Oxidation Fluorine Implantation into Si Process-Related MOS Characteristics," *J. Electrochem. Soc.* **139**, 238–241 (1992).

Wang, Y., Y. Nishioka, T.-P. Ma, and R. C. Barker, "Radiation and Hot-Electron Effects on SiO_2/Si Interfaces with Oxides Grown in O_2 Containing Small Amounts of Trichloroethane," *Appl. Phys. Lett.* **52**, 573–575 (1988).

Wang, Y., T.-P. Ma, and R. C. Barker, "Orientation Dependence of Interface-Trap Transformation," *IEEE Trans. Nucl. Sci.* **NS-36**, 1784–1791 (1989).

Wang, Z. Z., J. Suski, D. Collard, and E. Dubois, "Piezoresistivity Effects in *n*-MOSFET Devices," *Sens. Actuators A* **A34**, 59–65 (1992).

Wanlass, F. M. and C. T. Sah, "Nanowatt Logic Using Field-Effect MOS Triodes," *IEEE Tech. Dig. Int. Conf. Solid-State Circuits*, 32–33 (1963).

Watanabe, A., K. Fujimoto, M. Oda, T. Nakatsuka, and A. Tamura, "Rapid Degradation of WSi Self-Aligned Gate GaAs MESFET by Hot Carrier Effect," *Proc. Int. Reliab. Phys. Symp.*, 127–130 (1992).

Watanabe, D. S. and S. Slamet, "Numerical Simulation of Hot-Electron Phenomena," *IEEE Trans. Electron Devices* **30**, 1042–1049 (1983).

Watanabe, Y., K. Kaizu, and Y. Fukuda, "Threshold-Voltage Instability of MOS Transistors in an LSI Memory under Accelerated Operating Test Condition at 77 K," *Trans. Inst. Elect. Commun. Eng. Jpn.* **E66**, 397–398 (1983).

Watts, R. K., L. Manchanda, and R. L. Johnston, "Gate Current in 0.75 μm n-Channel MOSFETs with Doubly Diffused Drain," *Electron. Lett.* **23**, 468–469 (1987).

Weber, W., C. Werner, and G. Dorda, "Degradation of *n*-MOS-Transistors after Pulsed Stress," *IEEE Electron Devices Lett.* **5**, 518–520 (1984).

Weber, W., C. Werner, and A. V. Schwerin, "Lifetimes and Substrate Currents in Static and Dynamic Hot-Carrier Degradation," *Tech. Dig.—Int. Electron Devices Meet.*, 390–393 (1986).

Weber, W. and F. Lau, "Hot-Carrier Induced Degradation in Submicron *p*-MOSFETs," *IEE Colloq. Dig.* **15**, 6.1–6.3 (1987).

Weber, W., "Dynamic Stress Experiments for Understanding Hot-Carrier Degradation Phenomena," *IEEE Trans. Electron Devices* **35**, 1476–1486 (1988).

Weber, W., L. Risch, W. Krautschneider, and Q. Wang, "Hot-Carrier Degradation of CMOS-Inverters," *Tech. Dig.—Int. Electron Devices Meet.*, 208–211 (1988).

Weber, W. and I. Borchert, "Hot-Hole and -Electron Effects in Dynamically Stressed *n*-MOSFETs," *Proc. Eur. Solid State Devices Res. Conf., 19th*, 719–722 (1989).

Weber, W., M. Brox, F. Hofmann, H. Huber, D. Jager, and D. Rieger, "Spectroscopic Investigation of Oxide States Responsible for Hot-Carrier Degradation," *Appl. Phys. Lett.* **54**, 168–169 (1989).

Weber, W., Q. Wang, M. Brox, and D. Schmitt-Landsiedel, "Hot-Carrier Degradation Effects Relevant in Real Operation of MOSFETs," *Proc. Conf. Solid State Devices Mater., 22nd*, 295–298 (1990).

Weber, W. and M. Brox, "Dynamic Effects in Hot-Carrier Degradation Relevant for CMOS Operation," *Microelectron. Eng.* **19**, 453–460 (1992).

Wei, C.-Y., J. M. Pimbley, and Y. Nissan-Cohen, "Buried and Graded/Buried LDD Structures for Improved Hot-Electron Reliability," *IEEE Electron Devices Lett.* **7**, 380–382 (1986).

Wei, C.-Y., Y. Nissan-Cohen, and H. H. Woodbury, "Evaluation of 850°C Wet Oxide as the Gate Dielectric in a 0.8-μm CMOS Process," *IEEE Trans. Electron Devices* **38**, 2433–2441 (1991).

Wei, Y., Y. Loh, C. Wang, and C. Hu, "MOSFET Drain Engineering for ESD Performance," *Elect. Overstress/Electrostatic Discharge Symp. Proc.*, 1–6 (1992).

Weimer, P. K., "The TFT—A New Thin-Film Transistor," *Proc. IRE* **50**, 1462 (1962).

Weinberg, Z. A. and G. W. Rubloff, "Exciton Transport on SiO_2 as a Possible Cause of Surface State Generation in MOS Structures," *Appl. Phys. Lett.* **32**, 184–186 (1978).

Weiss, P. and L. Adams, "Development of Semiconductor Test Structures for Reliability Evaluation," *Proc. ESA Electron. Components Conf.*, 15–20 (1991).

Wei-Tsun S. and F. L. Terry, "Bias-Temperature Stability of Nitrided Oxides and Reoxidized Nitrided Oxides," *J. Electron. Mater.* **18**, 767–773 (1989).

Werner, C., R. Kuhnert, and L. Risch, "Optimization of Lightly-Doped Drain MOSFETs Using a New Quasiballistic Simulation Tool," *Tech. Dig.—Int. Electron Devices Meet.*, 770–773 (1984).

Werner, W. M., "The Work Function Difference of the MOS-System with Aluminum Field Plates and Polycystalline Silicon Field Plates," *Solid-State Electron.* **17**, 769 (1974).

White, M. H. and J. R. Cricchi, "Characterization of Thin-Oxide MNOS Memory Transistors," *IEEE Trans. Electron Devices* **19**, 1280 (1972).

Whitfield, J., "On Double Surface Conduction in SOI MOSFETs," *IEEE SOS/SOI Tech. Conf.*, 46–47 (1989).

Wijburg, R. C., G. J. Hemink, J. Middlehoek, H. Wallinga, and T. J. Mouthaan, "VIPMOS—A Novel Buried Injector Structure for EPROM Applications," *IEEE Trans. Electron Devices* **38**, 111–120 (1991).

Williams, R., "Photoemission of Electrons from Silicon into Silicon Dioxide," *Phys. Rev A* **140**, A569 (1965).

Wilmsen, C. W. and S. Szpak, "MOS Processing for III–V Compound Semiconductors: Overview and Bibliography," *Thin Solid Films* **46**, 17 (1977).

Wilson, A. H., "The Theory of Electronic Semiconductors. I," *Proc. R. Soc. London, Ser. A* **133**, 458–468 (1931).

Wilson, A. H., "The Theory of Electronic Semiconductors. II," *Proc. R. Soc. London, Ser. A* **134**, 277–287 (1931).

Wilson, C. L. and T. J. Russell, Two-Dimensional Modeling of Channel Hot-Electron Effects in Silicon MOSFETs," *Tech. Dig.—Int. Electron Devices Meet.*, 72–75 (1985).

Winnerl, J., A. Lill, D. Schmitt-Landsiedel, M. Orlowski, and F. Neppl, "Influence of Transistor Degradation on CMOS Performance and Impact on Life Time Criterion," *Tech. Dig.—Int. Electron Devices Meet.*, 204–207 (1988).

Wittmack, K., J. Maul, and F. Schulz, *Ion Implantation in Semiconductor and Other Materials*, Plenum, New York (1973).

Woerlee, P. H., C. A. H. Juffermans, H. Lifka, F. M. Oude Lansink, H. J. H. Merks-Eppingbroek, T. Poorter, and A. J. Walker, "Offset Diffused Drain Transistors for Half-Micron CMOS," *Proc. Eur. Solid State Devices Res. Conf., 17th*, 17–20 (1987).

Woerlee, P. H., A. H. Van Ommen, H. Lifka, C. A. H. Juffermans, L. Plaja, and F. M. Klaassen, "Half-Micron CMOS on Ultra-Thin Silicon on Insulator," *1989 Int. Electron Devices Meet. IEEE*, 821–824 (1989).

Woerlee, P. H., A. J. Walker, and A. L. J. Burgmans, "P-Channel Devices for Half-Micron CMOS: Advantages of High-Energy Channel Implantations," *Int. Symp. VLSI Technol. Syst. Appl. Proc. Tech. Pap.*, 213–216 (1989).

Woerlee, P. H., P. Damink, M. van Dort, C. Juffermans, C. deKort, H. Lifka, W. Manders, G. Paulzen, H. Pomp, J. Slotboom, G. Streutker, and R. Woltjer, "The Impact of Scaling on Hot-Carrier Degradation and Supply Voltage of Deep-Submicron NMOS Transistors," *Tech. Dig.—Int. Electron Devices Meet.*, 537–540 (1991).

Woerlee, P. H., C. A. H. Juffermans, H. Lifka, W. H. Manders, H. G. Pomp, G. M. Paulzen, A. J. Walker, and R. Woltjer, "Device Characterisation of a High-Performance 0.25 μm CMOS Technology," *Microelecton. Eng.* **19**, 21–24 (1992).

Wojtowicz, D. and T. Kammash, "The Effect of Alpha Particles on the Stability of the ELMO Bumpy Torus (EBT) Reactor," *Phys. Fluids* **28**, 1132–1138 (1985).

Wolters, D. R. and A. T. A. Zegers-Van Duynhoven, "Trapping of Hot Electrons," *Appl. Surf. Sci.* **39**, 565–577 (1989).

Woltjer, R., A. Hamada, and E. Takeda, "Time Dependence of *p*-MOSFET Hot-Carrier Degradation Measured and Interpreted Consistently over Ten Orders of Magnitude," *IEEE Trans. Electron Devices* **40**, 392–401 (1993).

Wong, H.-S., M. H. White, T. J. Krutsick, and R. V. Booth, "Modeling of Transconductance Degradation and Extraction of Threshold Voltage in Thin Oxide MOSFETs," *Solid-State Electron.* **30**, 953–968 (1987).

Wong, H.-S., and Y. C. Cheng, "Instabilities of Metal-Oxide-Semiconductor with High-Temperature Annealing of Its Gate Oxide in Ammonia," *J. Appl. Phys.* **67**, 7132–7138 (1990).

Wong, H.-S., and Y. C. Cheng, "On the Nitridation-Induced Enhancement and Degradation of MOSFET Characteristics," *Solid-State Electron.* **33**, 1107–1109 (1990).

Wong, H.-S., and Y. C. Cheng, "Study of the Electronic Trap Distribution at the SiO_2–Si Interface Utilizing the Low-Frequency Noise Measurement," *IEEE Trans. Electron Devices* **37**, 1743–1749 (1990).

Wong, H.-S., "Experimental Verification of the Mechanism of Hot-Carrier-Induced Photon Emission in *n*-MOSFETs with a CCD Gate Structure," *Tech. Dig.—Int. Electron Devices Meet.*, 549–552 (1991).

Wong, H.-S., "Experimental Verification of the Mechanism of Hot-Carrier-Induced Photon Emission in *n*-MOSFETs Using an Overlapping CCD Gate Structure," *IEEE Electron Devices Lett.* **13**, 389–391 (1992).

Wonshik, L., H. K. Lim, and B. W. Kim, "An Experimental Study on the Threshold Voltage and Punchthrough Voltage Reduction in Short-Channel n-MOS Transistors," *J. Korean Inst. Electron. Eng.* **20**, 1–6 (1983).

Woodruff, R. L. and J. R. Adams, "Radiation Effects in LDD MOS Devices," *IEEE Trans. Nucl. Sci.* **NS-34**, 1629–1634 (1987).

Woods, M. H. and B. L. Euzents, "Reliability in MOS Integrated Circuits," *Tech. Dig.—Int. Electron Devices Meet.*, 50–55 (1984).

Woods, M. H., "Scaling of VLSI; the Reliability Impact on Technology," *Proc. Electrochem. Soc. Meet.* **85-5**, 41–56 (1985).

Wright, P. J., N. Kasai, S. Inoue, and K. C. Saraswat, "Hot-Electron Immunity of SiO_2 Dielectrics with Fluorine Incorporation," *IEEE Electron Devices Lett.* **10**, 347–348 (1989).

Wright, P. J. and K. C. Saraswat, "Comments, with Reply, on 'Hot-Electron Hardened Si-Gate MOSFET Utilizing F Implantation' by Y. Nishioka *et al.*" *IEEE Electron Devices Lett.* **10**, 397–399 (1989).

Wright, P. J. and K. C. Saraswat, "The Effects of Fluorine in Silicon Dioxide Gate Dielectrics," *IEEE Trans. Electron Devices* **36**, 879–889 (1989).

Wright, P. J., N. Kasai, S. Inoue, and K. C. Saraswat, "Improvement in SiO_2 Gate Dielectrics with Fluorine Incorporation," *Tech. Dig. VLSI Tech. Symp.*, 51–52 (1989).

Wright, P. J., A. Kermani, and K. C. Saraswat, "Nitridation and Post-Nitridation Anneals of SiO_2 for Ultrathin Dielectrics," *IEEE Trans. Electron Devices* **37**, 1836–1841 (1990).

Wu, A. T., T. Y. Chan, P.-K. Ko, and C. Hu, "Novel High-Speed, 5-Volt Programming EPROM Structure with Source-Side Injection," *Tech. Dig.—Int. Electron Devices Meet.*, 584–587 (1986).

Wu, A. T., T. Y. Chan, P.-K. Ko, and C. Hu, "Source-Side Injection Erasable Programmable Read-Only-Memory (Si-EPROM) Device," *IEEE Electron Devices Lett.* **7**, 540–542 (1986).

Wu, A. T., S.-W. Lee, V. Murali, and M. Garner, "Off-State Gate Current in *n*-Channel MOSFETs with Nitrided Oxide Gate Dielectrics," *IEEE Electron Devices Lett.* **11**, 499–501 (1990).

Wu, N. R., S. Chiao, C. Wang, B. Bhushan, and C. Y. Yang, "Electron Trapping/Detrapping in Thin SiO_2 under High Fields," *Proc. Mater. Res. Soc. Symp.* **47**, 99–105 (1985).

Wu, N. R., S. Chiao, B. Bhushan, T. Batra, S. K. Fan, P. Pizzo, and C. Y. Yang, "Effect of Polysilicon Deposition and Thermal Cycling on Thin Oxide Quality," *Proc. Mater. Res. Soc. Symp.* **71**, 513–518 (1986).

Xiaolin, L., "Hot-Carrier Energy-Transfer Rate in GaAs–AlGaAs Heterostructures," *Chin. Phys. Lett.* **4**, 193–196 (1987).

Xu, J. and M. Shur, "Ballistic Transport in Hot-Electron Transistor," *J. Appl. Phys.* **62**, 3816–3820 (1987).

Xu, W., F. M. Peeters, and J. T. Devreese, "The Hot-Electron Distribution of Two-Dimensional Electrons in a Polar Semiconductor at Zero Temperature," *J. Phys.: Condens. Matter* **3**, 1783–1791 (1991).

Yamabe, K. and Y. Miura, "Discharge of Trapped Electrons from MOS Structures," *J. Appl. Phys.* **51**, 6258–6264 (1980).

Yamabe, K., Y. Ozawa, S. Nadahara, and K. Imai, "Thermally Grown Silicon Dioxide with High Reliability," *Proc.—Electrochem. Soc.* **90**, 349–363 (1990).

Yamada, H. and M. Abe, "Ultradry Annealing after Polysilicon Electrode Formation for Improving the TDDB Lifetime of Ultrathin Silicon Oxide Films in MOS Diodes," *IEEE Electron Devices Lett.* **EDL-12**, 536–538 (1991).

Yamada, M. T., "Hot-Electron Trapping of Short Channel 64 K Dynamic MOS RAM," *Jpn. J. Appl. Phys., Suppl.*, 59–62 (1982).

Yamada, M. T., T. Kobayashi, M. Kumanoya, M. Taniguchi, T. Nakano, and H. Matsumoto, "Hot-Electron Trapping Effects of Short Channel 64 K Dynamic MOS RAM," *Jpn. J. Appl. Phys.* **22**, 59–62 (1983).

Yamada, S., T. Suzuki, E. Obi, M. Oshikiri, K. Naruke, and M. Wade, "A Self-Convergence Erasing Scheme for a Simple Stacked Gate Flash EEPROM," *Tech. Dig.—Int. Electron Devices Meet.*, 307–310 (1991).

Yamaguchi, K., "Field-Dependent Mobility Model for Two-Dimensional Numerical Analysis of MOSFETs," *IEEE Trans. Electron Devices* **26**, 1068 (1979).

Yamaguchi, Y., M. Shimizu, Y. Inoue, T. Nishimura, and Y. Akasaka, "Hot Carrier Reliability of Submicron Ultra Thin SOI-MOSFETs," *Ext. Abstr., Int. Conf. Solid State Devices Mater., 1991*, 656–658 (1991).

Yamaguchi, Y., M. Shimizu, Y. Inoue, T. Nishimura, and K. Tsukamoto, "Hot-Carrier Reliability in Submicrometer Ultra-Thin SOI-MOSFETs," *IEICE Tran. Electron.* **E75-C**, 1465–1470 (1992).

Yamamoto, H. and H. Iwasawa, "Optimum Conditions for Second Harmonic Generation by Hot-Electron and Electron-Transfer Effects in Semiconductor," *Solid-State Electron.* **22**, 271–275 (1979).

Yamamoto, N., H. Kume, S. Iwata, K. Yagi, N. Kobayashi, N. Mori, and H. Miyazaki, "Characteristics of Tungsten Gate Devices for MOS VLSIs," *Proc. Workshop Tungsten Other Refractory Met. VLSI Appl.*, 297–311 (1986).

Yamamoto, N., H. Kume, S. Iwata, K. Yagi, N. Kobayashi, N. Mori, and H. Miyazaki, "Fabrication of Highly Reliable Tungsten Gate MOS VLSIs," *J. Electrochem. Soc.* **133**, 401–407 (1986).

Yamamoto, N., "Degradation of MOS Characteristics Caused by Internal Stresses in Gate Electrodes," *Ext. Abstr. Solid State Devices Mater.*, 415–418 (1987).

Yamamoto, N., S. Iwata, and H. Kume, "Influence of Internal Stresses in Tungsten-Gate Electrodes on Degradation of MOSFET Characteristics Caused by Hot Carriers," *IEEE Trans. Electron Devices* **34**, 607–614 (1987).

Yamamoto, T., Y. Nishimura, T. Iizuka, H. Matsumoto, and M. Fukuma, "Hot Carrier Effects in *n*MOSFET at 77 K and 300 K," *Ext. Abstr., Int. Conf. Solid State Devices Mater., 1991*, 8–10 (1991).

Yan, R. H., K. F. Lee, D. Y. Jeon, Y. O. Kim, B. G. Park, M. R. Pinto, C. S. Rafferty, D. M. Tennant, E. H. Westerwick, G. M. Chin, M. D. Moriss, K. Early, P. Mulgrew, W. M. Mansfield, R. K. Watts, A. M. Voshchenkov, J. Bokor, R. G. Swartz, and A. Ourmazd, "89-GHz f_T Room-Temperature Silicon MOSFETs," *IEEE Electron Devices Lett.* **EDL-13**, 259–258 (1992).

Yanagisawa, M., K. Nakamura, and M. Kikuchi, "Trench Transistor Cell with Self-Aligned Contact (TSAC) for Megabit MOS DRAM," *Tech. Dig.—Int. Electron Devices Meet.*, 132–135 (1986).

Yaney, D. S., J. C. Desko, M. M. Kelly, L. T. Lancaster, A. M. Lin, A. S. Manocha, T. K. McGuire, F. R. Peiffer, and H. C. Kirsch, "Technology for the Fabrication of a 1-Mb CMOS DRAM," *Tech. Dig.—Int. Electron Devices Meet.*, 698–701 (1985).

Yang, C. Y., C. M. Wang, G. Lin, and J. J. Tzou, "Time-Dependent Degradation in MOS Devices," *Proc. Symp. Reliab. Semicond. Devices/Interconnect.* **89-6**, 168–184 (1988).

Yang, C. Y., H. Inokawa, and F. E. Pagaduan, "Direct Determination of Interface Trapped Charges," *Jpn. J. Appl. Phys.* **30**, L888–L890 (1991a).

Yang, C. Y., E. M. Ajimine, and H. Inokawa, "Effects of Degradation and Recovery on MOS Device Reliability," *Proc. Symp. Reliab. Semicond. Devices/Interconnect. Dielectric Breakdown Laser Process Microelectron. Appl.* **92–94**, 177–195 (1991b).

Yang, K. S., H. S. Park, and B. R. Kim, "The Characteristics of MOSFET with Reoxidized Nitrided Oxide Gate Dielectrics," *J. Korean Inst. Telemat. Electron.* **28A**, 75–81 (1991).

Yang, K. S., J. T. Park, and B. R. Kim, "Analytical Model for Substrate and Gate Current of Stressed SC-PMOSFET in the Saturation Region," *Ext. Abstr., 1992 Int. Conf. Solid State Devices Mater.*, Tsukuba, Jpn., 164–166 (1992).

Yang, P. and S. Aur, "Modeling of Device Lifetime Due to Hot Carrier Effects," *Tech. Dig. VLSI Tech. Symp.*, 227–230 (1985).

Yang, Z. and J. Xu, "Study of Hot-Carrier Effect in Short Channel DD MOSFET," *Chin. J. Semicond.* **10**, 489–496 (1989).

Yano, K., M. Aoki, and T. Masuhara, "Hot Carrier Degradation Masking Due to Velocity Saturation in Low Temperature Operated MOSFETs," *Ext. Abstr., Solid State Devices Mater.*, 335–358 (1987).

Yano, K., K. Nakazato, M. Miyamoto, T. Onai, M. Aoki, and K. Shimohigashi, "A High-Current-Gain Low-Temperature Pseudo-Heterojunction Bipolar Transistor Utilizing Sidewall Base-Contact Structure (Sicos)," *IEEE Trans. Electron Devices* **38**, 555–565 (1991).

Yao, C., J. Tzou, R. Cheung, and H. Chan, "Temperature Dependence of CMOS Device Reliability," *Proc. Int. Reliab. Phys. Symp.*, 175–182 (1986).

Yao, C., J. Tzou, R. Cheung, H. Chan, and C. Y. Yang, "Structure and Frequency Dependence of Hot-Carrier-Induced Degradation in CMOS VLSI," *Proc. Int. Reliab. Phys. Symp.*, 195–200 (1987).

Yao, C. T., M. Peckerar, D. Friedman, and H. Hughes, "Hot Electron Effect on MOSFET Terminal Capacitances," *Proc. IEEE Custom Int. Circuits Conf.*, 208–211 (1986).

Yao, C. T., M. Peckerar, D. Friedman, and H. Hughes, "On the Effect of Hot-Carrier Stressing on MOSFET Terminal Capacitances," *IEEE Trans. Electron Devices* **35**, 384–386 (1988).

Yaron, G., W. A. Lukaszek, and D. Frohman-Benchkowsky, "E^2 FAMOS—An Electrically Erasable Reprogrammable Charge Storage Device," *IEEE Trans. Electron Devices* **26**, 1754–1759 (1979).

Yasuda, N., H. Nakamura, K. Taniguchi, C. Hamaguchi, and M. Kakumu, "Interface State Generation Mechanism in MOSFETs during Substrate Hot-Electron Injection," *Jpn. J. Appl. Phys., Part 2: Lett.* **27**, 2395–2397 (1988).

Yasuda, N., H. Nakamura, K. Taniguchi, and C. Hamaguchi, "Interface State Generation Mechanism in *n*-MOSFETs," *Solid-State Electron.* **32**, 1579–1583 (1989).

Yau, L. D., "A Simple Theory to Predict the Threshold Voltage of Short-Channel IGFETs," *Solid-State Electron.* **17**, 1059 (1974).

Yau, L. D., "Simple *I–V* Model for Short-Channel IGFETs in the Triode Region," *Electron Lett.* **11**, 44 (1975).

Yeow, Y. T., C. H. Ling, and L. K. Ah, "Evaluation and Modeling of MOSFET Degradation Due to Electrical Stressing through Gate-to-Source and Gate-to-Drain Capacitance Measurement," *Proc. Microelectron. Conf.*, 77–80 (1991).

Yeow, Y. T., C. H. Ling, and L. K. Ah, "Observation of MOSFET Degradation Due to Electrical Stressing through Gate-to-Source and Gate-to-Drain Capacitance Measurement," *IEEE Electron Devices Lett.* **12**, 366–368 (1991).

Yon, E., W. H. Ko, and A. B. Kuper, "Sodium Distribution in Thermal Oxide on Silicon by Radiochemical and MOS Analysis," *IEEE Trans. Electron Devices* **13**, 276 (1966).

Yoon, S., R. Siergiej, and M. H. White, "A Novel Substrate Hot Electron and Hole Injection Structure with a Double-Implanted Buried-Channel MOSFET," *IEEE Trans. Electron Devices* **38**, 2722 (1991).

Yoshida, A. and Y. Ushiku, "Hot Carrier Induced Degradation Mode Depending on the LDD Structure in *n*-MOSFETs," *Tech. Dig.—Int. Electron Devices Meet.*, 42–45 (1987).

Yoshida, M., M. Nakahara, M. Kimura, S. Taguchi, K. Maeguchi, and H. Tango, "Characteristics of MOSFETs on Double Solid-Phase Epitaxial SOS," *J. Microelectron.* **14**, 116–117 (1983).

Yoshida, M., M. Nakahara, M. Kimura, S. Taguchi, K.Maeguchi, and H. Tango, "High Speed 1 μm SOS CMOS Devices Using Double Solid-Phase Epitaxy," *Tech. Dig.—Int. Electron Devices Meet.*, 372–375 (1983).

Yoshida, M., D. Tohyama, K. Maeguchi, and K. Kanzaki, "Increase of Resistance to Hot Carriers in Thin Oxide MOSFETs," *Tech. Dig.—Int. Electron Devices Meet.*, 254–257 (1985).

Yoshida, S., K. Okuyama, F. Kanai, Y. Kawate, M. Motoyoshi, and H. Katto, "Improvement of Endurance to Hot Carrier Degradation by Hydrogen Blocking P-SiO," *Tech. Dig.—Int. Electron Devices Meet.*, 22–25 (1988).

Yoshida, S., T. Matsui, K. Okuyama, and K. Kubota, "AC Hot-Carrier Degradation in the Super-100 MHz Operation Range," *Tech. Dig. 1994 Symp. on VLSI Technology*, 145–146 (1994).

Yoshii, I., K. Hama, and K. Hashimoto, "Role of Hydrogen at Poly-Si/SiO_2 Interface in Trap Generation by Substrate Hot-Electron Injection," *Proc. Int. Reliab. Phys. Symp.*, 136–140 (1992).

Yoshikawa, K., M. Sato, S. Mori, Y. Mikata, T. Yanase, K. Kanzaki, and H. Nozawa, "Technology Requirements for Mega Bit CMOS EPROMS," *Tech. Dig. —Int. Electron Devices Meet.*, 456–459 (1984).

Yoshikawa, K., M. Sato, and Y. Ohshima, "A Reliable Profile Lightly-Doped Drain (PLD) Cell for High Density Submicron EPROMs and Flash EEPROMs," *Ext. Abstr., Solid State Devices Mater.*, 165–168 (1988).

Yoshikawa, K., N. Arai, S. Mori, Y. Kaneko, Y. Oshima, K. Narita, and H. Araki, "A New MOSFETs Degradation Induced by Gate Current in Off-State Condition," *1990 Symp. VLSI Technol., Dig. Tech. Pap.*, 73–74 (1990).

Yoshikawa, K., M. Sato, and Y. Oshima, "A Reliable Profiled Lightly Doped Drain (PLD) Cell for High-Density Submicrometer EPROMs," *IEEE Trans. Electron Devices* **37**, 999–1006 (1990).

Yoshimi, M., T. Wada, M. Takahashi, K. Numata, and K.Kawabuchi, "Characterization of Submicrometer Buried-Channel and Surface-Channel *p*-MOSFETs," *Tech. Dig. VLSI Tech. Symp.*, 76–77 (1984).

Yoshimi, M., M. Takahashi, S. Kambayashi, M. Kemmochi, T. Wada, and K. Natori, "Analysis of Drain Breakdown and Evaluation of Operation Speed in Ultra-Thin SOI MOSFETs," *Tech. Dig. VLSI Tech. Symp.*, 15–16 (1989).

Yoshimi, M., H. Hazama, M. Takahashi, S. Kambayashi, T. Wada, K. Kato, and H. Tango, "Two-Dimensional Simulation and Measurement on High-Performance MOSFETs Made on a Very Thin SOI Film," *IEEE Trans. Electron Devices* **36**, 493–503 (1989).

Yoshino, A., T.-P. Ma, and K. Okumura, "Hot-Carrier Effects in Fully Depleted Submicrometer NMOS/SIMOX as Influenced by Back Interface Degradation," *IEEE Electron Devices Lett.* **13**, 522–524 (1992).

Young, D. R. and D. P. Seraphim, "Surface Effects on Silicon: Introduction," *IBM J. Res. Dev.* **8**, 366–367 (1964).

Young, N. D., S. D. Brotherton, and A. Gill, "Drain Bias Instability in Polycrystalline Silicon Thin Film Transistors," *Proc. 7th Bienn. Eur. Conf.*, 231–234 (1991).

Young, N. D., A. Gill, and M. J. Edwards, "Hot Carrier Degradation in Low Temperature Processed Polycrystalline Silicon Thin Film Transistors," *Semicond. Sci. Technol.* **7**, 183–188 (1992).

Yu, H. N., A. Reisaman, C. M. Osburn, and D. L. Critchlow, "1 μm MOSFET VLSI Technology. I. An Overview," *IEEE J. Solid-State Circuits* **SC-14**, 240–246 (1979).

Yu, W., T.-P. Ma, and R. C. Barker, "Orientation Dependence of Interface-Trap Transformation," *Proc. 2nd Workshop Radiat.-Induced* and/or *Process-Relat. Electr. Active Defects Semicond.-Insul. Syst.*, 126–138 (1989).

Zeisler, P. and I. Ruge, "Device Performance Degradation of Short Channel MOS Transistors Due to Hot-Carrier Injection and Drain Profile Engineering," *NTZ Arch.* **8**, 191–197 (1986).

Zekeriya, V. and T.-P. Ma, "Dependence of X Rays Generation of Interface Traps on Gate Metal Induced Interfacial Stress in MOS Structures," *IEEE Trans. Nucl. Sci.* **NS-31**, 1261–1266 (1984).

Zhang, B. and T.-P. Ma, "Back-Channel Hot-Electron Effect on the Front-Channel Characteristics in Thin-Film SOI MOSFETs," *IEEE Electron Devices Lett.* **12**, 699–701 (1991).

Zhang, J. F., S. Taylor, and W. Eccleston, "Electron Trap Generation in Thermally Grown SiO_2 under Fowler–Nordheim Stress," *J. Appl. Phys.* **71**, 725–734 (1992).

Zhao, S. P., S. Taylor, W. Eccleston, and K. J. Barlow, "P-Well Bias Dependence of Electron Trapping in Gate Oxide of *n*-MOSFETs during Substrate Hot-Electron Injection," *Electron. Lett.* **28**, 2080–2082 (1992).

Zhou, B. and G. Ruan, "New Hot-Electron-Induced Threshold Voltage Degradation Model for *n*-MOSFETs," *Proc. Int. Conf. Solid State Devices Mater., 22nd*, 299–302 (1990).

Ziegler, K., E. Klausman, and S. Kar, "Determination of the Semiconductor Doping Profile Right Up to its Surface Using Thin MIS Capacitor," *Solid-State Electron.* **18**, 189–198 (1975a).

Ziegler, K. and E. Klausman, "Static Technique for Precise Measurement of Surface Potential and Interface State Density in MOS Structures," *Appl. Phys. Lett.* **26**, 400–402 (1975b).

Zimmer, G., "MOS-Technology," *Microprocess. Microprog.* **10**, 77–99 (1982).

Zsolt, G., G. Kovacs, T. Porjesz, T. Karman, and G. Gombos, "Avalanche Current Relaxation in *p*-Si MOSFETs," *Acta Physiol. Hung.* **62**, 19–22 (1987).

INDEX